IRMA Lectures in Mathematics

Edited by Vladimi

Institut de Recherche Mathématique Avancée
Université Louis Pasteur et CNRS
7 rue René Descartes
67084 Strasbourg Cedex
France

IRMA Lectures in Mathematics and Theoretical Physics
Edited by Vladimir G. Turaev

This series is devoted to the publication of research monographs, lecture notes, and other materials arising from programs of the Institut de Recherche Mathématique Avancée (Strasbourg, France). The goal is to promote recent advances in mathematics and theoretical physics and to make them accessible to wide circles of mathematicians, physicists, and students of these disciplines.

Previously published in this series:

1 Deformation Quantization, *Gilles Halbout* (Ed.)
2 Locally Compact Quantum Groups and Groupoids, *Leonid Vainerman* (Ed.)
3 From Combinatorics to Dynamical Systems, *Frédéric Fauvet and Claude Mitschi* (Eds.)
4 Three courses on Partial Differential Equations, *Eric Sonnendrücker* (Ed.)
5 Infinite Dimensional Groups and Manifolds, *Tilman Wurzbacher* (Ed.)
6 *Athanase Papadopoulos*, Metric Spaces, Convexity and Nonpositive Curvature

Volumes 1–5 are available from Walter de Gruyter (www.degruyter.de)

Numerical Methods for Hyperbolic and Kinetic Problems

CEMRACS 2003, Summer Research Center in Mathematics and Advances in Scientific Computing, July 21–August 29, 2003, CIRM, Marseille, France

Stéphane Cordier
Thierry Goudon
Michaël Gutnic
Eric Sonnendrücker
Editors

Editors:

Stéphane Cordier
Laboratoire MAPMO, UMR 6628
CNRS-Université d'Orléans
B.P. 6759
F-45067 Orléans Cedex 2
France

Michaël Gutnic
IRMA, UMR 7501
CNRS-Université Louis Pasteur
7 Rue René Descartes
F-67084 Strasbourg Cedex
France

Thierry Goudon
Laboratoire Paul Painlevé, UMR 8524
CNRS-Université Lille I
Cité scientifique
F-59655 Villeneuve d'Ascq Cedex
France

Eric Sonnendrücker
IRMA, UMR 7501
CNRS-Université Louis Pasteur
7 Rue René Descartes
F-67084 Strasbourg Cedex
France

2000 Mathematics Subject Classification 65M, 76-xx, 82-xx

ISBN 3-03719-012-4

Bibliographic information published by Die Deutsche Bibliothek
Die Deutsche Bibliothek lists this publication in the Deutsche Nationalbibliografie; detailed bibliographic data are available in the Internet at http://dnb.ddb.de.

Contact address:

European Mathematical Society Publishing House
Seminar for Applied Mathematics
ETH-Zentrum FLI C4
CH-8092 Zürich
Switzerland

Phone: +41 (0)1 632 34 36
Email: info@ems-ph.org
Homepage: www.ems-ph.org

Typeset using the author's TEX files: I. Zimmermann, Freiburg
Printed in Germany

9 8 7 6 5 4 3 2 1

Preface

Initiated from an original idea of Y. Maday in 1996, the CEMRACS is a yearly meeting for applied mathematicians interested in modeling and scientific computing. It is a special event of the SMAI, the French society for industrial and applied mathematics. Traditionally, the meeting is held at the Centre International de Rencontres Mathématiques (CIRM) on the campus of Luminy, Marseille. The eighth gathering was devoted to numerical methods for hyperbolic and kinetic problems.

The aim of the CEMRACS is two-fold. The first objective is to further fruitful interactions between academic laboratories and industrial centers of research. Secondly, the meeting provides a high level formation by offering a timely picture of a very active area in applied mathematics. Therefore, the CEMRACS aims at bringing together various groups of scientists: physicists, engineers, computer scientists, applied mathematicians, the combination of their skills and efforts being the condition of progress on the problems addressed by industrials. The applications include multi-phase flows, numerical resolution of plasma physics problems, simulations of non linear Schrödinger equations, diffusion approximation in radiative transfer.

The very heart of the CEMRACS is an intense research activity, organized by groups of two to six persons working together on the submitted subjects. Besides, the first week of the CEMRACS is devoted to lectures. This year the courses were given by Albert Cohen, Fréderic Coquel and Pierre Degond. These lectures survey the topics and offer the means to stay current with the cutting edges of the field.

More than one hundred people attended the summer school, with an average of fifty people present per week. This volume collects the progress performed during the summer school on the projects submitted by our industrial and academic partners.

Our thanks go to the scientific committee: Patrick Lascaux (Commissariat à l'Energie Atomique), Patrick Le Tallec (Ecole Polytechnique Paris), Pierre-Louis Lions (Membre de l'Institut, Université Paris IX and Ecole Polytechnique), Yvon Maday (Université Paris VI), Étienne Pardoux (Université Aix Marseille), Olivier Pironneau (Université Paris VI), Pierre-Arnaud Raviart (Ecole Polytechnique), Denis Talay (INRIA Sophia Antipolis).

We are grateful to our partners for their kind support – both in terms of funding, and in providing a true incentive in favor of interdisciplinary cooperation:

– the C.E.A. (Commissariat à l'Energie Atomique) and in particular the centers DAM (Ile de France, Bruyères), Cesta (Gironde), Cadarache (Bouches-du-Rhône), Saclay (Essonne);

– the I.F.P. (Institut Français du Pétrole);

– the D.G.A (Direction Générale de l'Armement);

– the academic laboratories: IRMA (UMR CNRS 7501, Université Louis Pasteur, Strasbourg), MAPMO (UMR CNRS 6628, Université d'Orléans), LMA (F.R.E. CNRS

2222, Université Lille 1), MIP (UMR CNRS 5640, Université Toulouse), Laboratoire Jacques-Louis Lions (UMR CNRS 7598 Université Pierre et Marie Curie);

– the INRIA (Institut National de Recherche en Informatique et en Automatique)

– the European Research Training Network HYKE HPRN-CT-2002-00282;

– the GdR CNRS 2250 GRIP.

Finally it is our pleasant duty to thank the staff of the CIRM in Luminy for its unforgettable warm hospitality.

Lille, Orléans, Strasbourg, February 2005 *The Editors*

Table of Contents

Hyperbolic problems

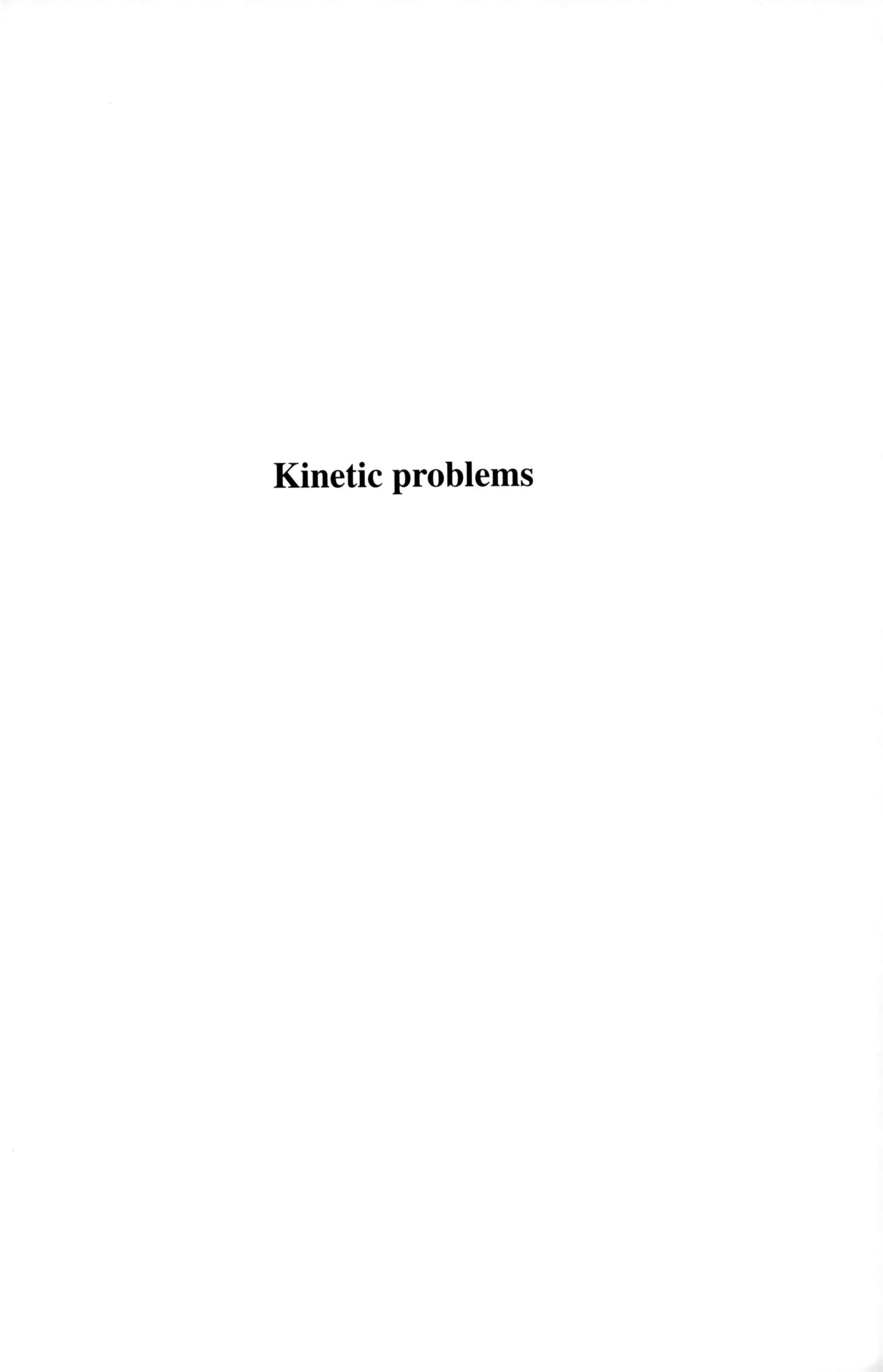

Kinetic problems

Introduction

by S. Cordier, T. Goudon and E. Sonnendrücker

Several projects were devoted to the numerical resolution of the Vlasov equation which is very important for the simulation of charged particle beams as well as globally neutral plasmas. The major difficulty involved is that the Vlasov equation is posed in phase space and thus the dimension is twice as high as for usual physical space simulations. For this reason Monte Carlo particle methods have had great success in this area. The coupling with the self-consistent electromagnetic field is done using the Particle-In-Cell (PIC) method. This technique is based on a Monte Carlo particle approximation of the Vlasov equation coupled with a grid based field solver in configuration space. Most simulation papers, in the last twenty years, involving charged particles have been performed using variants of this technique. Still, there are many problems, for which the method needs to be better understood and improved such as the coupling of the particle solution with a Maxwell solver, or the limitation of numerical noise in the simulations. The COCPIC project addresses the first of these issues and the PICONU project the second.

Other techniques for the numerical resolution of the Vlasov equation have been introduced for 1D problems in the seventies and have regained a new interest for multi-dimensional simulations recently with the increase of computer power. However, when going to 4D, 5D or 6D, phase space methods based on a uniform phase-space grid cannot compete with PIC methods except in some cases where very high precision is needed because of their computational complexity. In order to overcome this problem, adaptive phase-space grid methods have been introduced in the last years. Two such techniques were studied at CEMRACS 2003, one based on hierarchical finite elements and one based on interpolating wavelets. The first one is introduced in this book.

The work by Régine Barthelmé and Céline Parzani deals with the coupling of a Particle-In-Cell Vlasov solver with an FDTD Maxwell solver based on the Yee method. The underlying problem comes from the fact that the approximate charge and current densities calculated from the particles normally do not satisfy a discrete version of the continuity equation $\partial_t \rho + \nabla \cdot J = 0$. Because of this the computed electric field does not satisfy Gauss' law well enough on long time scales, which leads to very unphysical solutions. Several remedies have been proposed. They fall into two categories. The first consists in correcting the electric field from time to time in order that the error in Gauss' law does not become too large, and the second consists in computing the current density J from the particles such that the continuity equation remains satisfied on the discrete level. This of course depends fundamentally on the field solver. Up to now two such methods had been introduced. The first one consists in computing the current using the actual fluxes of the numerical particles through the grid faces. This was introduced for the first order by Morse and Nielson and for the second order by Villasenor and Buneman. The other technique consists in computing J directly form the continuity equation by splitting between the different directions. This was introduced by Ezyrkepov at any order. The main contribution of the work in this book is first to give a better and clearer understanding of the Morse–Nielson and Villasenor–Buneman method and then to generalize the method to any order.

The paper by Chehab, Cohen, Jennequin, Nieto, Roland and Roche deals with the very important issue of noise reduction in Particle-In-Cell simulations. Many efforts have been made in this direction since this is the major problem of the method. These efforts include statistical techniques like variance reduction methods as in the variant which is called the δf method and which consists in applying the Particle-In-Cell method only to a perturbation of an equilibrium. This technique is very efficient but is only applicable in a limited set of problems where a known equilibrium, to which the solution stays close, exists. Apart from that, many filtering techniques applied generally in Fourier space have been introduced and are now available in most of the codes. In this work an original approach which consists in computing the charge density ρ in a well chosen wavelet basis and applying a thresholding procedure is introduced. This method is based on ideas introduced by Donoho in non parametric statistics and looks very promising for PIC simulations.

The work of Campos Pinto and Mehrenberger introduces adaptive grid based Vlasov solvers. The underlying numerical method is the semi-Lagrangian method which is based on the fact that the distribution function is conserved along characteristics. The method then consists in updating the grid values of the distribution function in two steps: 1) Find the origin of the characteristic curve ending at the different grid points. 2) Interpolate the distribution function at the origin of the characteristic curve from the known grid values. The adaptivity is introduced using non linear approximation. Indeed, using a hierarchical decomposition of the distribution function, it is possible to obtain an approximation, depending on the function being approximated, based on considerably fewer points. This technique allows to reduce the number of needed interpolation points compared to a uniform fine grid while increasing the error only slightly and in a controlled manner.

The paper by Campos Pinto and Mehrenberger investigates non linear approximation based on hierarchical bi-quadratic finite elements. This approach has the advantage of being cell based and completely local, which makes it far easier to parallelize.

The contribution of Crouseilles and Filbet is devoted to the derivation of a high order scheme for the simulation of a collisional plasma. The model they consider is the Vlasov equation with an additional Fokker–Planck–Landau collision operator. The numerical scheme decouples the transport part from the collision part. A new second order method for the Vlasov equation is derived which conserves mass and energy. Slope limiters are introduced to ensure positivity of the distribution function. However, when those are needed, energy conservation is no more exactly verified. On the other hand the collision operator is discretized using a spectral method. The numerical scheme is validated in particular on the Landau damping problem which is thoroughly discussed and which should be of great importance to people interested in code validation examples.

The project by C. Besse, N. Mauser and H. P. Stimming is concerned with numerical experiments of a time splitting strategy to solve two non linear models arising in different physical situations. The first is the so-called Schrödinger–Poisson-X_α (SPX_α) model. This model arises in the modeling of quantum particle dynamics and is intended to be an approximation of the time dependent Hartree–Fock equation. The second is the Davey–Stewardson (DS) model, which is motivated by the description of water waves. On the one hand, SPX_α involves two nonlinearities: one is purely local and is nothing but the usual focusing cubic nonlinearity in the Schrödinger equation, while the other is non local, since it is related to self-consistent interactions. The behavior of the solutions, in the semi classical limit $\varepsilon \to 0$, ε being defined from the Planck constant, highly depends on the relative strength of both nonlinearities, scaled in an ε-dependent way. Addressing the problem from a numerical viewpoint is a really challenging issue: using a classical finite difference approach would require unrealistic restrictions on the mesh size. On the other hand, DS involves a set of parameters, and depending on the range of the values of these parameters, the equation changes type. In turn, the solutions of DS can exhibit a very rich behavior, like formation of soliton structures, or blowup phenomena. Therefore, the numerical challenge is the precise evaluation of time and space localization of the blowup and the description of the profiles with sharp accuracy. Here the authors adapt a time splitting spectral method introduced by Bao, Jin and Markowich. As usual, the idea is to separate the problem by first solving a linear PDE, and then dealing with a nonlinear ODE, possibly stiff, which can be solved exactly. A proof of convergence of the semi-discrete scheme is proposed, and simulations illustrate the behavior of the solutions and the accuracy of the method by various multidimensional examples.

The contribution by C. Besse, P. Degond, F. Deluzet, J. Claudel, G. Gallice, and T. Tessieras is devoted to modeling issues for ionospheric plasmas, a subject of crucial importance for communication satellites. The models in this field can exhibit instability phenomena and strong heterogeneities can occur, a situation comparable to Rayleigh–Taylor instabilities in fluid mechanics. The basic model which describes ionospheric plasmas couples Euler equations for ions and electrons, and Maxwell equations for the evolution of electric and magnetic fields. The computational cost to solve numerically such a complicated system is prohibitive when dealing with physical values of the parameters, which motivates that one seeks for simplified models. Hence, having identified a set of dimensionless relevant parameters, a hierarchy of models is presented, which ends with the so-called striation model. This model involves a $3D$ convection diffusion equation coupled with a $2D$ elliptic equation. However, the stability analysis reveals that the striation does not reproduce the main features of the physics: dissipation phenomena have been destroyed in the asymptotics. Then, the idea consists in taking into account turbulence effects in the fluid evolution, which introduces an additional eddy viscosity and restores nice dissipation properties. In turn, the new "turbulent striation model" presents better stability properties. Numerical experiments validate the modeling discussion.

The paper by L. Gosse is a review of different numerical and asymptotic methods for high frequency asymptotics of the 1D Schrödinger equation.

Four techniques for dealing with such asymptotics are briefly presented: stationary phase, Wigner measure, WKB, ray tracing. Two numerical methods based on moment closure are described (the classical moment system and the K branch entropy solution) and are illustrated by a classical example.

Numerical charge conservation in Particle-In-Cell codes

R. Barthelmé and C. Parzani

Institut de Recherche Mathématique Avancée
Université Louis Pasteur, 7 rue René Descartes
67084 Strasbourg Cedex, France
email: `barthelm@math.u-strasbg.fr`

Laboratoire MIP, Université Paul Sabatier
118 route de Narbonne, 31062 Toulouse Cedex, France
email: `parzani@mip.ups-tlse.fr`

Abstract. A difficulty in electromagnetic PIC simulations of plasmas modeled by the Vlasov–Maxwell system is that the continuity equation may not be satisfied and that then the Maxwell system isn't any more well-posed. Indeed, in PIC codes, fields and densities are defined at spatial grid points while charged particles can take arbitrary positions. Therefore, the current mapped from particles to the mesh may not satisfy the continuity equation at the discrete level. In this paper, we are interested in charge conserving current deposition algorithms. We study the efficiency of two methods ensuring that the continuity equation is satisfied at the discrete level: an extension of the Villasenor–Buneman method and the Esirkepov method.

1 Introduction

Numerical simulations of plasmas or charged particle beams modeled by the Vlasov–Maxwell system are often performed using the Particle-In-Cell method. Indeed, in physical cases where the independent variables (position and velocity) are defined in the six dimensional phase-space, PIC approximations of this kinetic model have a lower numerical cost than grid based methods. The PIC method consists in following the trajectories of a large set of macro-particles describing the plasma in the phase space. Next, the charge and current densities associated to these particles are deposited onto a mesh on which the self-field is computed. On the other hand, the electromagnetic field at the position of a particle is interpolated from those at adjacent grid points in order to evaluate the particle motion. The assignment process, first from the particle cloud to the mesh and next from the grid to each particle, is obtained using shape factors e.g. splines. However, this method may fail in simulations of some realistic cases.

Indeed, let us consider a noncollisional plasma constituted of charged particles. At a kinetic level, the evolution of these particles is described by a distribution function $f(\boldsymbol{x}, \boldsymbol{u}, t)$ depending on space $\boldsymbol{x} \in \mathbb{R}^3$, momentum $\boldsymbol{u} \in \mathbb{R}^3$ and time $t > 0$ which satisfies the Vlasov equation

$$\frac{\partial f}{\partial t} + \boldsymbol{v} \cdot \nabla_{\boldsymbol{x}} f + q(\boldsymbol{E} + \boldsymbol{v} \times \boldsymbol{B}) \cdot \nabla_{\boldsymbol{u}} f = 0. \tag{1.1}$$

In the relativistic case, the momentum $\boldsymbol{u}$ is defined by $\boldsymbol{u} = m\boldsymbol{v}\gamma(\boldsymbol{v})$ with

$$\gamma(\boldsymbol{v}) = (1 - |\boldsymbol{v}|^2/c^2)^{-1/2} = (1 + |\boldsymbol{u}|^2/(m^2c^2))^{1/2}$$

where $\boldsymbol{v}$ is velocity and c is the speed of light in vacuum, while in the non-relativistic case we have $\boldsymbol{u} = m\boldsymbol{v}$. Charge and current densities are then deduced from this distribution function by setting

$$\rho(\boldsymbol{x}, t) = \int q f(\boldsymbol{x}, \boldsymbol{u}, t)\, d\boldsymbol{u}, \quad \boldsymbol{J}(\boldsymbol{x}, t) = \int q \boldsymbol{v} f(\boldsymbol{x}, \boldsymbol{u}, t)\, d\boldsymbol{u}. \tag{1.2}$$

The electromagnetic field $(\boldsymbol{E}, \boldsymbol{B})$ is described by the Maxwell's equations and the coupling with the Vlasov equation is done through the source terms such that

$$\frac{\partial \boldsymbol{E}}{\partial t} = c^2 \textbf{rot}\, \boldsymbol{B} - \frac{\boldsymbol{J}}{\varepsilon_0}, \quad \frac{\partial \boldsymbol{B}}{\partial t} = -\,\textbf{rot}\, \boldsymbol{E}, \tag{1.3}$$

$$\text{div}\, \boldsymbol{E} = \frac{\rho}{\varepsilon_0}, \quad \text{div}\, \boldsymbol{B} = 0, \tag{1.4}$$

where ε_0 is the vacuum permittivity. By integrating the Vlasov equation for all $\boldsymbol{u} \in \mathbb{R}^3$, we get

$$\frac{\partial \rho}{\partial t} + \text{div}\, \boldsymbol{J} = 0. \tag{1.5}$$

This last relation, called the continuity equation, is crucial. Indeed, first it ensures that the Vlasov–Maxwell system is well-posed. Next, if it is right, relations (1.4) are verified at any time as soon as they are initially satisfied so that it only remains to solve evolution equations (1.3). This last property represents a great advantage in numerical simulations because Vlasov–Maxwell system can then be implemented without solving equations (1.4) and especially the Poisson equation for the electric potential. Unfortunately, classical PIC codes using Cloud-In-Cell algorithm do not satisfy the continuity equation locally and as a consequence, errors may appear in the irrotational part of the electric field in Gauss's law.

In order to keep these errors small enough such that the physics is correctly described, two main directions have been explored. The first deals with different formulations of the Maxwell system. Among the main ones we find the Boris formulation (hyperbolic-elliptic), which consists in discarding the irrotational part of the electric field and replacing it by the solution of the Poisson equation [1]; the Marder formulation (hyperbolic-parabolic), which consists in introducing a pseudo-current in Ampere's law [5]; and the hyperbolic formulation proposed by Munz, Schneider, Sonnendrücker and Voss in [6].

In this paper we rather focus on charge conserving current algorithms. The key point is then to ensure the continuity equation at a discrete level and locally i.e. for the charge and current density associated to each macro-particle. Several methods have been developed in order to satisfy it. They use an appropriate definition of the current density, naturally related to the variation of charge resulting from the particle motion. Charge conserving schemes were developed by Villasenor and Buneman [9] for simple shape of particles (a recent faster algorithm is given by [8]) and by Esirkepov [3] for a wider class of form factors.

The main goal of this work is to give an overview of this series of charge conserving current methods. To do this, we give a new viewpoint on Villasenor–Buneman's method and develop it for higher order spline form-factors. Then we give a detailed presentation of the Esirkepov scheme [3] noting that this algorithm is not a simple extension of the Villasenor–Buneman method.

The paper is organized as follows. In Section 2 we recall the usual PIC algorithm. Then, in Section 3, we detail the different charge conserving methods implemented in PIC codes. Finally, the last section is devoted to the study of computational efficiency and accuracy of these techniques. We study three test cases: the first one deals with order in form-factors and the two other ones show that the Esirkepov and Villasenor–Buneman methods well cure the unphysical effects appearing in the solution using the classical Cloud-In-Cell method.

2 The Particle-In-Cell method

In this section we present the numerical resolution of the Vlasov–Maxwell system (1.1)–(1.4) using a PIC method.

A typical cycle for one time step in Particle-In-Cell simulations is composed as follows.

Let us assume that the initial distribution function f_0 is represented by a collection of N macro-particles with positions $\boldsymbol{x}_\alpha^0$, momenta $\boldsymbol{u}_\alpha^0$ and weights p_α, such that

$$f_0(\boldsymbol{x}, \boldsymbol{u}) \approx f_0^h(\boldsymbol{x}, \boldsymbol{u}) = \sum_{\alpha=1}^{N} p_\alpha \delta(\boldsymbol{x} - \boldsymbol{x}_\alpha^0)\delta(\boldsymbol{u} - \boldsymbol{u}_\alpha^0).$$

Then (see [7]) the particle approximation f^h of f, solution of the Vlasov equation with initial data f_0^h, is given for all $t > 0$ by

$$f^h(\boldsymbol{x}, \boldsymbol{u}, t) = \sum_{\alpha=1}^{N} p_\alpha \delta(\boldsymbol{x} - \boldsymbol{x}_\alpha(t))\delta(\boldsymbol{u} - \boldsymbol{u}_\alpha(t)),$$

where the position and momentum of the αth particle $(\boldsymbol{x}_\alpha(t), \boldsymbol{u}_\alpha(t))$ are solutions of

$$\frac{d\boldsymbol{x}_\alpha}{dt} = \boldsymbol{v}_\alpha, \quad \boldsymbol{x}_\alpha(0) = \boldsymbol{x}_\alpha^0, \tag{2.1}$$

$$\frac{d\boldsymbol{u}_\alpha}{dt} = q\,(\boldsymbol{E}_\alpha(t) + \boldsymbol{v}_\alpha \times \boldsymbol{B}_\alpha(t)), \quad \boldsymbol{u}_\alpha(0) = \boldsymbol{u}_\alpha^0, \tag{2.2}$$

which are the characteristics of the Vlasov equation. $(\boldsymbol{E}_\alpha(t), \boldsymbol{B}_\alpha(t))$ is the approximation of the electromagnetic field acting on the αth particle.

Hence, the particle approximation of the charge and current densities are given by

$$\left.\begin{aligned} \rho^h(\boldsymbol{x}, t) &= q \sum_{\alpha=1}^{N} p_\alpha \delta(\boldsymbol{x} - \boldsymbol{x}_\alpha(t)), \\ \boldsymbol{J}^h(\boldsymbol{x}, t) &= q \sum_{\alpha=1}^{N} p_\alpha \boldsymbol{v}_\alpha(t) \delta(\boldsymbol{x} - \boldsymbol{x}_\alpha(t)). \end{aligned}\right\} \tag{2.3}$$

where q is the elementary charge.

The system (2.1)–(2.2) is numerically solved using a leapfrog scheme

$$\left.\begin{aligned} \frac{\boldsymbol{x}_\alpha^{n+1} - \boldsymbol{x}_\alpha^n}{\Delta t} &= \boldsymbol{v}_\alpha^{n+\frac{1}{2}}, \\ \frac{\boldsymbol{u}_\alpha^{n+\frac{1}{2}} - \boldsymbol{u}_\alpha^{n-\frac{1}{2}}}{\Delta t} &= q(\boldsymbol{E}_\alpha^n + \boldsymbol{v}_\alpha^n \times \boldsymbol{B}_\alpha^n), \end{aligned}\right\} \tag{2.4}$$

where $(\boldsymbol{E}_\alpha^n, \boldsymbol{B}_\alpha^n)$ is the field acting on the αth particle at time t^n. In order to compute it, we start solving Maxwell's equations (1.3) on a mesh.

Let us assume that we are looking for an approximation of the Vlasov–Maxwell system on a bounded domain $\Omega \in \mathbb{R}^2$. In two space dimensions, the equations uncouple in TM (tranverse magnetic) and TE (tranverse electric) modes. We are interested in the TE mode. Let us consider a grid cell of size $\Delta x \times \Delta y$. The components of electromagnetic field $(\boldsymbol{E}, \boldsymbol{B})$, charge density ρ and current density $\boldsymbol{J}$ are located at positions shown in Figure 1 and are taken at different times.

$$\boldsymbol{E}^n = \left((E_x)^n_{i+\frac{1}{2},j}, (E_y)^n_{i,j+\frac{1}{2}}\right), \qquad \boldsymbol{B}^{n+\frac{1}{2}} = (B_z)^{n+\frac{1}{2}}_{i+\frac{1}{2},j+\frac{1}{2}},$$

$$\rho^n = \rho^n_{i,j}, \qquad \boldsymbol{J}^{n+\frac{1}{2}} = \left((J_x)^{n+\frac{1}{2}}_{i+\frac{1}{2},j}, (J_y)^{n+\frac{1}{2}}_{i,j+\frac{1}{2}}\right).$$

We can now write the discrete version of system (1.3) given by the standard, second order in time and space, finite difference Yee scheme [10].

$$\frac{(E_x)^{n+1}_{i+\frac{1}{2},j} - (E_x)^{n}_{i+\frac{1}{2},j}}{\Delta t} = -\frac{1}{\varepsilon_0}(J_x)^{n+\frac{1}{2}}_{i+\frac{1}{2},j} + c^2 \frac{(B_z)^{n+\frac{1}{2}}_{i+\frac{1}{2},j+\frac{1}{2}} - (B_z)^{n+\frac{1}{2}}_{i+\frac{1}{2},j-\frac{1}{2}}}{\Delta y},$$

$$\frac{(E_y)^{n+1}_{i,j+\frac{1}{2}} - (E_y)^{n}_{i,j+\frac{1}{2}}}{\Delta t} = -\frac{1}{\varepsilon_0}(J_y)^{n+\frac{1}{2}}_{i,j+\frac{1}{2}} - c^2 \frac{(B_z)^{n+\frac{1}{2}}_{i+\frac{1}{2},j+\frac{1}{2}} - (B_z)^{n+\frac{1}{2}}_{i-\frac{1}{2},j+\frac{1}{2}}}{\Delta x},$$

$$\frac{(B_z)^{n+\frac{1}{2}}_{i+\frac{1}{2},j+\frac{1}{2}} - (B_z)^{n-\frac{1}{2}}_{i+\frac{1}{2},j+\frac{1}{2}}}{\Delta t} = -\frac{(E_y)^{n}_{i+1,j+\frac{1}{2}} - (E_y)^{n}_{i,j+\frac{1}{2}}}{\Delta x} + \frac{(E_x)^{n}_{i+\frac{1}{2},j+1} - (E_x)^{n}_{i+\frac{1}{2},j}}{\Delta y}.$$

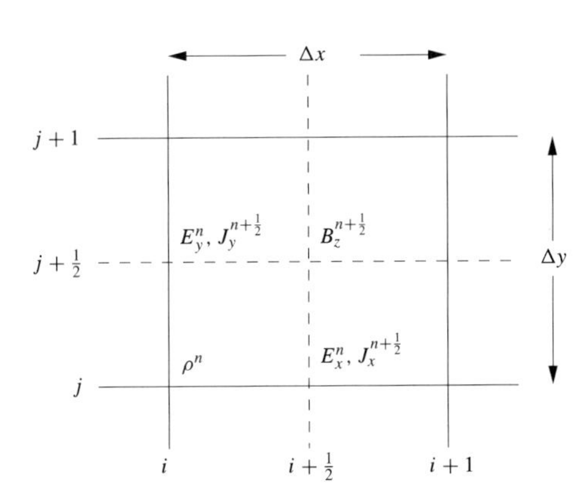

Figure 1. Locations of the field components on the spatial grid.

The current densities J_x, J_y need to be computed at the grid points $(X_{i+1/2}, Y_j)$ and $(X_i, Y_{j+1/2})$. They result from the superposition of the charge flux contributed by each particle. One of the most natural assignment of particle charge is the Cloud-In-Cell one also called "area weighting". This is a first order assignment consisting in a linear weighting to the nearest four grid points. The idea consists in representing each macro-particle by a square cloud of size $\Delta x \times \Delta y$, centered at the particle location and such that its total charge q is uniformly distributed. Then the charge and current densities $\boldsymbol{J}$ at each grid point are computed from velocities and positions of charged particles with the following charge assignment procedure:

$$
\begin{aligned}
(\rho)^n_{i,j} &= \sum_{\alpha=1}^{N} q\, S^1(X_i - x^n_\alpha, Y_j - y^n_\alpha),\\
(J_x)^{n+\frac{1}{2}}_{i,j} &= \sum_{\alpha=1}^{N} q\, (v_x)^{n+\frac{1}{2}}_\alpha\, S^1\big(X_i - x^{n+\frac{1}{2}}_\alpha, Y_j - y^{n+\frac{1}{2}}_\alpha\big),\\
(J_y)^{n+\frac{1}{2}}_{i,j} &= \sum_{\alpha=1}^{N} q\, (v_y)^{n+\frac{1}{2}}_\alpha\, S^1\big(X_i - x^{n+\frac{1}{2}}_\alpha, Y_j - y^{n+\frac{1}{2}}_\alpha\big),
\end{aligned}
\tag{2.5}
$$

where S^1 is the first order form factor. J_x and J_y are then averaged at their right location. In the $2D$ case, the form-factors we use are the tensor product of one dimensional splines such that

$$S^1(x, y) = S^1_{\Delta x}(x) S^1_{\Delta y}(y)$$

where

$$
S^1_k(x) = \begin{cases} \frac{1}{k}\left(1 - \left|\frac{x}{k}\right|\right) & \text{if } |x| < k,\\ 0 & \text{else.} \end{cases}
$$

Remark 2.1. Let us note that the first-order form-factor S^1_k can be expressed as the convolution of zero-order shape-factors. Indeed, we have $S^1_k(x) = S^0_k(x) * S^0_k(x)$ and, more generally, for all $m \in \mathbb{N}^*$,

$$S^m_k(x) = S^0_k(x) * S^{m-1}_k(x) = \frac{1}{k}\int_{x-\frac{k}{2}}^{x+\frac{k}{2}} S^{m-1}_k(u)du$$

with

$$
S^0_k(x) = \begin{cases} \frac{1}{k} & \text{if } |x| < \frac{k}{2},\\ 0 & \text{else.} \end{cases}
$$

Once the fields are known at the grid points, we can deduce the electromagnetic field acting on the αth particle by the following interpolation procedure:

$$
\begin{aligned}
E^n_\alpha &= \Delta x \Delta y \sum_{\alpha=1}^{N} E^n_{i,j} S^1(X_i - x^n_\alpha, Y_j - y^n_\alpha),\\
B^n_\alpha &= \Delta x \Delta y \sum_{\alpha=1}^{N} B^n_{i,j} S^1(X_i - x^n_\alpha, Y_j - y^n_\alpha),
\end{aligned}
\tag{2.6}
$$

where $E^n_{i,j}$ and $B^n_{i,j}$ are obtained by space averaging, and finally, we solve the motion equations of the particles.

Let us denote that the same shape-factor is used both for the assignment and interpolation step in the PIC method in order to avoid a self-force, i.e. a particle accelerating itself. Nevertheless, the current densities obtained with this area-weighting scheme do not satisfy the charge conservation equation exactly. In order to remedy this, we are looking for a scheme such that if $\boldsymbol{E}$ is generated time step after time step from the finite-difference version of (1.3) and if we calculate $\nabla \cdot \boldsymbol{E}$ then the resulting ρ should satisfy the charge conservation law (1.5) exactly in this finite difference version.

Therefore, several numerical techniques for solving the continuity equation have been developed. These methods are based on the following remark: a charge flux of a particle can be computed from the start and end points of the particle motion. Thus, when a particle moves across the mesh cells, the particle motion is split into several straight line motions, each one assigned to a cell. From the superposition of charge flux, charge conservation can be realized for any particle trajectories made up of straight line segments between any start and end points.

In the next section we detail the two charge conservation schemes we are interested in.

3 Charge conserving schemes in Particle-In-Cell methods

3.1 The Villasenor–Buneman method

In the Villasenor–Buneman method, a particle trajectory over one time step is assumed to be a straight line such that for all $t \in [t^n, t^{n+1}]$, we have

$$x_\alpha(t) = x^n_\alpha + (t - t^n)(v_x^{n+1/2})_\alpha,$$

$$y_\alpha(t) = y^n_\alpha + (t - t^n)(v_y^{n+1/2})_\alpha,$$

where $((v_x^{n+1/2})_\alpha, (v_y^{n+1/2})_\alpha)$ is the αth particle velocity. Moreover, from the C.F.L. condition (coming from the Maxwell solver) we have that

$$\Delta t < \frac{\sqrt{\Delta x^2 + \Delta y^2}}{c},$$

and thus a particle cannot cover a distance longer than one cell size. As in the area weighting method, each macro-particle is represented by a rectangular cloud of size $\Delta x \times \Delta y$, centered at the particle, with its total charge q uniformly distributed. This charged cloud is in fact the graph of the function $(x, y) \in \Omega \mapsto q S^0_{\Delta x}(x - x(t)) S^0_{\Delta y}(y - y(t))$. Following the area weighting method, the charge density computed at the grid point (X_i, Y_j) is

$$\rho^n_{i,j} = q \sum_{\alpha=1}^{N} p_\alpha S^1_{\Delta x}(X_i - x^n_\alpha) S^1_{\Delta y}(Y_j - y^n_\alpha) \tag{3.1}$$

which can be rewritten using Remark (2.1) as

$$\rho_{i,j}^n = q \sum_{\alpha=1}^{N} p_\alpha \frac{1}{\Delta x \Delta y} \int_{X_{i-\frac{1}{2}}}^{X_{i+\frac{1}{2}}} \int_{Y_{j-\frac{1}{2}}}^{Y_{j+\frac{1}{2}}} S_{\Delta x}^0(x - x_\alpha^n) S_{\Delta y}^0(y - y_\alpha^n)\, dx\, dy. \tag{3.2}$$

As the current density at a point is given by the sum of all particle contributions to this density, let us consider the current density created by a single particle α. In the Villasenor–Buneman algorithm, three kinds of particle motions are considered. Let us start with the simplest one, called "four boundaries motion" which occurs when the rectangular cloud of the αth particle crosses four interfaces of the dual mesh as shown in Figure 2. Now, let us focus on the current $(J_x)_{i+1/2,j}^n$. It results from the flux of

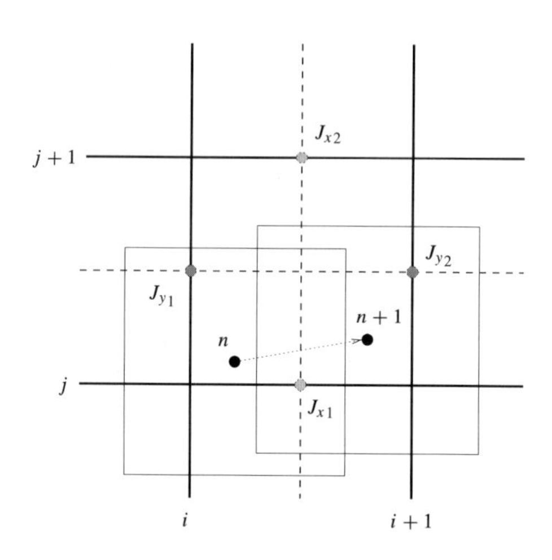

Figure 2. 4 boundaries motion.

current through the interface $\Gamma = \{X_{i+1/2}\} \times [Y_{j-1/2}, Y_{j+1/2}]$ between t^n and t^{n+1}, such that

$$\begin{aligned} J_{x\,i+\frac{1}{2},j}^{\;n+\frac{1}{2}} &= \frac{qp_\alpha}{\Delta t} \int_{t^n}^{t^{n+1}} \frac{1}{|\Gamma|} \int_\Gamma \boldsymbol{v}_\alpha(t) \cdot \boldsymbol{n}\, S_{\Delta x}^0(x - x_\alpha(t)) S_{\Delta y}^0(y - y_\alpha(t))\, d\sigma(x,y)\, dt, \\ &= \frac{qp_\alpha}{\Delta t \Delta y} \int_{t^n}^{t^{n+1}} \int_{Y_{j-\frac{1}{2}}}^{Y_{j+\frac{1}{2}}} (v_x^{n+\frac{1}{2}})_\alpha S_{\Delta x}^0(X_{i+\frac{1}{2}} - x_\alpha(t)) S_{\Delta y}^0(y - y_\alpha(t))\, dy\, dt, \\ &= \frac{qp_\alpha (v_x^{n+\frac{1}{2}})_\alpha}{\Delta t} \int_{t^n}^{t^{n+1}} S_{\Delta x}^0(X_{i+\frac{1}{2}} - x_\alpha(t)) S_{\Delta y}^1(Y_j - y_\alpha(t))\, dt. \end{aligned} \tag{3.3}$$

In this case of motion, both the start and end points of the αth particle motion are located in the same cell, that is $X_i \le x(t) \le X_{i+1}$ and $Y_j \le y(t) \le Y_{j+1}$, for all

$t^n \leq t \leq t^{n+1}$. Then we have

$$\begin{aligned}
J_{x\,i+\frac{1}{2},j}^{\;n+\frac{1}{2}} &= \frac{qp_\alpha \left(v_x^{n+\frac{1}{2}}\right)_\alpha}{\Delta t \Delta x} \int_{t^n}^{t^{n+1}} S^1_{\Delta y}(Y_j - y_\alpha(t))\, dt \\
&= \frac{qp_\alpha \left(v_x^{n+\frac{1}{2}}\right)_\alpha}{\Delta t \Delta x \Delta y} \int_{t^n}^{t^{n+1}} \left(1 + j - \frac{y_\alpha(t)}{\Delta y}\right) dt \\
&= \frac{qp_\alpha \left(v_x^{n+\frac{1}{2}}\right)_\alpha}{\Delta t \Delta x \Delta y} \int_0^{\Delta t} \left(1 + j - \frac{y_\alpha^n}{\Delta y} - t\frac{\left(v_y^{n+\frac{1}{2}}\right)_\alpha}{\Delta y}\right) dt \\
&= \frac{qp_\alpha \left(v_x^{n+\frac{1}{2}}\right)_\alpha}{\Delta x \Delta y} \left(1 + j - \frac{y_\alpha^n}{\Delta y} - \frac{\Delta t}{2}\frac{\left(v_y^{n+\frac{1}{2}}\right)_\alpha}{\Delta y}\right) \\
&= \frac{qp_\alpha}{\Delta x \Delta y}\left(\frac{x_\alpha^{n+1} - x_\alpha^n}{\Delta t}\right)\left(1 + j - \frac{y_\alpha^{n+1} + y_\alpha^n}{2\Delta y}\right) \\
&= \frac{qp_\alpha}{\Delta t \Delta y}\mathcal{S},
\end{aligned}$$

where $\mathcal{S} = \frac{x_\alpha^{n+1} - x_\alpha^n}{\Delta x}\left(1 + j - \frac{y_\alpha^{n+1}+y_\alpha^n}{2\Delta y}\right)$ is the hatched area in Figure 3 representing the quantity of charge having crossed the border Γ during time $[t^n, t^{n+1}]$. We find

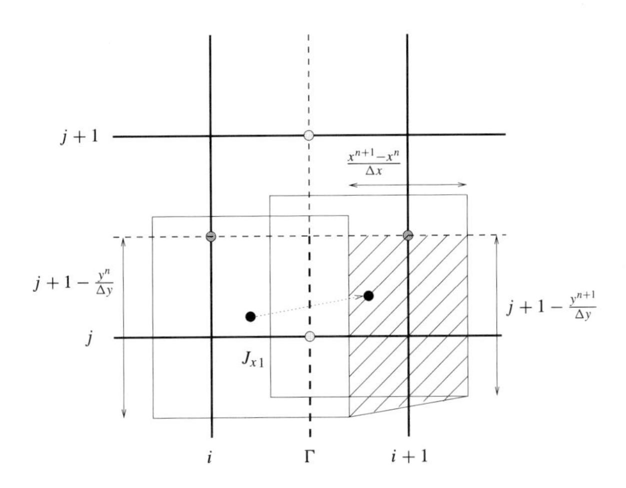

Figure 3. Computation of $J_{x\,i+\frac{1}{2},j}^{\;n+\frac{1}{2}}$.

thus the relation given in [9]. A similar computation gives

$$J_{x_{i+\frac{1}{2},j+1}}^{n+\frac{1}{2}} = \frac{qp_\alpha}{\Delta x \Delta y}\left(\frac{x^{n+1}-x^n}{\Delta t}\right)\left(\frac{y^n+y^{n+1}}{2\Delta y}-j\right),$$

$$J_{y_{i,j+\frac{1}{2}}}^{n+\frac{1}{2}} = \frac{qp_\alpha}{\Delta x \Delta y}\left(\frac{y^{n+1}-y^n}{\Delta t}\right)\left(i+1-\frac{x^n+x^{n+1}}{2\Delta x}\right),$$

$$J_{y_{i+1,j+\frac{1}{2}}}^{n+\frac{1}{2}} = \frac{qp_\alpha}{\Delta x \Delta y}\left(\frac{y^{n+1}-y^n}{\Delta t}\right)\left(\frac{x^n+x^{n+1}}{2\Delta x}-i\right).$$

The second kind of motion is the "seven boundaries motion" which occurs when the αth particle crosses one interface of the mesh, and the cloud crosses seven interfaces of the dual mesh. Four different cases of "seven boundaries motion" depending on the direction can occur. We give an example in Figure 4.

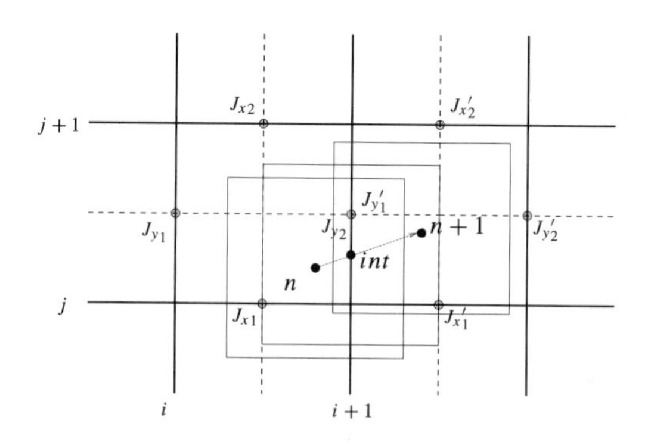

Figure 4. 7 boundaries motion.

The seven boundaries motion can be treated as two four boundaries ones. Let $\boldsymbol{x}^{\text{int}} = (x^{\text{int}}, y^{\text{int}})$ be the intersection between the interface of the mesh and the αth particle trajectory. The first four boundaries motion from $\boldsymbol{x}^n$ to $\boldsymbol{x}^{\text{int}}$ produces 4 currents denoted by J_{x1}, J_{x2}, J_{y1}, J_{y2} in Figure 4. The second four boundaries motion from $\boldsymbol{x}^{\text{int}}$ to $\boldsymbol{x}^{n+1}$ produces 4 others currents denoted by J'_{x1}, J'_{x2}, J'_{y1}, J'_{y2}. They satisfy

$$J_{x_{i+\frac{1}{2},j}}^{n+\frac{1}{2}} = J_{x1}, \quad J_{x_{i+\frac{1}{2},j+1}}^{n+\frac{1}{2}} = J_{x2},$$

$$J_{x_{i+\frac{3}{2},j}}^{n+\frac{1}{2}} = J'_{x1}, \quad J_{x_{i+\frac{3}{2},j+1}}^{n+\frac{1}{2}} = J'_{x2},$$

and moreover

$$J_{y_{i,j+\frac{1}{2}}}^{n+\frac{1}{2}} = J_{y1}, \quad J_{y_{i+1,j+\frac{1}{2}}}^{n+\frac{1}{2}} = J_{y2}+J'_{y1}, \quad J_{y_{i+2,j+\frac{1}{2}}}^{n+\frac{1}{2}} = J'_{y2}$$

where J comes again from the flux of current through the associated interface. In the example given in Figure 4, we have $X_i \leq x^n < X_{i+1}$, $X_{i+1} \leq x^{n+1} < X_{i+2}$

and $Y_j \leq y(t) < Y_{j+1}$ for all $t^n \leq t < t^{n+1}$. Then let $t^n \leq t_0 < t^{n+1}$ such that $x(t_0) = X_{i+1}$, and let us set $y^{\text{int}} = y(t_0)$. Then $X_i \leq x(t) < X_{i+1}$ for all $t^n \leq t < t_0$ and $X_i \leq x(t) < X_{i+1}$ for all $t_0 \leq t < t^{n+1}$. Therefore, on the one hand we get

$$\begin{aligned}
J_{x_{i+\frac{1}{2},j}}^{n+\frac{1}{2}} &= \frac{qp_\alpha}{\Delta t}\int_{t^n}^{t^{n+1}} \frac{\left(v_x^{n+\frac{1}{2}}\right)_\alpha}{\Delta y}\int_{Y_{j-\frac{1}{2}}}^{Y_{j+\frac{1}{2}}} S_{\Delta x}^0(X_{i+\frac{1}{2}} - x_\alpha(t))S_{\Delta y}^0(y - y_\alpha(t))\,dy\,dt \\
&= qp_\alpha\frac{1}{\Delta t}\int_{t^n}^{t^{n+1}} \left(v_x^{n+\frac{1}{2}}\right)_\alpha S_{\Delta x}^0(X_{i+\frac{1}{2}} - x_\alpha(t))S_{\Delta y}^1(Y_j - y_\alpha(t))\,dt \\
&= qp_\alpha\frac{1}{\Delta t}\int_{t^n}^{t_0} v_x^{n+\frac{1}{2}} S_{\Delta y}^1(Y_j - y_\alpha(t))\,dt = J_{x1},
\end{aligned}$$

and on the other hand

$$\begin{aligned}
J_{y_{i+1,j+\frac{1}{2}}}^{n+\frac{1}{2}} &= \frac{qp_\alpha}{\Delta t}\int_{t^n}^{t^{n+1}} \frac{\left(v_y^{n+\frac{1}{2}}\right)_\alpha}{\Delta x}\int_{X_{i+\frac{1}{2}}}^{X_{i+\frac{3}{2}}} S_{\Delta x}^0(x - x_\alpha(t))S_{\Delta y}^0(Y_{j+\frac{1}{2}} - y_\alpha(t))\,dx\,dt \\
&= qp_\alpha\frac{1}{\Delta t}\int_{t^n}^{t^{n+1}} \left(v_y^{n+\frac{1}{2}}\right)_\alpha S_{\Delta x}^1(X_{i+1} - x_\alpha(t))S_{\Delta y}^0(Y_{j+\frac{1}{2}} - y_\alpha(t))\,dt \\
&= \frac{qp_\alpha}{\Delta t}\int_{t^n}^{t_0} \left(v_y^{n+\frac{1}{2}}\right)_\alpha S_{\Delta x}^1(X_{i+1} - x_\alpha(t))\,dt \\
&\quad + \frac{qp_\alpha}{\Delta t}\int_{t_0}^{t^{n+1}} \left(v_y^{n+\frac{1}{2}}\right)_\alpha S_{\Delta x}^1(X_{i+1} - x_\alpha(t))\,dt \\
&= J_{y2} + J'_{y1}.
\end{aligned}$$

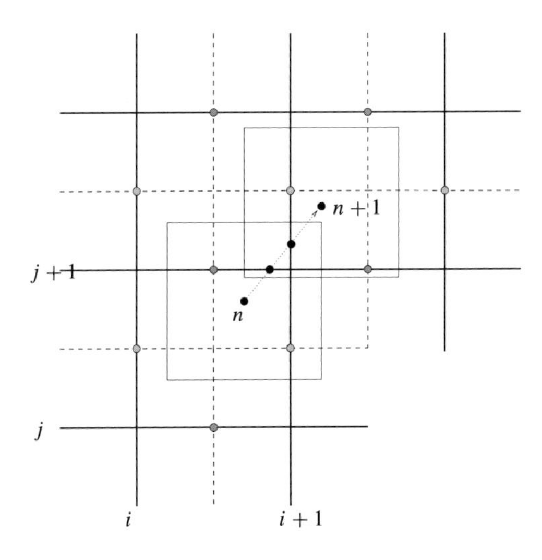

Figure 5. 10 boundaries motion.

The last case is called the "ten boundaries motion" occurring when two interfaces of the mesh are crossed by the αth particle (see Figure 5 for instance). The trajectory is then decomposed in three four boundaries motions and the current J is computed in the same way as in the two previous cases.

Since we are interested in high order charge conserving algorithms, we propose an extension of the Villasenor–Buneman method to higher order spline shape factors. Using Remark (2.1) and the relation (3.3), we get the following result.

Proposition 3.1. *For all $m \in \mathbb{N}^*$, the charge and current densities defined for all i, j by*

$$\rho_{i,j}^n = q \sum_{\alpha=1}^{N} p_\alpha S_{\Delta x}^m (X_i - x_\alpha^n) S_{\Delta y}^m (Y_j - y_\alpha^n), \tag{3.4}$$

$$J_{x\,i+\frac{1}{2},j}^{\,n+\frac{1}{2}} = q \sum_{\alpha=1}^{N} p_\alpha \frac{(v_x^{n+\frac{1}{2}})_\alpha}{\Delta t} \int_{t^n}^{t^{n+1}} S_{\Delta x}^{m-1} (X_{i+\frac{1}{2}} - x_\alpha(t)) S_{\Delta y}^m (Y_j - y_\alpha(t))\, dt, \tag{3.5}$$

$$J_{y\,i,j+\frac{1}{2}}^{\,n+\frac{1}{2}} = q \sum_{\alpha=1}^{N} p_\alpha \frac{(v_y^{n+\frac{1}{2}})_\alpha}{\Delta t} \int_{t^n}^{t^{n+1}} S_{\Delta x}^{m} (X_i - x_\alpha(t)) S_{\Delta y}^{m-1} (Y_{j+\frac{1}{2}} - y_\alpha(t))\, dt, \tag{3.6}$$

satisfy the discrete equation of charge conservation.

Proof. By additivity, we only have to show the charge conservation for a single particle α.

$$\begin{aligned}
&\rho_{i,j}^{n+1} - \rho_{i,j}^n \\
&\quad = \int_{t^n}^{t^{n+1}} \frac{d}{dt} \rho_{i,j}(t)\, dt \\
&\quad = \frac{q\, p_\alpha}{\Delta x \Delta y} \int_{t^n}^{t^{n+1}} \frac{d}{dt} \int_{X_{i-\frac{1}{2}}}^{X_{i+\frac{1}{2}}} \int_{Y_{j-\frac{1}{2}}}^{Y_{j+\frac{1}{2}}} S_{\Delta x}^{m-1}(x - x_\alpha(t)) S_{\Delta y}^{m-1}(y - y_\alpha(t))\, dx\, dy\, dt \\
&\quad = \frac{q\, p_\alpha}{\Delta x \Delta y} \int_{t^n}^{t^{n+1}} \int_{X_{i-\frac{1}{2}}}^{X_{i+\frac{1}{2}}} \int_{Y_{j-\frac{1}{2}}}^{Y_{j+\frac{1}{2}}} \frac{d}{dt} \left(S_{\Delta x}^{m-1}(x - x_\alpha(t)) S_{\Delta y}^{m-1}(y - y_\alpha(t)) \right) dx\, dy\, dt \\
&\quad = \frac{q\, p_\alpha}{\Delta x \Delta y} \int_{t^n}^{t^{n+1}} \int_{X_{i-\frac{1}{2}}}^{X_{i+\frac{1}{2}}} \int_{Y_{j-\frac{1}{2}}}^{Y_{j+\frac{1}{2}}} \Big(-(v_x^{n+\frac{1}{2}})_\alpha \frac{d S_{\Delta x}^{m-1}}{dx}(x - x_\alpha(t)) S_{\Delta y}^{m-1}(y - y_\alpha(t)) \\
&\qquad - \left(v_y^{n+\frac{1}{2}}\right)_\alpha S_{\Delta x}^{m-1}(x - x_\alpha(t)) \frac{d S_{\Delta y}^{m-1}}{dx}(y - y_\alpha(t)) \Big)\, dx\, dy\, dt
\end{aligned}$$

$$\rho_{i,j}^{n+1} - \rho_{i,j}^{n} = -q\, p_\alpha \frac{\left(v_x^{n+\frac{1}{2}}\right)_\alpha}{\Delta x \Delta y} \int_{t^n}^{t^{n+1}} \int_{Y_{j-\frac{1}{2}}}^{Y_{j+\frac{1}{2}}} S_{\Delta y}^{m-1}(y - y_\alpha(t))$$

$$\left(S_{\Delta x}^{m-1}(X_{i+\frac{1}{2}} - x_\alpha(t)) - S_{\Delta x}^{m-1}(X_{i-\frac{1}{2}} - x_\alpha(t))\right) dy\, dt$$

$$- q\, p_\alpha \frac{\left(v_y^{n+\frac{1}{2}}\right)_\alpha}{\Delta x \Delta y} \int_{t^n}^{t^{n+1}} \int_{X_{i-\frac{1}{2}}}^{X_{i+\frac{1}{2}}} S_{\Delta x}^{m-1}(x - x_\alpha(t))$$

$$\left(S_{\Delta y}^{m-1}(Y_{j+\frac{1}{2}} - y_\alpha(t)) - S_{\Delta y}^{m-1}(Y_{j-\frac{1}{2}} - y(t))\right) dx\, dt$$

$$= -q\, p_\alpha \frac{\left(v_x^{n+\frac{1}{2}}\right)_\alpha}{\Delta x} \int_{t^n}^{t^{n+1}} S_{\Delta y}^{m}(Y_j - y_\alpha(t))\big(S_{\Delta x}^{m-1}(X_{i+\frac{1}{2}} - x_\alpha(t))$$

$$- S_{\Delta x}^{m-1}(X_{i-\frac{1}{2}} - x_\alpha(t))\big)\, dt$$

$$- q\, p_\alpha \frac{\left(v_y^{n+\frac{1}{2}}\right)_\alpha}{\Delta y} \int_{t^n}^{t^{n+1}} S_{\Delta x}^{m}(X_i - x_\alpha(t))\big(S_{\Delta y}^{m-1}(Y_{j+\frac{1}{2}} - y_\alpha(t))$$

$$- S_{\Delta y}^{m-1}(Y_{j-\frac{1}{2}} - y_\alpha(t))\big)\, dt$$

$$= -\Delta t \left(\frac{1}{\Delta x}(J_{x\,i+\frac{1}{2},j}^{\,n} - J_{x\,i-\frac{1}{2},j}^{\,n}) + \frac{1}{\Delta y}(J_{y\,i,j+\frac{1}{2}}^{\,n} - J_{y\,i,j-\frac{1}{2}}^{\,n}) \right). \qquad \square$$

We can now describe the second order Villasenor–Buneman method. Let us represent the particle as a charged cloud whose charge distribution is given by the first order shape factor (see Figure 6). Then the charge density is given by

$$\rho_{i,j}^{n} = q p_\alpha\, S_{\Delta x}^{2}(X_i - x_\alpha^n) S_{\Delta y}^{2}(Y_j - y_\alpha^n)$$

$$= q\, \frac{1}{\Delta x\, \Delta y} \int_{X_{i-\frac{1}{2}}}^{X_{i+\frac{1}{2}}} \int_{Y_{j-\frac{1}{2}}}^{Y_{j+\frac{1}{2}}} S_{\Delta x}^{1}(x - x_\alpha^n) S_{\Delta y}^{1}(y - y_\alpha^n)\, dx\, dy$$

that is, the integral of the charged cloud over the dual grid cell, centered in (X_i, Y_j). A particle contributes now to charge density at the nine nearest grid points. Like in the first order case, there will be three types of deplacements: 12, 17 and 22 boundaries motions, which reduce respectively to one, two and three 12 boundaries motions (see Figure 7).

Let us treat this “12 boundaries” motion. By denoting

$$a = i + \frac{1}{2} - \frac{x^n}{\Delta x}, \quad b = j + \frac{1}{2} - \frac{y^n}{\Delta y},$$

$$c = \frac{1}{2} - i + \frac{x^n}{\Delta x}, \quad d = \frac{1}{2} - j + \frac{y^n}{\Delta y},$$

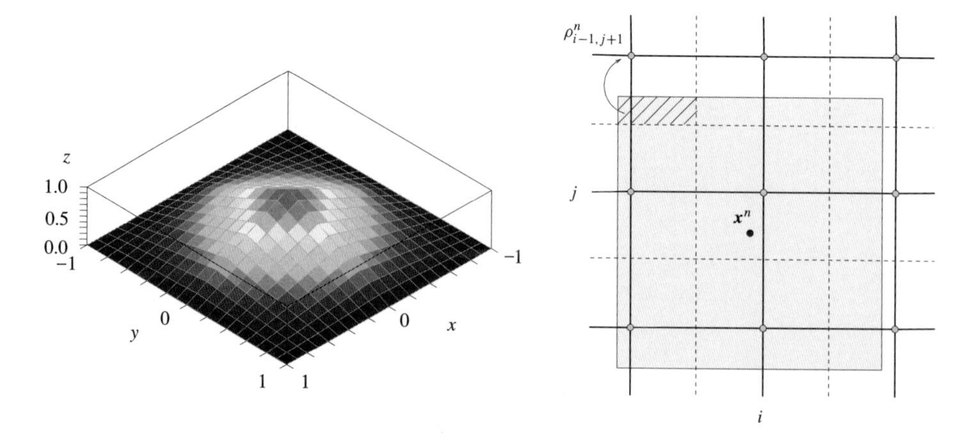

(a) Charged cloud ($m = 2$) or first order shape factor.

(b) The particle contributes to charge density at the nine nearest gridpoints.

Figure 6. Second order cloud.

and

$$\delta x = \frac{x_\alpha^{n+1} - x_\alpha^n}{\Delta x}, \qquad \delta y = \frac{y_\alpha^{n+1} - y_\alpha^n}{\Delta y},$$

we obtain after calculation (see Figure 7)

$$J_{x1} = \frac{q p_\alpha (v_x^{n+\frac{1}{2}})_\alpha}{2\Delta x \Delta y} \left(\left(a - \frac{\delta x}{2}\right) b^2 + \left(-a + \frac{2\delta x}{3}\right) \delta y\, b + \left(\frac{a}{3} - \frac{\delta x}{4}\right) \delta y^2 \right),$$

$$J_{x2} = \frac{q p_\alpha (v_x^{n+\frac{1}{2}})_\alpha}{2\Delta x \Delta y} \left(\left(c + \frac{\delta x}{2}\right) b^2 + \left(-c - \frac{2\delta x}{3}\right) \delta y\, b + \left(\frac{c}{3} + \frac{\delta x}{4}\right) \delta y^2 \right),$$

$$J_{x3} = \frac{q p_\alpha (v_x^{n+\frac{1}{2}})_\alpha}{\Delta x \Delta y} \left(\left(a - \frac{\delta x}{2}\right) \left(\frac{1}{2} + bd\right) + \left(a - \frac{2\delta x}{3}\right) \delta y \frac{b-d}{2} + \left(-\frac{a}{3} + \frac{\delta x}{4}\right) \delta y^2 \right),$$

$$J_{x4} = \frac{q p_\alpha (v_x^{n+\frac{1}{2}})_\alpha}{\Delta x \Delta y} \left(\left(c + \frac{\delta x}{2}\right) \left(\frac{1}{2} + bd\right) + \left(c + \frac{2\delta x}{3}\right) \delta y \frac{b-d}{2} + \left(-\frac{c}{3} - \frac{\delta x}{4}\right) \delta y^2 \right),$$

$$J_{x5} = \frac{q p_\alpha (v_x^{n+\frac{1}{2}})_\alpha}{2\Delta x \Delta y} \left(\left(a - \frac{\delta x}{2}\right) d^2 + \left(a - \frac{2\delta x}{3}\right) \delta y\, d + \left(\frac{a}{3} - \frac{\delta x}{4}\right) \delta y^2 \right),$$

$$J_{x6} = \frac{q p_\alpha (v_x^{n+\frac{1}{2}})_\alpha}{2\Delta x \Delta y} \left(\left(c + \frac{\delta x}{2}\right) d^2 + \left(c + \frac{2\delta x}{3}\right) \delta y\, d + \left(\frac{c}{3} + \frac{\delta x}{4}\right) \delta y^2 \right).$$

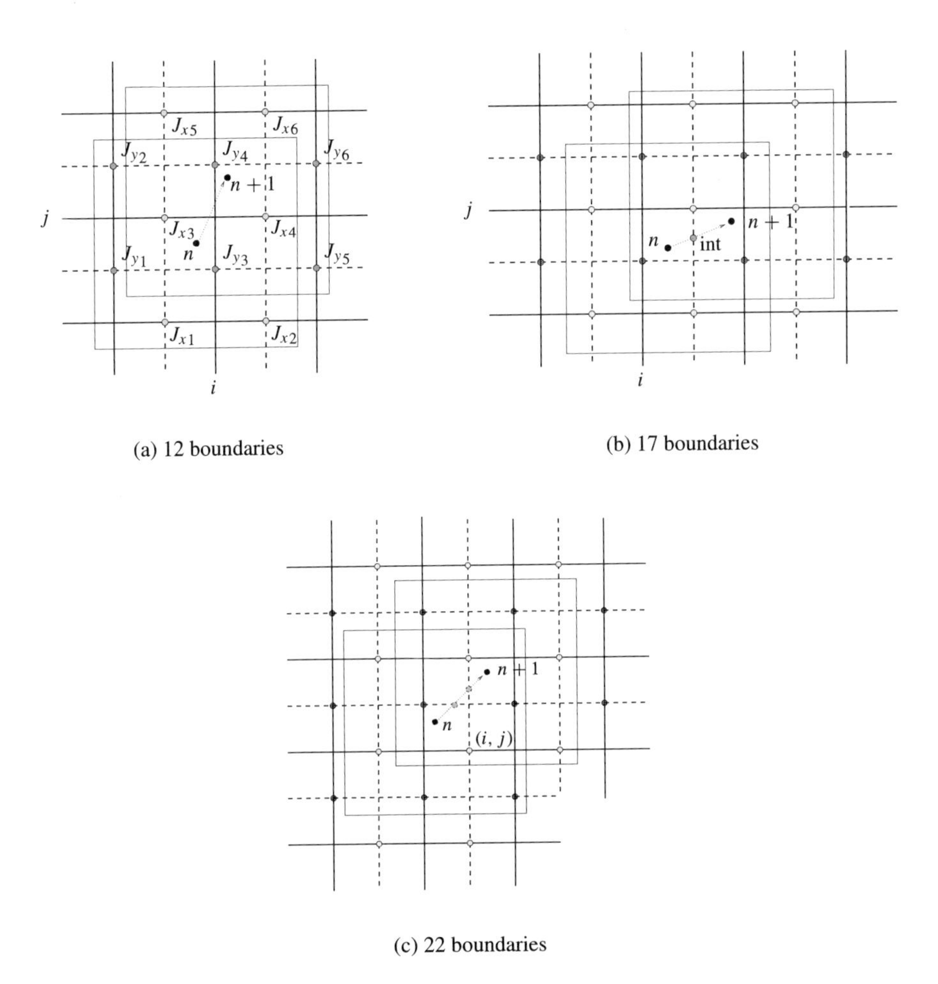

(a) 12 boundaries

(b) 17 boundaries

(c) 22 boundaries

Figure 7. Different possible motion types for second order shape factors.

The results in J_y are symmetric. We are going to compare this method to Esirkepov's one.

3.2 The Esirkepov method

This scheme also called "Density Decomposition scheme" is valid for an arbitrary form-factor S and is limited to Cartesian geometry. As in the Villasenor–Buneman method, the macro particle trajectory over one time step is again assumed to be a straight line but will be decomposed into four trajectories along the x and y axes (see

Figure 8). Let us consider the continuity equation (1.5) in finite differences, we get

$$\frac{\rho_{i,j}^{n+1} - \rho_{i,j}^{n}}{\Delta t} + \frac{(J_x)_{i+\frac{1}{2},j}^{n+\frac{1}{2}} - (J_x)_{i-\frac{1}{2},j}^{n+\frac{1}{2}}}{\Delta x} + \frac{(J_y)_{i,j+\frac{1}{2}}^{n+\frac{1}{2}} - (J_y)_{i,j-\frac{1}{2}}^{n+\frac{1}{2}}}{\Delta y} = 0.$$

Due to the linearity of this equation, it is sufficient to define the current density associated with the motion of a single macro particle. Therefore, let us consider a single macro particle located at (x^n, y^n) and let us introduce the vector $\boldsymbol{W}$ of density decomposition which satisfies:

$$\frac{(J_x)_{i+\frac{1}{2},j} - (J_x)_{i-\frac{1}{2},j}}{\Delta x} = -\frac{q}{\Delta t}(W_x)_{i,j}, \tag{3.7}$$

$$\frac{(J_y)_{i,j+\frac{1}{2}} - (J_y)_{i,j-\frac{1}{2}}}{\Delta y} = -\frac{q}{\Delta t}(W_y)_{i,j}. \tag{3.8}$$

Then the discrete charge conservation is satisfied if and only if

$$q\,((W_x)_{i,j} + (W_y)_{i,j}) = \rho_{i,j}^{n+1} - \rho_{i,j}^{n},$$

that is

$$(W_x)_{i,j} + (W_y)_{i,j} = S(X_i - x^{n+1}, Y_j - y^{n+1}) - S(X_i - x^n, Y_j - y^n). \tag{3.9}$$

Since (see Figure 8)

$$\begin{aligned} S(X_i - x^{n+1}, Y_j - y^{n+1}) &- S(X_i - x^n, Y_j - y^n) \\ &= S(X_i - x^{n+1}, Y_j - y^{n+1}) - S(X_i - x^{n+1}, Y_j - y^n) \\ &\quad + S(X_i - x^{n+1}, Y_j - y^n) - S(X_i - x^n, Y_j - y^n) \\ &= S(X_i - x^{n+1}, Y_j - y^{n+1}) - S(X_i - x^n, Y_j - y^{n+1}) \\ &\quad + S(X_i - x^n, Y_j - y^{n+1}) - S(X_i - x^n, Y_j - y^n) \end{aligned}$$

we assume that the vector $\boldsymbol{W}$ satisfies the following properties:

- $\boldsymbol{W}$ is a linear combination of functions

$$\begin{aligned} &S(X_i - x^n, Y_j - y^n), &&S(X_i - x^{n+1}, Y_j - y^n), \\ &S(X_i - x^n, Y_j - y^{n+1}), &&S(X_i - x^{n+1}, Y_j - y^{n+1}), \end{aligned} \tag{3.10}$$

- The sum of components of the vector $\boldsymbol{W}$ satisfies the relation (3.9),

- If the motion is equal to zero in direction x, then $W_x = 0$ and respectively in the y direction then $W_y = 0$,

- If $S(x, y)$ is symmetrical with respect to the permutation (x, y) and the equality $(x^{n+1} - x^n)/\Delta x = (y^{n+1} - y^n)/\Delta y$ holds, then $W_x = W_y$.

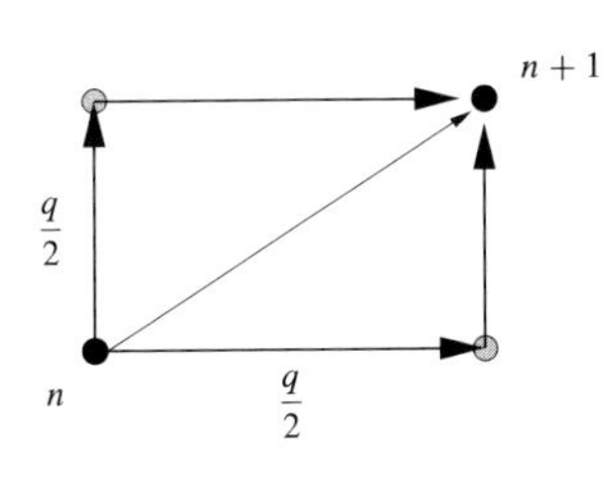

Figure 8. Esirkepov's motion decomposition.

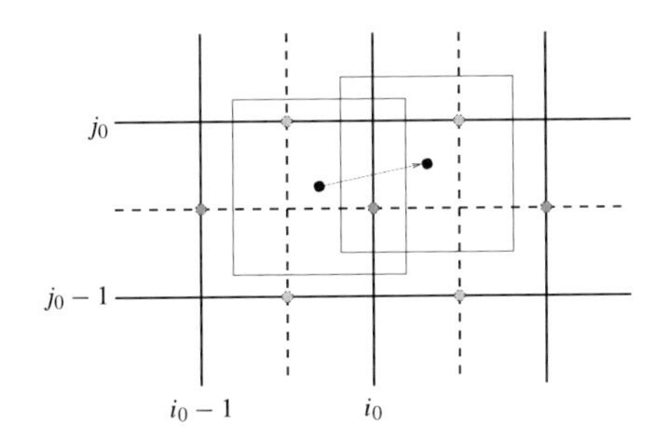

Figure 9. Example of motion.

Then (see [3]) only one linear combination of the functions defined by (3.10) obeys these conditions, which is

$$\begin{aligned}(W_x)_{i,j} &= \frac{1}{2}(S(X_i - x^{n+1}, Y_j - y^{n+1}) - S(X_i - x^n, Y_j - y^{n+1})) \\ &\quad + \frac{1}{2}(S(X_i - x^{n+1}, Y_j - y^n) - S(X_i - x^n, Y_j - y^n)), \\ (W_y)_{i,j} &= \frac{1}{2}(S(X_i - x^{n+1}, Y_j - y^{n+1}) - S(X_i - x^{n+1}, Y_j - y^n)) \\ &\quad + \frac{1}{2}(S(X_i - x^n, Y_j - y^{n+1}) - S(X_i - x^n, Y_j - y^n)).\end{aligned}$$

The current $\boldsymbol{J}^{n+\frac{1}{2}}$ is finally reconstructed from relations (3.7), (3.8) ($\boldsymbol{J}$ is zero far from the particle).

In particular, for the first order spline shape-factor

$$S(x, y) = S^1_{\Delta x}(x) S^1_{\Delta y}(y),$$

we have the following expression of the vector $\boldsymbol{W}$:

$$\begin{aligned}(W_x)_{i,j} &= \frac{1}{2}\big(S^1_{\Delta x}(X_i - x^{n+1}) - S^1_{\Delta x}(X_i - x^n)\big)\big(S^1_{\Delta y}(Y_j - y^n) + S^1_{\Delta y}(Y_j - y^{n+1})\big), \\ (W_y)_{i,j} &= \frac{1}{2}\big(S^1_{\Delta y}(Y_j - y^{n+1}) - S^1_{\Delta y}(Y_j - y^n)\big)\big(S^1_{\Delta x}(X_i - x^n) + S^1_{\Delta x}(X_i - x^{n+1})\big).\end{aligned}$$

We now show that Esirkepov's method is not a direct extension of Villasenor–Buneman's one by considering first order spline shape factors. When the particle stays in the same cell during the time $[t^n, t^{n+1}]$, we find the same current as in the Villasenor–Buneman "four boundaries motion". Let us go back to the "seven boundaries motion" case illustrated in Figure 9. For clarity, we assume $\Delta x = \Delta y = 1$.

Then, applying the previous Esirkepov algorithm, we get

$$J_{x\,i_0-\frac{1}{2},j_0-1} = -q\frac{1}{\Delta t}(x^n - i_0)\left(j_0 - \frac{y^n + y^{n+1}}{2}\right),$$
$$J_{x\,i_0+\frac{1}{2},j_0-1} = -q\frac{1}{\Delta t}(i_0 - x^{n+1})\left(j_0 - \frac{y^n + y^{n+1}}{2}\right),$$
$$J_{x\,i_0-\frac{1}{2},j_0} = -q\frac{1}{\Delta t}(x^n - i_0)\left(\frac{y^n + y^{n+1}}{2} - j_0 + 1\right),$$
$$J_{x\,i_0+\frac{1}{2},j_0} = -q\frac{1}{\Delta t}(i_0 - x^{n+1})\left(\frac{y^n + y^{n+1}}{2} - j_0 + 1\right)$$

and

$$J_{y\,i_0-1,j_0-\frac{1}{2}} = -q\frac{1}{\Delta t}\frac{i_0 - x^n}{2}(y^n - y^{n+1}),$$
$$J_{y\,i_0,j_0-\frac{1}{2}} = -q\frac{1}{\Delta t}(y^n - y^{n+1})\left(1 - \frac{x^{n+1} - x^n}{2}\right),$$
$$J_{y\,i_0+1,j_0-\frac{1}{2}} = -q\frac{1}{\Delta t}(y^n - y^{n+1})\frac{x^{n+1} - i_0}{2}$$

with the other terms equal to zero, while the Villasenor–Buneman method gives

$$J_{x\,i_0-\frac{1}{2},j_0-1} = \frac{q}{\Delta t}(i_0 - x^n)\left(j_0 - \frac{y^n + y^{\text{int}}}{2}\right),$$
$$J_{x\,i_0-\frac{1}{2},j_0} = \frac{q}{\Delta t}(i_0 - x^n)\left(\frac{y^n + y^{\text{int}}}{2} - (j_0 - 1)\right),$$
$$J_{x\,i_0+\frac{1}{2},j_0-1} = \frac{q}{\Delta t}(x^{n+1} - i_0)\left(j_0 - \frac{y^{\text{int}} + y^{n+1}}{2}\right),$$
$$J_{x\,i_0+\frac{1}{2},j_0} = \frac{q}{\Delta t}(x^{n+1} - i_0)\left(\frac{y^{\text{int}} + y^{n+1}}{2} - (j_0 - 1)\right)$$

and

$$J_{y\,i_0-1,j_0-\frac{1}{2}} = \frac{q}{\Delta t}(y^{\text{int}} - y^n)\left(\frac{i_0 - x^n}{2}\right),$$
$$J_{y\,i_0,j_0-\frac{1}{2}} = \frac{q}{\Delta t}(y^{\text{int}} - y^n)\left(\frac{x^n - i_0}{2} + 1\right) + \frac{q}{\Delta t}(y^{n+1} - y^{\text{int}})\left(\frac{i_0 - x^{n+1}}{2} + 1\right),$$
$$J_{y\,i_0+1,j_0-\frac{1}{2}} = \frac{q}{\Delta t}(y^{n+1} - y^{\text{int}})\left(\frac{x^{n+1} - i_0}{2}\right),$$

where $y^{\text{int}} = y^n + \frac{y^{n+1} - y^n}{x^{n+1} - x^n}(i_0 - x^n)$ is the intermediate position on the y-axis.

Finally, we can verify that although the expression of the currents are different, both methods satisfy the charge conservation equation at the discrete level. In the next section, we propose numerical simulations illustrating these results.

4 Numerical results

We implemented the traditional PIC, the Villasenor–Buneman and the Esirkepov methods for spline shape factors of order one and two. All the results are given with dimensionless units (we fixed $c = \epsilon_0 = 0$).

Our first test case will show the advantage to take second order shape factors compared to first order shape factors. We consider a plasma with stationary electron distribution function

$$f(\boldsymbol{x}, \boldsymbol{v}, t) = \frac{1}{2\pi v_{th}^2} \exp\Big(- \frac{v_x^2 + v_y^2}{2 v_{th}} \Big)$$

where v_{th} is the thermal velocity, and uniform neutralizing background. We take $v_{th} = 0.1$ and initialize this distribution with a quiet start [1]. Exact charge and current densities are identically zero. It is well known that PIC methods introduce noise, as can be seen in Figure 10 where we plotted the error on the first component of the current density. We note that this noise level is lower with second order shape

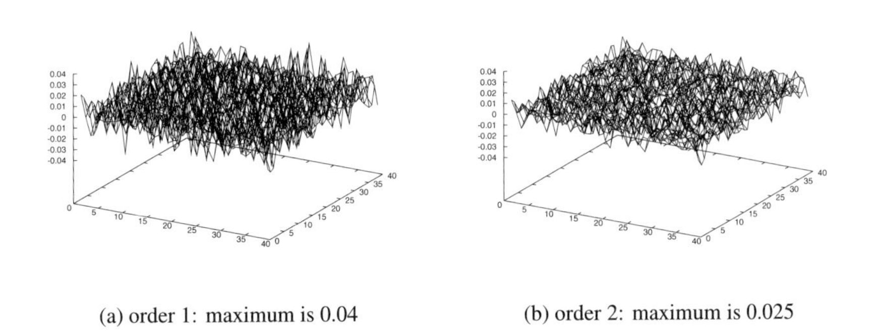

(a) order 1: maximum is 0.04

(b) order 2: maximum is 0.025

Figure 10. First test case: error on the first component of the current density with 50 particles per cell (on a 40×40 grid).

factors (the maximum is 0.025) than with first order shape factors (the maximum is 0.04). Figure 11 shows the L^2-norm of computed charge and current densities for different number of particles per cell. Here we show the results obtained with the Villasenor Buneman method, but those given by the CIC and Esirkepov methods are similar. We see that we have a lower noise level by taking 40 particles per cell and

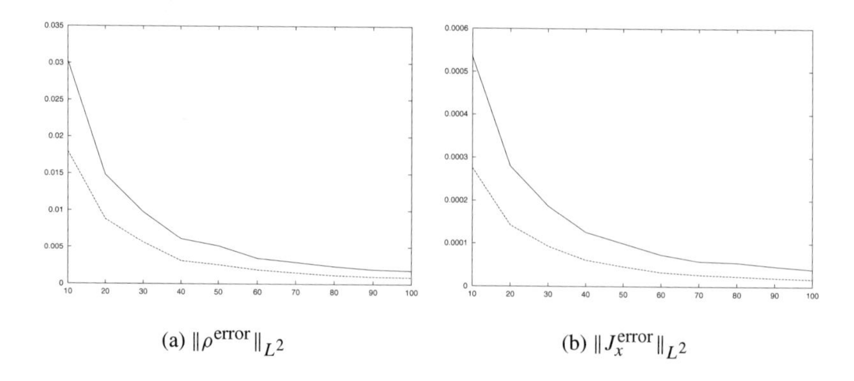

(a) $\|\rho^{\text{error}}\|_{L^2}$ (b) $\|J_x^{\text{error}}\|_{L^2}$

Figure 11. First test case: L^2-norm of the errors on ρ and J_x in function of the number of particles per cell (on a 40×40 grid). Order 1 is represented with full lines, order 2 with hatched lines.

second order shape factors, than by taking 60 particles per cell and first order shape factors.

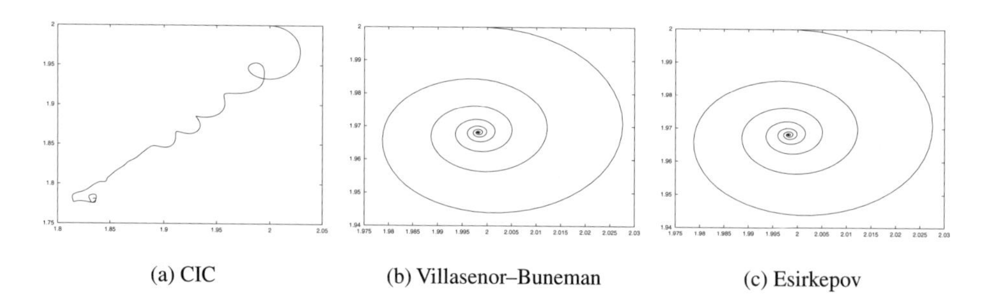

(a) CIC (b) Villasenor–Buneman (c) Esirkepov

Figure 12. Second test case: trajectories using first order shape factors.

In our second test case we study the trajectory of a relativistic charged particle in a constant uniform magnetic field. We place the particle at $t = 0$ in the center of a 4×4 square. We take $q = 1$, $B_z = 10$ and $\boldsymbol{v}_0 = (0.5, 0)$. Figure 12 represents the particle trajectory in the (x, y) plane during time $[0, 10]$. We obtain exactly the same results taking second order shape factors (with another deviation in the CIC case).

This figure shows that non conservation of charge, and then violation of Gauss's law, gives rise to non physical behaviours. But it doesn't allow to decide between Villasenor–Buneman and Esirkepov, because we obtain exactly the same trajectories. we compare computation times and notice that the Esirkepov method is the fastest.

Our third test case is the simulation of a charged particle beam. The particles are injected at a boundary with same speed so as to keep a constant intensity at

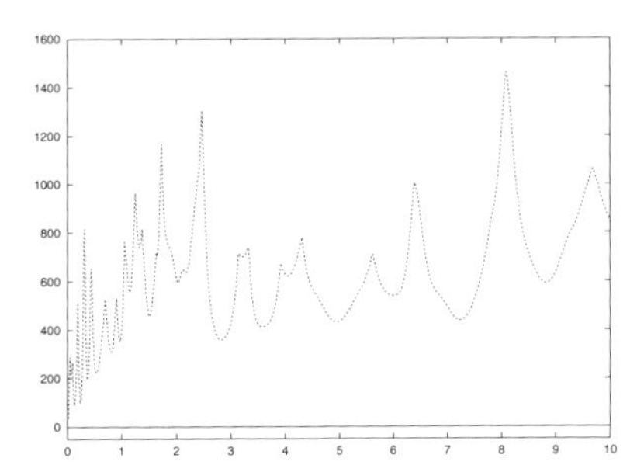

Figure 13. Second test case: $\| \operatorname{div} \boldsymbol{E} - \rho \|_{L^2}$ in function of time (with first order shape factors). The non charge conserving method (CIC) is represented with hatched line, while the Esirkepov and Villasenor–Buneman results are identically zero.

this boundary. We apply an external electric field, calculated to avoid the Child Langmuir phenomenon. In Figure 14 we represent the resulting beam at time 3 (after 206 iterations). There are about 3774 particles in a 40×40 grid. The intensity is maintained at 3.51 at the injection boundary. We have performed the computation until time 1000 (68942 iterations), and noticed that in the two charge conserving cases we have a stationary state, while the CIC simulation gives a non physical beam, where particles of same charge aren't repulsed.

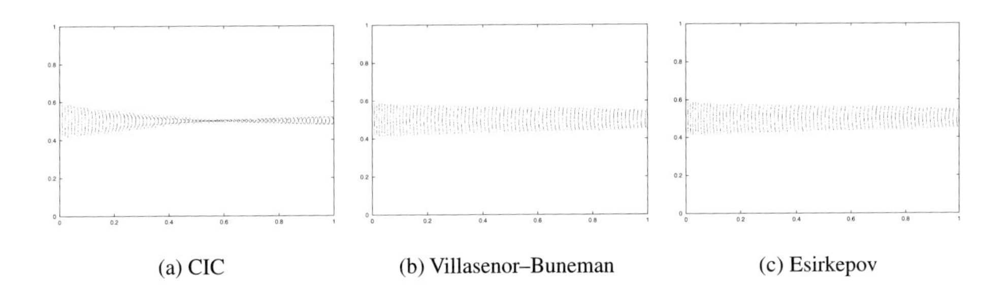

(a) CIC (b) Villasenor–Buneman (c) Esirkepov

Figure 14. Third test case with first order shape factors at time 3.

To conclude, we have illustrated the necessity to satisfy the discrete charge conservation law. We have developed the existing Villasenor–Buneman method for higher order shape factors and proposed a new viewpoint on the Esirkepov method. We noted that the current computed for one particle is different for each one of the algorithms. We are not still able to prove that a method is better than the other. Therefore we are elaborating a new test case which could enhance some difference.

Acknowledgments. The authors thank Mihai Bostan, Yves Elskens and Pierre Navaro, who contributed to the beginning of this work, Eric Sonnendrücker and the CEMRACS organizers.

References

[1] O. Birdsall, A. Langdon, *Plasma physics via computer simulation*, MacGraw-Hill, New York 1985.

[2] G. H Cottet, P. A. Raviart, Particle methods for the one-dimensional Vlasov-Poisson equations, *SIAM J. Numer. Anal.* 21 (1) (1984), 53–76.

[3] T. Zh. Esirkepov, *Exact charge conservation scheme for Particle-in-Cell simulation with an arbitrary form-factor, Comput. Phys. Comm.* 135 (2001) 144–153.

[4] R. W. Hockney, J. W. Eastwood, *Computer simulations using particles*, McGraw Hill, New York 1981; reprinted, IOPP, 1988.

[5] B. Marder, A method for incorporating Gauss's law into electromagnetic PIC codes, *J. Comput. Phys.* 68 (1987), 48–55.

[6] C. D. Munz, R. Schneider, E. Sonnendrücker, U. Voss, *Maxwell's equations when the charge conservation is not satisfied, C. R. Acad. Sci. Paris Sér. I* 328 (1999), 431–436.

[7] P. A. Raviart, An analysis of particle methods, Numerical Methods in Fluid Dynamics, Lecture Notes in Math. 1127, Springer-Verlag, Berlin, New York 1985, 243–324.

[8] T. Umeda, Y. Omura, T. Tominaga, H. Matsumoto, A new charge conservation method in electromagnetic particle-in-cell simulations, *Comput. Phys. Comm.* 156 (2003), 73–85.

[9] J. Villasenor, O. Buneman, Rigorous charge conservation for electromagnetic field solvers, *J. Comput. Phys.* 69 (1992), 306–316.

[10] K. S. Yee, Numerical solution of initial boundary value problems involving Maxwell's equations in isotropic media-, *IEE Trans. Antennas and Propagation* 14 (1966), 302–307.

An adaptive Particle-In-Cell method using multi-resolution analysis

J.-P. Chehab[1], *A. Cohen, D. Jennequin, J. J. Nieto, Ch. Roland*[1] *and J. R. Roche*

[1]*Laboratoire de Mathématiques Paul Painlevé, UMR 8524*
Université de Lille 1, 59655 Villeneuve d'Ascq Cédex, France
email: chehab@math.univ-lille1.fr, roland@math.univ-lille1.fr

Laboratoire Jacques-Louis Lions, Université Paris VI
Boîte courrier 187, 75252 Paris Cedex 05, France
email: cohen@ann.jussieu.fr

Laboratoire J. Liouville, Centre Universitaire de la Mi-Voix
Maison de la Recherche Blaise Pascal, 50 rue F. Buisson, B.P. 699
62228 Calais Cedex, France
email: jennequin@lmpa.univ-littoral.fr

Universidad de Granada, Facultad de Ciencias
Campus Universitario de Fuentenueva
Avda. Fuentenueva s/n, 18071 Granada, Spain
email: jjmnieto@ugr.es

Institut de Mathématiques Élie Cartan, Université de Nancy I
B.P. 239, 54506 Vandoeuvre-lès-Nancy Cedex, France
email: roche@iecn.u-nancy.fr

Abstract. In this paper we introduce a new PIC method based on an adaptive multi-resolution scheme for solving the one dimensional Vlasov–Poisson equation. Our approach is based on a description of the solution by particles of unit weight and on a reconstruction of the density at each time step of the numerical scheme by an adaptive wavelet technique: the density is firstly estimated in a proper wavelet basis as a distribution function from the current empirical data and then "de-noised" by a thresholding procedure. The so-called Landau damping problem is considered for validating our method. The numerical results agree with those obtained by the classical PIC scheme, suggesting that this multi-resolution procedure could be extended with success to plasma dynamics in higher dimensions.

1 Introduction

The kinetic motion of a physical plasma of charged particles in which the collisions between particles are neglected is usually modeled by the Vlasov equation [4],

$$\frac{\partial f}{\partial t} + v \cdot \nabla_x f - (E \cdot \nabla_v) f = 0, \quad (x, v) \in \mathbb{R}^d \times \mathbb{R}^d, \tag{1.1}$$

$f(x, v, t)$ being the distribution function and E the electrostatic field. The self-consistent field produced by the charge of the particles is

$$E_{\text{self}}(x, t) = -\nabla_x \phi_{\text{self}}(x, t), \tag{1.2}$$

where

$$-\Delta_x \phi_{\text{self}} = \rho(x, t), \quad \rho(x, t) = \int f(x, v, t)\, dv. \tag{1.3}$$

The system is closed with an initial data $f(x, v, 0) = f_0(x, v)$ and some decay conditions for the Poisson equation.

If $E = E_{\text{self}}$, this model is clearly dispersive due to the repulsive forces and then, in order to confine the particles in a bounded domain, as usual, we consider an additional given external potential $\phi_{\text{ext}}(x)$ and rewrite E as

$$E(x, t) = E_{\text{self}} + E_{\text{ext}} := -(\nabla_x \phi_{\text{self}} + \nabla_x \phi_{\text{ext}}). \tag{1.4}$$

In this project we are interested in the numerical resolution of the repulsive VP system (1.1)–(1.2)–(1.4) endowed with an appropriate initial data by means of the *Particle-In-Cell* (PIC) method. In the classical PIC method (see [6], [12]) the initial data is approximated by a set of particles, and the method aims to follow the trajectories (the characteristic curves) of these particles. In order to reflect the distribution function f_0, the initial set of particles can either be uniformly distributed and weighted with the value of f_0 at the corresponding point, or distributed randomly according to the distribution function f_0 and identically weighted. In this paper we follow the second approach.

The main difficulty in the PIC method lies in the construction of the characteristic curves

$$\begin{cases} \dfrac{dX(t)}{dt} = V(t), \\ \\ \dfrac{dV(t)}{dt} = E(X(t), t), \end{cases}$$

because of the nonlinearity due to the self-consistent potential. This requires to rebuild the charge density ρ at each time step in order to solve the Poisson equation and to obtain the electric field. Generally, this method gives good results with a relatively small number of particles but produces some numerical noise which prevent from describing precisely the tail of the distribution function. To overcome this drawback, it has been proposed to solve the problem by Eulerian methods [1] or semi Lagrangian methods [2]. Here we propose to combine the PIC method with density estimation techniques based on wavelet thresholding [8] in order to reduce the noise level: the density $\rho(x, t)$ is estimated at time $t > 0$ by an expansion of the type

$$\sum_k \hat{c}_{J_1,k}\, \varphi_{J_1,k}(x) + \sum_{j=J_1}^{J_0} \Big(\sum_{k=0}^{2^j-1} T_\eta(\hat{d}_{j,k})\, \psi_{j,k}(x) \Big). \tag{1.5}$$

Here $\varphi_{J_1,k}$ and $\psi_{j,k}$ are the scaling functions and the wavelets, respectively. The scaling and detail coefficients $\hat{c}_{J_1,k}$ and $\hat{d}_{j,k}$ are estimated from the particle distribution at time t, and T_η is a thresholding operator at level η. Both the threshold level η and the finest resolution level J_0 are chosen depending on the number of particles.

The paper is organized as follows. In Section 2 we describe how the density is estimated by wavelets, in particular we present several thresholding strategies. Then, in Section 3, we present the new numerical scheme, making a comparison with the classical PIC method. In Section 4 we give a numerical illustration with the simulation of the so-called Landau damping.

2 Density estimation by wavelet thresholding

Wavelet decompositions have been widely studied since the last two decades both from the theoretical and practical point of view. In a nutshell, these decompositions are based on a hierarchy of nested approximation spaces $(V_j)_{j\geq 0}$ which should be thought as finite element spaces of mesh size $h \sim 2^{-j}$, endowed with a nodal basis of the form $\varphi_{j,k} := 2^{j/2}\varphi(2^j \cdot -k)$. The functions $\varphi_{j,k}$ are often referred to as primal scaling functions. A projector onto V_j is of the form

$$P_j f := \sum_k c_{j,k}\varphi_{j,k} \quad \text{with } c_{j,k} := \langle f, \tilde{\varphi}_{j,k}\rangle, \tag{2.1}$$

where $\tilde{\varphi}_{j,k}$ are dual scaling functions. The primal and dual wavelets $\psi_{j,k}$ and $\tilde{\psi}_{j,k}$ characterize the update between two successive level of approximation in the sense that

$$P_{j+1}f - P_j f := \sum_k d_{j,k}\psi_{j,k} \quad \text{with } d_{j,k} := \langle f, \tilde{\psi}_{j,k}\rangle. \tag{2.2}$$

We refer to [7] for a classical introduction on wavelets, [5] for more information on their application to numerical simulation of PDE's.

In the particular context of PIC methods, we are interested in the reconstruction of the density $\rho(x,t)$ from the locations $(x_i)_{i=1,\dots,N}$ of the particles at time t. As explained in the introduction, this reconstruction has the form (1.5) where $\hat{c}_{J_1,k}$ and $\hat{d}_{j,k}$ are estimators of the exact coefficients $c_{J_1,k} := \int \rho(x,t)\tilde{\varphi}_{J_1,k}(x)\,dx$ and $d_{J_1,k} := \int \rho(x,t)\tilde{\psi}_{j,k}(x)\,dx$ from the empirical distribution according to

$$\hat{c}_{J_1,k} := \frac{1}{N}\sum_{i=1}^{N}\varphi_{J_1,k}(x_i), \tag{2.3}$$

and

$$\hat{d}_{j,k} := \frac{1}{N} \sum_{i=1}^{N} \psi_{j,k}(x_i). \tag{2.4}$$

The interest of the thresholding procedure, in contrast to a simple projection or regularization at a fixed scale j which would compute $\sum_k \hat{c}_{j,k}\varphi_{j,k}$ is twofold: (i) the regularization level j is allowed to vary locally in the sense that the procedure might retain coefficients $d_{j,k}$ at scale j only for some k which typically corresponds to the regions where the density has sharp transitions and requires more resolution, and (ii) the local regularization level automatically adapts to the unknown amount of smoothness of the density through the thresholding procedure which only depends on the number N of samples.

Following Donoho *et al.* [8], the maximal scale level J_0 and the threshold η depend on the number of samples according to

$$2^{J_0} \sim N^{1/2} \tag{2.5}$$

and

$$\eta \sim \sqrt{\log(N)/N}. \tag{2.6}$$

Another choice proposed in [8] is a threshold parameter which also depends on the scale level j according to $\eta = \eta_j = K\sqrt{j/N}$. Two techniques are generally used to threshold the details: "hard" thresholding defined by $T_\eta(y) = y\chi_{\{|y|\geq\eta\}}$ and "soft" thresholding defined by $T_\eta(y) = \mathrm{Sign}(y)\max\{0, |y| - \eta\}$. We shall precise thresholding strategies that we choose for our applications in Section 4.

The density reconstruction method varies with the choice of the wavelet basis. This choice is dictated by two constraints:

1. Numerical simplicity: according to (2.3) and (2.4), the coefficients are estimated through the evaluation of dual scaling functions $\tilde{\varphi}_{J_1,k}$ and dual wavelets $\tilde{\psi}_{j,k}$ at the points x_i. It is therefore useful that these functions have a simple analytical form. In particular, high order compactly supported orthonormal wavelets cannot be used since they do not have an explicit analytical expression.

2. High order accuracy and smoothness: the primal wavelet system should have high order accuracy and smoothness in order to ensure the quality of the approximation of $\rho(x,t)$ by the expansion (1.5).

The choice of the Haar system is good with respect to the first constraint, since in this case the scaling function $\tilde{\varphi} = \varphi$ is simply the box function $\chi_{[0,1]}$, so that the estimation of a scaling coefficient $c_{j,k} = \langle \rho, \tilde{\varphi}_{j,k}\rangle$ simply amounts in counting the points falling in the interval $I_{j,k} = [2^{-j}k, 2^{-j}(k+1)[$:

$$\hat{c}_{j,k} := 2^{j/2}\frac{1}{N}\#\{i;\ x_i \in I_{j,k}\}. \tag{2.7}$$

In particular, we can compute the $\hat{c}_{J_0,k}$ at the finest scale level and use the Haar transform algorithm to compute the $\hat{d}_{j,k}$ according to the classical relations:

$$\hat{c}_{j,k} = \frac{\hat{c}_{j+1,2k} + \hat{c}_{j+1,2k+1}}{\sqrt{2}} \quad \text{and} \quad \hat{d}_{j,k} = \frac{\hat{c}_{j+1,2k} - \hat{c}_{j+1,2k+1}}{\sqrt{2}}. \tag{2.8}$$

However, this choice is not good with respect to the second constraint since piecewise constant functions are low order accurate. In order to fix this defect, while preserving the numerical simplicity of the method, we propose to use a higher order (third order) reconstruction still based on the box function $\chi_{[0,1]}$ as $\tilde{\varphi}$, as proposed by Ami Harten in [11]. This means that the coefficients $\hat{c}_{j,k}$ are still defined by (2.7), but the relation between the approximation and detail coefficients $\hat{c}_{j,k}$ and $\hat{d}_{j,k}$ is modified according to

$$\hat{d}_{j,k} = \frac{1}{\sqrt{2}}\hat{c}_{j+1,2k} - \hat{c}_{j,k} - \frac{1}{8}(\hat{c}_{j,k-1} - \hat{c}_{j,k+1}). \tag{2.9}$$

This is an instance of the so-called *lifting scheme* introduced in [14]. Using this relation, we estimate all the coefficients $c_{j,k}$ and $d_{j,k}$ for $J = J_1, \dots, J_0 - 1$ and we apply the thresholding operator T_η to the estimated coefficients $\hat{d}_{j,k}$.

It should be remarked that the primal scaling functions $\varphi_{J_1,k}$ and wavelets $\psi_{j,k}$ do not have an explicit analytical expression, in contrast to the dual scaling functions and wavelets. However, we can reconstruct the estimator (1.5) at arbitrarily fine resolution by applying the reconstruction formulae

$$\hat{c}_{j+1,2k} = \sqrt{2}\Big[\hat{c}_{j,k} - \frac{1}{8}(\hat{c}_{j,k-1} - \hat{c}_{j,k+1}) + T_\eta(\hat{d}_{j,k})\Big], \tag{2.10}$$

and

$$\hat{c}_{j+1,2k+1} = \sqrt{2}\hat{c}_{j,k} - \hat{c}_{j+1,2k}. \tag{2.11}$$

It is also possible to construct wavelet-like multiscale decompositions where both the dual and primal functions have a simple analytical expression, based on the quasi-interpolation operator

$$P_j f := \sum_k c_{j,k}\varphi_{j,k}, \quad c_{j,k} = \langle f, \varphi_{j,k} \rangle, \tag{2.12}$$

where $\varphi = (1 - |x|)_+$ is the classical hat function. We therefore estimate the coefficients by

$$\hat{c}_{j,k} := \frac{1}{N}\sum_{i=1}^{N} \varphi_{j,k}(X_i), \tag{2.13}$$

at the finest scale $j = J_0$ and derive them recursively at coarser levels by the formula

$$\hat{c}_{j,k} = \frac{1}{\sqrt{2}}\hat{c}_{j+1,2k} + \frac{1}{2\sqrt{2}}(\hat{c}_{j+1,2k+1} + \hat{c}_{j+1,2k-1}).$$

In this case, the detail components at level j reads

$$(P_{j+1} - P_j)f = \sum_k [\hat{d}^0_{j,k}\varphi_{j+1,2k} + \hat{d}^+_{j,k}\varphi_{j+1,2k+1} + \hat{d}^-_{j,k}\varphi_{j+1,2k-1}], \tag{2.14}$$

with

$$\begin{aligned}
\hat{d}^0_{j,k} &:= \frac{\sqrt{2}-1}{\sqrt{2}}\hat{c}_{j+1,2k} - \frac{1}{2\sqrt{2}}\hat{c}_{j+1,2k+1} - \frac{1}{2\sqrt{2}}\hat{c}_{j+1,2k-1},\\
\hat{d}^+_{j,k} &:= \frac{2\sqrt{2}-1}{4\sqrt{2}}\hat{c}_{j+1,2k+1} - \frac{1}{2\sqrt{2}}\hat{c}_{j+1,2k} - \frac{1}{4\sqrt{2}}\hat{c}_{j+1,2k-1},\\
\hat{d}^-_{j,k} &:= \frac{2\sqrt{2}-1}{4\sqrt{2}}\hat{c}_{j+1,2k-1} - \frac{1}{2\sqrt{2}}\hat{c}_{j+1,2k} - \frac{1}{4\sqrt{2}}\hat{c}_{j+1,2k}.
\end{aligned}$$

The triplet $(d^0_{j,k}, d^+_{j,k}, d^-_{j,k})$ plays the role of the wavelet coefficient and it is jointly thresholded in order to preserve the density mass.

In the sequel of the paper, we shall denote W0 for the first algorithm based on the lifting scheme we have described and W1 for the second algorithm based on the quasi-interpolation operator. In the numerical scheme, we apply these algorithms and reconstruct the denoised density at the finest level J_0 on which we apply the Poisson solver to derive the electric field.

3 Numerical schemes

We present here the new scheme we introduce in the paper (PICONU[1]) as a modification of the classical PIC method which will be used to compare the numerical results. Of course, the considered Vlasov–Poisson equation is one dimensional in space and in velocity, so we can write a formal expression using the fundamental solution of the Poisson equation. Indeed, we have

$$-\Delta\Phi_{\text{self}} = \rho \iff \phi_{\text{self}} = \frac{-1}{2}|x| \iff E_{\text{self}} = \frac{1}{2}\frac{x}{|x|} * \rho.$$

On the other hand, if we denote $X_i(t)$ the position of the ith particle at time t for $i = 1 \dots N$, the density is

$$\rho(x,t) = \frac{1}{N}\Big(\sum_{i=1}^{N} \delta_{X_i(t)}\Big),$$

[1] which stands in French for PIC Ondelettes NUmérique

where δ_ξ stands for the Dirac measure at point ξ. Hence, the self-consistent field E_{self} can be computed by the following formula (written in general dimension d)

$$E_{\text{self}}(x,t) = \frac{1}{2N}\Big(\sum_{i=1}^{N} \frac{x - X_i}{|x - X_i|^d}\Big).$$

However, we underline that, in the practical point of view, we can not proceed in such a way in higher dimensions ($d > 1$) due to the singularity of the Green kernel, and that the adaptive method proposed in this paper is aimed at being extended, e.g., to the 2-D Vlasov–Poisson problem.

3.1 The PIC method

The PIC method consists in following the track to particles with position X_i and velocity V_i along the characteristic curves

$$\begin{aligned}
\frac{dX_i(t)}{dt} &= V_i(t),\\
\frac{dV_i(t)}{dt} &= E(X_i(t),t),\\
X_i(0) &= x_i,\ \ V_i(0) = v_i.
\end{aligned}$$

Let f_0 be the initial distribution, the distribution at time $t = T$ is computed as follows:

- Initialization: build (x_i, v_i), N pair of random variables drawn of the initial distribution f_0.
- Time marching scheme: the (nonlinear) characteristic equation is split and integrated as follows: set $\delta t = \frac{T}{N_{\max}}$ where $N_{\max}$ is the number of time steps, then, for $n = 0, \dots, N_{\max}$,

$$V^{n+1/2} = V^n + \frac{\delta t}{2} E^n(X^n) \tag{3.1a}$$

$$X^{n+1} = X^n + \delta t\, V^{n+1/2} \tag{3.1b}$$

$$\texttt{Build}\ \rho^{n+1} \tag{3.1c}$$

$$\texttt{Solve}\ -\Delta_h \phi^{n+1} = \rho^{n+1} \tag{3.1d}$$

$$E^{n+1}(X^{n+1}) = \nabla_h \phi^{n+1}(X^{n+1}) \tag{3.1e}$$

$$V^{n+1} = V^{n+1/2} + \frac{\delta t}{2} E^{n+1}(X^{n+1}). \tag{3.1f}$$

- Build $f^{N_{\max}}$ by interpolation.

In step 3.1c we must compute the charge density ρ on the discrete grid points. Two classical methods are Nearest Grid Point (NGP) and Cloud In Cell (CIC): they consist on $P0$ and $P1$ interpolations respectively. According to Birdsall and Langdon in [3, p. 19–23], CIC reduces the noise relative to the NGP. Higher order techniques could be used too, some of them consist on quadratic and cubic spline interpolations. In our numerical results, we will compare our schemes to a PIC method with CIC and NGP density reconstruction.

3.2 The adaptive scheme (PICONU)

The PICONU scheme differs from the classical PIC method in the step (3.1c). The density is computed by Donoho's technique described in Section 2. For this method, we have to select the finest and the coarsest resolution level (J_0 and J_1), the threshold and the mesh size for the Poisson equation (3.1d). In the classical density estimation, the noise appears when the mesh size is locally too small. We expect to find the "good threshold" which refines the density mesh only in the region where there are a lot of particles.

4 Numerical results

The numerical results presented hereafter were obtained with SCILAB, the (free) numerical software of the INRIA [13]. As a validation of our scheme, we consider the simulation of the so-called Landau damping. This is indeed a significant numerical test, due to its difficulty in simulating, and it has been considered by several authors for validating a code, see, e.g., [3], [9] and references therein. This test consists in the observation of the decay rate of the electrostatic energy obtained when the initial distribution is the perturbed Maxwellian distribution defined by

$$f_0(x,v) = \frac{1}{\sqrt{2\pi}} \exp\left(-v^2/2v_{\text{th}}{}^2\right)(1+\alpha\cos(kx)) \quad \text{for all } (x,v) \in \left[-\frac{L}{2},\frac{L}{2}\right] \times \mathbb{R},$$

where v_{th} is the thermal mean velocity, $L = \frac{2\pi}{k}$ and k, α are positive constants with $\alpha \ll 1$. Let us recall that if E denotes the electric field, the electrostatic energy is defined by $\int_{-L/2}^{L/2} E^2\,dx$. We consider that the tail of the distribution does not contribute to the problem for $|v| > v_{\max}$ for some $v_{\max}$ large enough. We will choose $v_{\text{th}} = 1$ and $v_{\max} = 6$.

More precisely, one must observe numerically that

- the decay rate of electrostatic energy defines a line of director coefficient

$$\gamma_L = \sqrt{\frac{\pi}{8}} k^{-3} \exp\left(-\frac{1}{2k^2} - \frac{3}{2}\right),$$

- the oscillation frequency of the electrostatic energy must be

$$\omega^2 = 1 + 3k^2.$$

The Landau damping is very sensitive to the initial distribution and we follow [3] using for that purpose the so-called “quiet start” initialization for α small enough. For all the tests, we do vary only α taking as fixed value $k = 0.5$.

4.1 Comparison between NGP and W0

As a reference for validating our new scheme, we shall compare the results obtained with the classical PIC method where the charge density is computed using the NGP technique described in [3, pp. 21, 22] to the wavelets build from Haar system. In the finest resolution level J_0 is equal to the coarsest one J_1, these two methods are equivalent. In Figures 1, 2 and 3 we plotted the charge density, the electrostatic field and the discrete electrostatic energy of the plasma. The graph of the electrostatic energy is in a log-scale and the line corresponds to the theoretical decay of director coefficient γ_L.

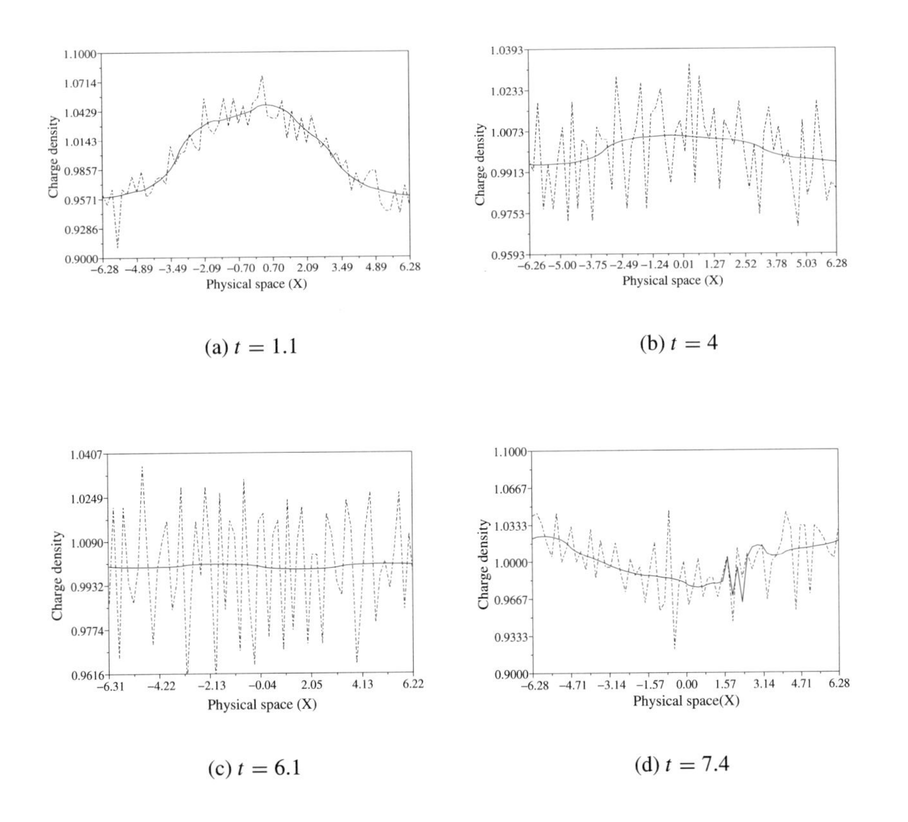

(a) $t = 1.1$ (b) $t = 4$

(c) $t = 6.1$ (d) $t = 7.4$

Figure 1. Landau damping with $\alpha = 0.1$. Charge density computed with 26000 particles, in solid line: the PICONU method (W0), in dashed line: the classical PIC (NGP).

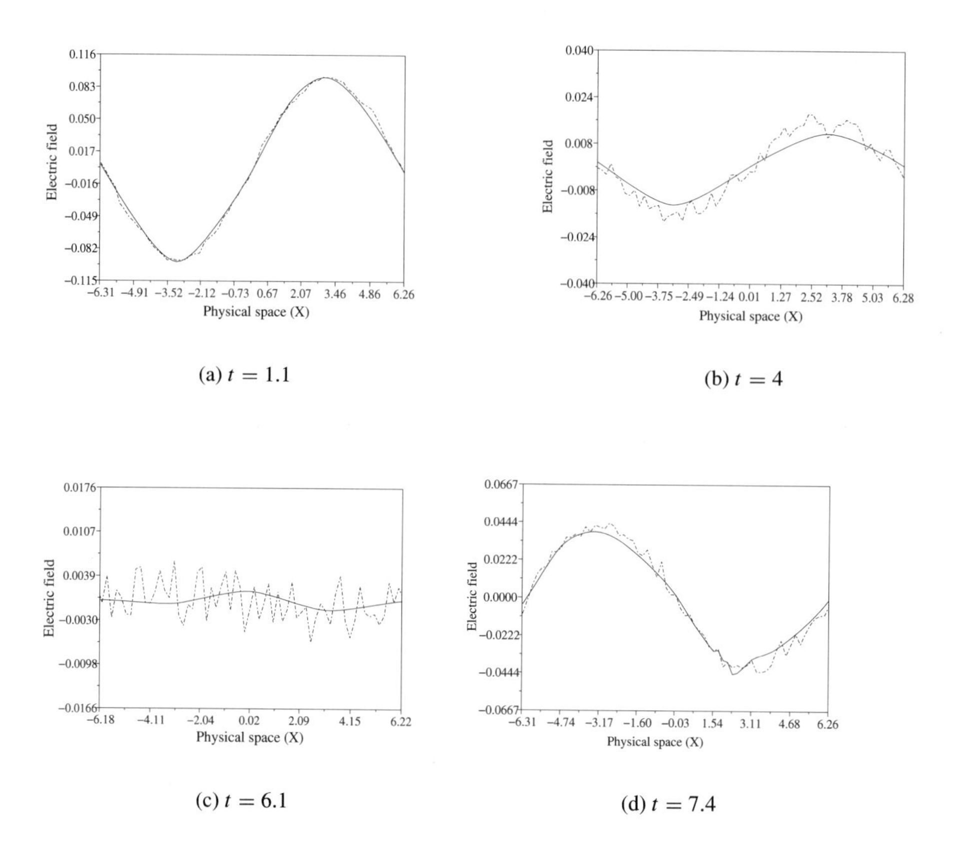

Figure 2. Landau damping with $\alpha = 0.1$. Electrostatic field computed with 26 000 particles, in solid line: the PICONU method (W0), in dashed line: the classical PIC (NGP).

We used the following parameters: the time step δt is chosen equal to 0.1. The threshold is the one given in (2.6) and more precisely, we choose $\eta = 0.5 \times \sqrt{j/N}$. The finest and coarsest resolution levels are 6 and 2, this corresponds to about 800 particles per cell. Then we use NGP on a grid of 2^6 intervals.

The NGP method requires a high number of particles per cell. The only way to reduce it is to increase the accuracy of the interpolation. The wavelets, based on the same interpolation, inherit the same problem. However, we observe that wavelets allow to reduce efficiently the noise of the method. This is obvious in Figure 3 and 4 looking at the minima of electrostatic energy: the local minima are obtained theoretically when the electrostatic field vanishes and numerically, these minima have the same magnitude than the discrete L^2-norm of the noise. Furthermore, we observe that the charge density and the electrostatic field are smoother than in the simple $P0$ interpolation case. Whatever noisy is the charge density, the electrostatic field is nearly smooth. It is due to the particular case of the one dimensional integration which has a

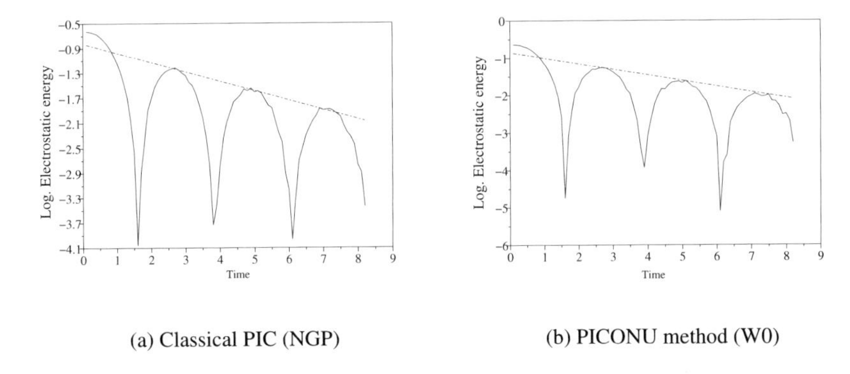

(a) Classical PIC (NGP)

(b) PICONU method (W0)

Figure 3. Landau damping with $\alpha = 0.1$, $k = 0.5$. Electrostatic energy.

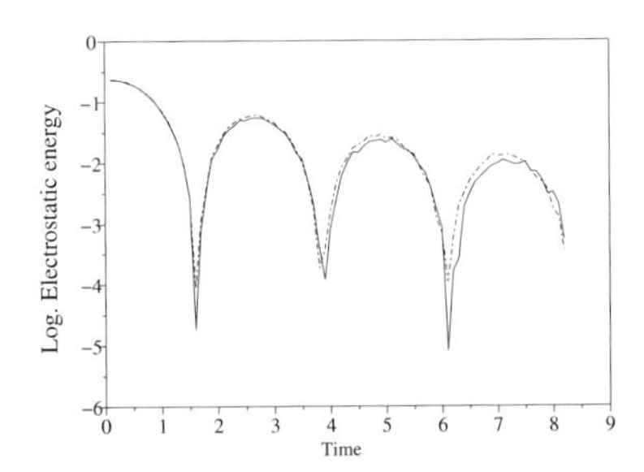

Figure 4. Superposition of electrostatic energy with 26 000 particles. In solid line: damping obtained with PICONU method (W0); in dashed line: classical PIC method (NGP).

smoothing effect. In higher dimension, the smoothness will take a greater importance and the use of wavelets should be more pertinent.

4.2 Comparison between CIC and W1

The simulation parameters are chosen as follows: the highest level of resolution equals 7 and the number of particles equals 10 000. It implies that there is about 80 particles per cell. The time step δt equals 0.1. For the density estimation using wavelets, the coarsest resolution level is equal to 3 and the threshold is this one prescribed by Donoho, that is $K \times \sqrt{j/N}$. The coefficient R_T gives the rate of thresholded coefficients that is the ratio of the mean value of thresholded coefficients at each time step. Figure 5 gives the electrostatic energy computed with the classical PIC (CIC). We observe that the classical PIC fails when α becomes small. On the contrary, the adap-

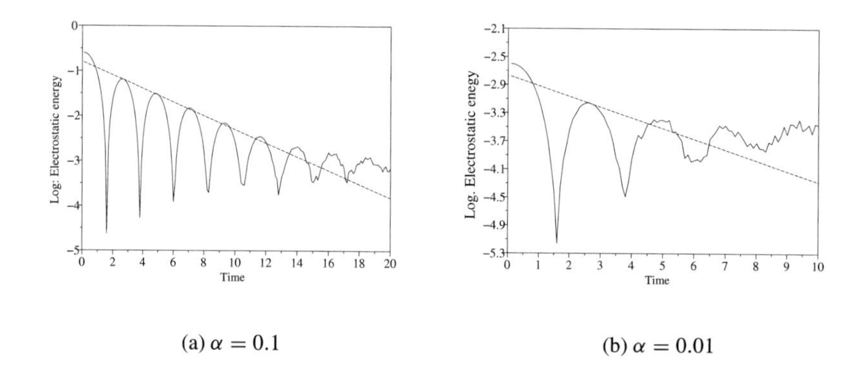

(a) $\alpha = 0.1$

(b) $\alpha = 0.01$

Figure 5. Linear Landau damping with classical PIC (CIC). 10 000 particles, 128 cells.

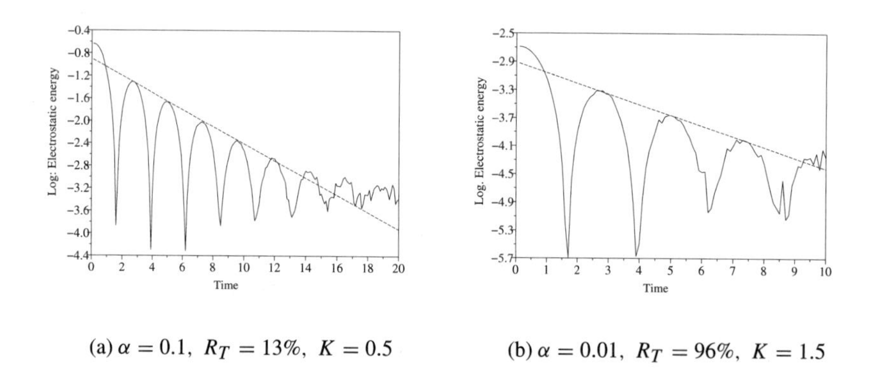

(a) $\alpha = 0.1,\ R_T = 13\%,\ K = 0.5$

(b) $\alpha = 0.01,\ R_T = 96\%,\ K = 1.5$

Figure 6. Linear Landau damping, PICONU (W1), 10 000 particles.

tive method gives some better results (see Figure 6). A finer analysis of the threshold shows that all the coefficients are thresholded on the two finest grid. This is a natural consequence of the landau test: the distribution of particles corresponds to a small perturbation of the uniform distribution. Since we use only a first order reconstruction, the charge density is less smooth than in the case of W0.

5 Concluding remarks and perspectives

The results presented in this paper show that the adaptive wavelet reconstruction of the density for the Vlasov–Poisson equation is a promising approach to solve such plasma dynamics in a Lagrangian framework even though the choice of appropriated wavelets must still be discussed. The W0 wavelets are not completely satisfying because they require a high number of particles. Moreover, these methods are unable to verify the landau damping test for small perturbation magnitude α. We are interested in the numerical simulation where there are less than 100 particles per cell. The W1 wavelets satisfy this condition but do not smooth the density. The results proved that threshold helps to find the appropriate (adaptive) mesh and it should become crucial in tests where particles are very dispersed. Moreover, highest order reconstruction is particularly efficient to reduce the noise. A compromise has to be found between the reconstruction and accuracy order which minimize the computational time.

We have considered here one dimensional Vlasov–Poisson but our approach will be extended in the near future to higher dimensional problems for which the Eulerian framework becomes more costly in terms of CPU time since large numbers of grid points must be used in that case.

Acknowledgements. The authors are very grateful to Prof. Eric Sonnendrücker for the numerous discussions we had with him and for the precious advice he gave us.

References

[1] Arber, T. D., Vann, R. G. L., A critical comparison of Eulerian-Grid-Based Vlasov solvers, *J. Comput. Phys.* 180 (2002), 339–357.

[2] Besse, N., Sonnendrücker, E., Semi-Lagrangian schemes for the Vlasov equation on an unstructured mesh of phase space. *J. Comput. Phys.* 191 (2) (2003), 341–376.

[3] Birdsall, C. K., Langdon, A. B., *Plasma Physics via Computer Simulation*, Institute of Physics Publishing, Bristol and Philadelphia 1991.

[4] Chapman, S., Cowling, T. G., *The mathematical theory of non-uniform gases*, An account of the kinetic theory of viscosity, thermal conduction and diffusion in gases, third edition, prepared in co-operation with D. Burnett, Cambridge University Press, London 1970.

[5] Cohen, A., *Numerical Analysis of Wavelet Methods*, Stud. Math. Appl. 32, Elsevier, North-Holland, Amsterdam 2003.

[6] Cottet, G.-H., Raviart, P.-A., Particle methods for the one-dimensional Vlasov-Poisson equations, *Siam J. Numer. Anal.* 21 (1) (1984), 52–76.

[7] Daubechies, I., *Ten Lectures on Wavelets*, CBMS-NSF Reg. Conf. Ser. Appl. Math. 61, SIAM, Philadelphia, PA, 1992.

[8] Donoho D. L., Johnstone I. M., Kerkyacharian G., Picard, D., Density estimation by wavelet thresholding, *Ann. Statist.* 24 (2) (1996), 508–539.

[9] Filbet, F., Contribution à l'analyse et la simulation numérique de l'équation de Vlasov, Thèse, Université Henri Poincaré, Nancy, France, 2001.

[10] Filbet, F., Sonnendrücker, E., Bertrand, P., Conservative numerical schemes for the Vlasov equation, *J. Comput. Phys.* 172 (2001), 166–187.

[11] Harten, A., Discrete multi-resolution analysis and generalized wavelets, *Appl. Numer. Math.* 12 (1993), 153–193.

[12] Raviart, P. A., An analysis of particle methods, in *Numerical methods in fluid dynamics*, Lecture Notes in Math. 1127, Springer-Verlag, Berlin 1985, 243–324.

[13] Scilab, `http://scilabsoft.inria.fr`

[14] Sweldens, W., The lifting scheme: a custom design construction of biorthogonal wavelets, *Appl. Comput. Harmon. Anal.* 3 (1996), 186–200.

Adaptive numerical resolution of the Vlasov equation*

M. Campos Pinto and M. Mehrenberger

Laboratoire Jacques-Louis Lions, UMR CNRS 7598
Université Pierre et Marie Curie, Boîte courrier 187
75252 Paris Cedex 05, France
email: `campos@ann.jussieu.fr`

Institut de Recherche Mathématique Avancée, UMR CNRS 7501
Université Louis Pasteur, 7 rue René Descartes
67084 Strasbourg Cedex, France
email: `mehrenbe@math.u-strasbg.fr`

Abstract. A fully adaptive scheme (based on hierarchical continuous finite element decomposition) is derived from a semi-Lagrangian method for solving a periodic Vlasov–Poisson system. The first numerical results establish the validity of such a scheme.

1 Introduction

Most Vlasov solvers in use today are based on the Particle In Cell method which consists in solving the Vlasov equation with a gridless particle method coupled with a grid based field solver (see e.g. [4]). For some problems in plasma physics or beam physics, particle methods are too noisy and it is of advantage to solve the Vlasov equation on a grid of phase space. This has proven very efficient on uniform meshes in the two-dimensional phase space (see e.g. [14], [10] for structured meshes and [3] for unstructured meshes). However, when the dimensionality increases, the number of points on a uniform grid becomes too important for the method to be efficient, and it is essential to regain optimality by keeping only the 'necessary' grid points. Such adaptive methods have recently been developed, like in [15] and [2]–[11] where the authors use moving distribution function grids or interpolatory wavelets of Deslaurier and Dubuc. We refer also to [9] for a summary of many Vlasov solvers.

In this project, our first objective was to write a simpler algorithm based on hierarchical finite element decompositions. Compared to [2]–[11], this leads to less regularity and a heavier data structure, but the underlying partitions of dyadic tensor-product cells allow simpler memory space management and parallel implementations.

*This work was done in collaboration with our PhD advisors, Albert Cohen and Eric Sonnendrücker.

After describing such hierarchical decompositions, we recall in Section 3 the semi-Lagrangian scheme. The actual implementation of the scheme is given in Section 4, and numerical results are presented in Section 5 for two classical test cases, the linear Landau damping and the semi-Gaussian beam.

2 Adaptive representation of a phase space density

The *adaptivity* of our scheme relies on the representation, for each time step n, of the numerical solution f^n on a non-uniform moving mesh $\mathcal{M}^n$. But before introducing such adaptive meshes, let us first describe (with a few notations) the

Uniform meshes. Considering the unit square $[0,1[^2$ as the computational domain, we denote by $\mathcal{M}_j$ the set of all phase space dyadic cells of resolution $j \in \mathbb{N}$:

$$\mathcal{M}_j := \{\alpha^j_{k,l} := [k\,2^{-j}, (k+1)\,2^{-j}] \times [l\,2^{-j}, (l+1)\,2^{-j}] : k, l \in \mathbb{Z}\},$$

and the corresponding approximation space by

$$V_j := \{f \in \mathcal{C}^0 : f_{|\alpha} \in \Pi_{1,1} \text{ for all } \alpha \in \mathcal{M}_j\},$$

consisting of all continuous functions that are bilinear on each cell. In the sequel, we shall use the short notation $|\alpha| := j$ to mean that $\alpha \in \mathcal{M}_j$, and always consider that j stands between two reference resolutions

$$j_0 \leq j \leq J, \tag{2.1}$$

or in other words, that $\mathcal{M}_j := \emptyset$ for any $j \notin \{j_0, \ldots, J\}$.

In a very classical manner, we denote by $\Gamma(\alpha)$ the vertices of a cell α, and for any node a in $\Gamma_j := \bigcup_{\alpha \in \mathcal{M}_j} \Gamma(\alpha)$, we define the nodal function φ^j_a as the unique element of V_j that satisfies

$$\varphi^j_a(b) = \delta_{a,b}, \quad b \in \Gamma_j.$$

The family $\{\varphi^j_a\}_{a \in \Gamma_j}$ is then a natural basis for V_j (normalized in L^∞) and a natural interpolatory projection on V_j is given by

$$P_j : f \in \mathcal{C}^0 \to P_j f = \sum_{a \in \Gamma_j} f(a) \varphi^j_a \in V_j.$$

Adaptive meshes. To gain some flexibility in discretizing the numerical solutions, we construct meshes of variable resolution, and a natural way for doing so is to define *cell trees*. For that purpose, we first denote for any dyadic phase space cell α its children and parents by

$$\mathcal{D}(\alpha) := \{\beta \in \mathcal{M}_{|\alpha|+1} : \beta \subset \alpha\} \quad \text{and} \quad \mathcal{P}(\alpha) := \Big\{\beta \in \bigcup_{j \leq |\alpha|-1} \mathcal{M}_j : \beta \supset \alpha\Big\},$$

and then define an *adaptive tree* Λ as a set of cells satisfying the two following properties:

1. $\mathcal{M}_{j_0} \subset \Lambda$.
2. For all $\alpha \in \Lambda$, $\bigcup_{\beta \in \mathcal{P}(\alpha)} \mathcal{D}(\beta) \subset \Lambda$.

This last property implies that no cell of Λ is *partially* refined, so that the leaves (*i.e.* the unrefined cells) form a partition of the phase space. Such a set $\mathcal{M} = \mathcal{M}(\Lambda)$ will be referred to as an *adaptive mesh*. Accordingly to the uniform case, we denote by

$$V_{\mathcal{M}} := \{f \in \mathcal{C}^0 : f_{|\alpha} \in \Pi_{1,1} \text{ for all } \alpha \in \mathcal{M}\}$$

the set of all continuous functions that are bilinear on each cell of $\mathcal{M}$, and by

$$\Gamma(\mathcal{M}) := \bigcup_{\alpha \in \mathcal{M}} \Gamma(\alpha),$$

the set of all nodes of $\mathcal{M}$.

Now if we want to define a projection $f_{\mathcal{M}} \in V_{\mathcal{M}}$ from a continuous function f, it is readily seen that the interpolatory equations

$$f_{\mathcal{M}}(a) = f(a), \quad a \in \Gamma(\mathcal{M})$$

are incompatible (unless $\mathcal{M}$ is uniform). Instead, we can define $f_{\mathcal{M}}$ as the unique element of $V_{\mathcal{M}}$ that satisfies

$$f_{\mathcal{M}}(a) = f(a), \quad a \in \Gamma_c(\mathcal{M}),$$

where

$$\Gamma_c(\mathcal{M}) := \{a \in \Gamma(\mathcal{M}) : \text{ if } \beta \in \mathcal{M} \text{ satisfies } a \in \beta, \text{ then } a \in \Gamma(\beta)\}$$

denotes the *conforming nodes* of $\mathcal{M}$.

Obviously, $P_{\mathcal{M}} : f \to f_{\mathcal{M}}$ is a linear mapping, and with the additional requirement that $\mathcal{M}$ is *graded* (that is, two neighboring cells α and β satisfy $\big||\alpha| - |\beta|\big| \leq 1$), $P_{\mathcal{M}}$ is locally stable in the sense that for any $\alpha \in \mathcal{M}$, we have

$$\|P_{\mathcal{M}} f\|_{L^\infty(\alpha)} \leq \|f\|_{L^\infty(\mathcal{P}^1(\alpha))},$$

where $\mathcal{P}^1(\alpha) := \mathcal{P}(\alpha) \cap \mathcal{M}_{|\alpha|-1}$ is the mother cell of α.

Wavelets. This data structure can also be described in the spirit of multiresolution analysis (see, for instance, [6] or [12]), since we see that the spaces V_j are nested. Thus for each j, we may define the *detail space* W_j as the complement of V_j with respect to V_{j+1} whose basis is given by $\{\varphi_a^{j+1}\}_{a \in \nabla_j}$, where $\nabla_j := \Gamma_{j+1} \setminus \Gamma_j$. It is indeed easily verified that

$$P_{j+1} f = P_j f + \sum_{a \in \nabla_j} d_a(f) \varphi_a^{j+1},$$

where the details $d_a(f)$ are defined by

$$d_a(f) := [P_{j+1}f - P_j f](a) = \sum_{b\in\Gamma_j} [f(a) - f(b)]\varphi_b^j(a), \tag{2.2}$$

where j is such that $a \in \nabla_j$. Following the usual wavelet notation, we write ψ_a instead of φ_a^{j+1}, and when a is a *coarse node* of $\nabla_{j_0-1} := \Gamma_{j_0}$, write for short $(d_a(f), \psi_a)$ instead of $(f(a), \varphi_a^{j_0})$. Hence the wavelet decomposition of f reads

$$P_J f = \sum_{a\in\nabla^J} d_a(f)\psi_a, \tag{2.3}$$

where $\nabla^J := \bigcup_{j=j_0-1}^{J-1} \nabla_j$.

Now defining the truncation operator

$$\mathcal{T}_B f := \sum_{a\in B} d_a(f)\psi_a, \quad B \subset \nabla^J,$$

we can verify that $\mathcal{P}_{\mathcal{M}} = \mathcal{T}_{\Gamma_c(\mathcal{M})}$.

3 The semi-Lagrangian scheme

We briefly recall here the principle of the semi-Lagrangian scheme for solving the periodic one-dimensional Vlasov–Poisson system, and refer the reader to [1] for a more detailed presentation and a proof of convergence in the uniform case.

Denoting by $f(t, x, v)$ the distribution function in phase space (with $x, v \in \mathbb{R}$), and by $E(t, x)$ the self consistent electric field, the Vlasov–Poisson system reads

$$\partial_t f + v \cdot \partial_x f + E \cdot \partial_v f = 0 \tag{3.1}$$

$$\partial_x E = \rho, \tag{3.2}$$

where the charge density is given by

$$\rho(t, x) = \int_{-\infty}^{\infty} f(t, x, v)\, dv - 1. \tag{3.3}$$

If we consider that f and E are 1-periodic in space, this problem becomes well posed with the additional zero-mean electrostatic condition

$$\int_0^1 E(t, x)\, dx = 0, \quad t \geq 0, \tag{3.4}$$

and an initial data

$$f(0, x, v) = f^0(x, v), \quad x \in [0, 1],\ v \in \mathbb{R}. \tag{3.5}$$

The semi-Lagrangian method consists in taking advantage of the conservation of the density f along the characteristics, so let us first consider for a time step Δt some approximation $\mathcal{B}(x, v)$ of the backward characteristics $(X(0), V(0))$ of (3.1)–(3.3) defined by

$$\begin{cases} \dot{X}(s) = V(s) \\ \dot{V}(s) = E(s, X(s)) \end{cases} \quad \text{for } s \in [0, \Delta t], \quad \text{and} \quad \begin{cases} X(\Delta t) = x \\ V(\Delta t) = v. \end{cases} \tag{3.6}$$

In the uniform case, for $n = 0, 1 \ldots$, we approximate on each time step the solution of (3.1)–(3.3) with initial value f^n by

$$f^{n+1} := P_J \mathcal{A} f^n, \tag{3.7}$$

where the nonlinear *advection operator* $\mathcal{A}$ is defined by

$$\mathcal{A}\colon f^n \to f^n \circ \mathcal{B}. \tag{3.8}$$

The adaptive version of this scheme only differs by the use of an adaptive (and moving) mesh $\mathcal{M}^n$ on which f^n is represented (in the sense that $f^n \in V_{\mathcal{M}^n}$). We thus basically have to add a *mesh moving* step in the scheme, that is a prediction of the new mesh $\mathcal{M}^{n+1}$ from the data $(f^n, \mathcal{M}^n)$. The *adaptive semi-Lagrangian scheme* reads then

$$(f^n, \mathcal{M}^n) \to \mathcal{M}^{n+1}, \tag{3.9}$$

$$f^{n+1} := P_{\mathcal{M}^{n+1}} \mathcal{A} f^n \tag{3.10}$$

for $n = 0, 1, \ldots$.

Remark (practical mesh prediction). The problem of constructing an adaptive mesh $\mathcal{M}_\varepsilon(f)$ well fitted to a *given* f is well understood (provided that f has some smoothness, see [8] or [6] for a general survey on nonlinear approximation) and near optimal algorithms can be obtained by using adaptive splitting, or wavelet-based *thresholding operators*

$$\mathcal{T}_\varepsilon\colon f \to P_{\mathcal{M}_\varepsilon(f)} f$$

for which $\| f - \mathcal{T}_\varepsilon f \|$ is lower than a prescribed tolerance $\varepsilon > 0$. On the other hand, it is a much more difficult task to predict a mesh $\mathcal{M}^{n+1}$ that is well adapted to the *unknown* solution f^{n+1}. Hence, like most adaptive methods, we first construct a 'pessimistic' guess $\tilde{\mathcal{M}}^{n+1}$ from f^n (by a heuristic procedure), compute then a temporary solution $\tilde{f}^{n+1} \in V_{\tilde{\mathcal{M}}^{n+1}}$ following (3.10), and eventually discard the small coefficients by a compression step, computing $f^{n+1} := P_{\mathcal{M}^{n+1}} \tilde{f}^{n+1}$ on $\mathcal{M}^{n+1} := \mathcal{M}_\varepsilon(\tilde{f}^{n+1})$. In the context of finite volume schemes for solving conservation laws, such a refinement strategy is precisely described in [7], together with a proof of convergence.

4 Biquadratic second order in time adaptive scheme

It is well known that linear reconstruction is too diffusive in the numerical resolution of the Vlasov equation. We now detail our practical implementation of the scheme, based on biquadratic finite elements instead of bilinear ones.

Adaptive biquadratic representation. We denote by $\Gamma^2(\alpha)$ the 9 equidistant nodes of a biquadratic cell α, and by $\Gamma^2(\mathcal{M})$ the nodes of all the cells of a given adaptive mesh $\mathcal{M}$. So, our numerical solution at time n is the data

$$\mathcal{F}_n := \big(\mathcal{M}^n,\ (f^n(a))_{a\in\Gamma^2(\mathcal{M}^n)}\big), \tag{4.1}$$

where $\mathcal{M}^n$ is an adaptive mesh. The evaluation $f^n(c)$ of the solution at a point $c \in [0,1[^2$ is obtained by searching the unique cell α of the adaptive mesh $\mathcal{M}$ where the point lives, using the values $f^n(\alpha) := (f^n(a))_{a\in\Gamma^2(\alpha)}$ and computing the local biquadratic interpolation on that cell, say $I(c, \alpha, f^n(\alpha))$.

Characteristics. Let us denote by $(X, V)(s; x, v, t)$ the characteristics of the Vlasov equation, i.e. the solutions of the following system of ordinary differential equations:

$$\begin{cases} \dot{X}(s) = V(s) \\ \dot{V}(s) = E(s, X(s)), \end{cases}$$

with initial conditions $X(t) = x$, $V(t) = v$. The advection is classically performed by time-splitting (see e.g [13]). A general two time step method has been introduced in [14] to solve the characteristics, by a direct $2D$ computation (see also [15] for an efficient new method). We have used the following completely local algorithm. Knowing the final position $a = (X^{n+1}, V^{n+1})$ at time $(n+1)\Delta t$, we compute a second order approximation (X^n, V^n) of the backward advected position $(X, V)(n\Delta t; a, (n+1)\Delta t)$.

So, with the notations $t^n = n\Delta t$, $t^{n+\frac{1}{2}} = (n+\frac{1}{2})\Delta t$ and $X^{n+\frac{1}{2}} = \frac{X^n+X^{n+1}}{2}$, we can write to second order accuracy:

$$\frac{X^{n+1}-X^n}{\Delta t} = \frac{V^n+V^{n+1}}{2} + O(\Delta t^2).$$

On the other hand, again to second order accuracy:

$$\begin{aligned} \frac{V^{n+1}-V^n}{\Delta t} &= E(t^{n+\frac{1}{2}}, X^{n+\frac{1}{2}}) + O(\Delta t^2) \\ &= \frac{1}{2}[E(t^n, X^{n+\frac{1}{2}}) + E(t^{n+1}, X^{n+\frac{1}{2}})] + O(\Delta t^2) \end{aligned}$$

and

$$\frac{1}{2}[E(t^{n+1}, X^{n+\frac{1}{2}}) + E(t^{n-1}, X^{n+\frac{1}{2}})] = E(t^n, X^{n+\frac{1}{2}}) + O(\Delta t^2).$$

So, after some more computations, we obtain a second order accurate formula,

$$\frac{V^{n+1} - V^n}{\Delta t} = 2E\left(t^n, \frac{X^n + X^{n+1}}{2}\right) - \frac{E(t^{n-1}, X^n) + E(t^n, X^{n+1})}{2} + O(\Delta t^2). \tag{4.2}$$

Poisson electric field. From the data $\mathcal{F}_n$ (defined in (4.1)) we can derive the density of charge $\rho^n = \int f^n(x, v)\,\mathrm{dv}$, and then obtain the electric field E^n from the Poisson equation

$$-\partial_x E^n = \rho^n - 1$$

(in the axisymmetric case, we will take $-\frac{1}{r}\partial_r(rE_r) = \rho$).

Backward/forward advection. We denote by $\mathrm{B}_n^{\mathrm{wd}}(a)$ the backward advected position of a node a. We obtain it here by (4.2) with $a = (X^{n+1}, V^{n+1})$ and $\mathrm{B}_n^{\mathrm{wd}}(a) = (X^n, V^n)$. This system can be solved iteratively for the unknown V^n. On the other hand $\mathrm{F}_n^{\mathrm{wd}}(a)$ denotes the forward advected position and is also given by (4.2), writing $a = (X^n, V^n)$ and $\mathrm{F}_n^{\mathrm{wd}}(a) = (X^{n+1}, V^{n+1})$.

The algorithm is now guided by two quantities: the *finest resolution level* J and a *thresholding tolerance number* ε .

Time marching algorithm

• Initialization.

The initial function is given by f_0 on the unit square. Let $\mathcal{M}^0$ be at first empty. Starting from $j = J - 1$ to j_0, for each cell $\alpha \in \mathcal{M}_j$,

▷ compute: $d(b) := f_0(b) - I(b, \alpha, f_0(\alpha))$ for $b \in \Gamma^2(\mathcal{D}(\alpha))\backslash\Gamma^2(\alpha)$

▷ add α in $\mathcal{M}^0$ if the following compression test is false:

$$\sum_{b\in\Gamma^2(\mathcal{D}(\alpha))\backslash\Gamma^2(\alpha)} \omega_{|\alpha|}|d(b)| \le \varepsilon, \tag{4.3}$$

where $|\alpha|$ is the level of the cell α and ω_j stands for 1, 2^{-dj}, $2^{-dj/2}$ depending on the norm on which we want to compress (respectively L_∞, L_1 or L_2), with d the dimension of the code (here $d = 2$). In our simulations, we will always take the L_2 norm.

▷ Add the necessary cells in order to have an adaptive mesh (i.e., so that the tree structure is respected).

For the sequel of the algorithm, we loop on n.

• Prediction of $\tilde{\mathcal{M}}^{n+1}$.

Let $\tilde{\mathcal{M}}^{n+1}$ be at first empty. For each center c of cell α of the adaptive mesh $\mathcal{M}^n$,

▷ compute the forward advected point $\mathrm{F}_n^{\mathrm{wd}}(c)$,

▷ add the unique cell of level $|\alpha|$ which fits at that place in $\tilde{\mathcal{M}}^{n+1}$.

Finally,

▷ add the necessary cells so that $\tilde{\mathcal{M}}^{n+1}$ is an adaptive mesh,

▷ refine the cells (which are not of level J) of one level, that is, replace each cell by its daughters.

• Semi-Lagrangian advection.

For each node $a \in \Gamma^2(\tilde{\mathcal{M}}^{n+1})$,

▷ compute the backward advected point $\mathrm{B}_n^{\mathrm{wd}}(a)$,

▷ set $\tilde{f}^{n+1}(a)$ to $f^n(\mathrm{B}_n^{\mathrm{wd}}(a))$.

Now, we have a first $\tilde{\mathcal{F}}_{n+1}$.

• Data compression: $\tilde{\mathcal{F}}_{n+1} \to \mathcal{F}_{n+1}$.

Starting from $j = J - 1$ to $j_0 + 1$, for each cell $\alpha \in \tilde{\mathcal{M}}^{n+1}$ of level j,

▷ compute: $d(b) := \tilde{f}^{n+1}(b) - I(b, \alpha, \tilde{f}^{n+1}(\alpha))$ for $b \in \Gamma^2(\mathcal{D}(\alpha))\backslash\Gamma^2(\alpha)$,

▷ remove α in $\tilde{\mathcal{M}}^{n+1}$ if the compression test (4.3) is true.

The remaining data is $\mathcal{F}_{n+1}$.

5 Test cases

Linear Landau damping. In order to test the precision of the numerical scheme, the linear Landau damping is very classical. The initial condition is given by

$$f(0, x, v) = \frac{1}{\sqrt{2\pi}} e^{-v^2/2}(1 + \alpha \cos(kx))$$

with $\alpha = 0.01$, the period equals $L = 4\pi$ and $k = 0.5$. For the time discretization, we choose $\Delta t = 1/8$. We restrain our computational domain in velocity to an interval $[-v_{\max}, v_{\max}]$, with a number $v_{\max}$ big enough. The electric field decays exponentially with a theoretical decay rate of $\gamma = 0.1533$ in the L_2 norm (see e.g. [13]). In fact, the numerical solution cannot decay all the time and the solution should restitute its energy at a "recurrence time" (see e.g. [13]). However, by taking more points in the velocity direction, we can push this phenomenon away and thus have a better description of the electric field for longer times.

In the adaptive case (see Figure 1), we take $J = 6$ and $v_{\max} = 7.15$ since these parameters give good results in the uniform case (see Figure 2): we obtain about 20 periods of oscillations. As expected, the accuracy of the damping increases, as the tolerance decreases and we reach the accuracy of the underlying uniform case with $\varepsilon = 10^{-8}$. The L_1 and L_2 norm are well conserved: the L_2 norm loses less than $4 \cdot 10^{-6}$ relative weight for $\varepsilon = 10^{-6}$ and the relative total mass error is less than 10^{-6}

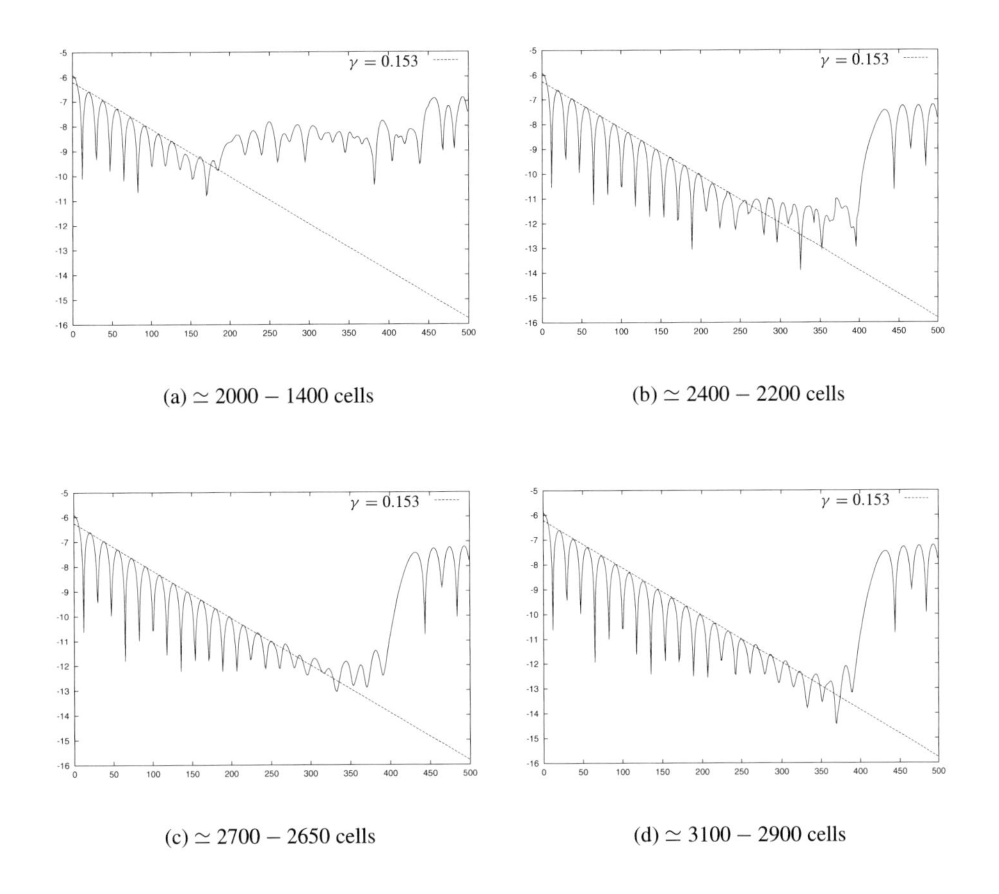

(a) $\simeq 2000 - 1400$ cells

(b) $\simeq 2400 - 2200$ cells

(c) $\simeq 2700 - 2650$ cells

(d) $\simeq 3100 - 2900$ cells

Figure 1. Evolution of $\log(\int E(x,t)^2\,dx)$ in terms of the number of iterations ($\Delta t = 0.125$) with $\Delta v = 0.112$ in adaptive case (with finest resolution level $J = 6$ and thresholding tolerance $\varepsilon = 10^{-5}, 10^{-6}, 10^{-7}, 10^{-8}$) for the linear Landau damping.

(however, if the tolerance is too large, the results are worse: if $\varepsilon = 10^{-5}$, the L_2 norm increases of 10^{-4} after 500 iterations).

On the other hand, the proportion of cells saved is about $1/4$ (for $\varepsilon = 10^{-8}$, about 3100 cells are used at the beginning and 2900 after the 100 first iterations; for $\varepsilon = 10^{-6}$, we go from 2400 to 2200 cells): the compression is occurring where the solution is almost null.

Thus, if the use of such an adaptive grid seems not to be the most natural way to treat this test case since the solution does not develop small scales, we are now convinced that, according to this example, the adaptive scheme can go to the accuracy of the uniform solution if we choose a tolerance small enough. We have also pointed out that, with such methods, we can increase the domain of calculation with a very small additional cost.

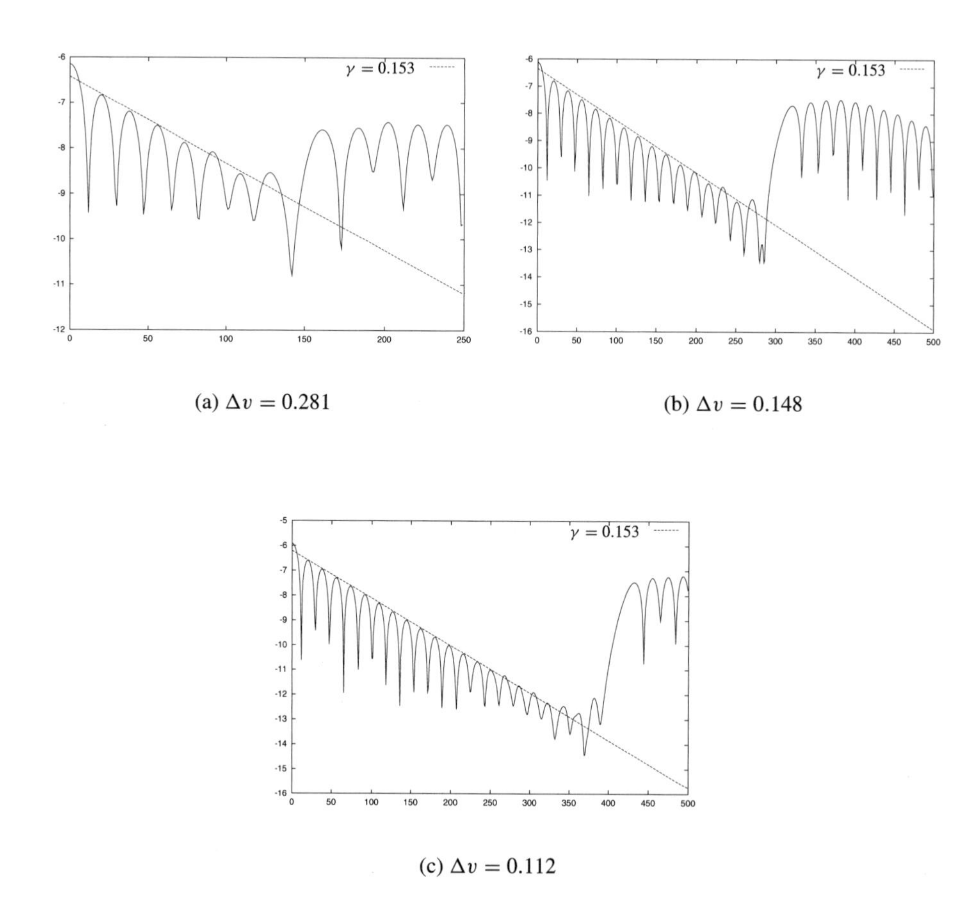

Figure 2. Evolution of $\log(\int E(x,t)^2\,dx)$ in terms of the number of iterations ($\Delta t = 0.125$) in the uniform case with resolution level $J = 4, 5$ and 6 (that is, respectively 256, 1024 and 4096 cells) for the linear Landau damping.

Semi-Gaussian beam. We now consider the semi-Gaussian beam defined by the initial condition

$$f_0(r,v) = \frac{1}{\pi a^2\sqrt{2\pi}} b e^{-\frac{1}{2}(v^2/b^2)}, \quad \text{if} \quad r < a,$$

and $f_0(r, v) = 0$ elsewhere. Here $a = 4/\sqrt{15}$ and $b = 1/(2\sqrt{15})$. The time step is $\pi/16$ which corresponds to $1/32$ of a period. We make 480 iterations, that is 15 periods.

The numerical solution develops small scales which disappear after a while, by diffusion. In our simulation, we have set $\varepsilon = 10^{-3}$ and $J = 7$ for the adaptive case.

We compare our adaptive solution to a coarse uniform solution (with the coarser resolution $J = 6$), and to a fine uniform one (with the same resolution $J = 7$). We see in Figures 4, 5 and 6 that the grid follows the development of the small scales, while the ratio $\#(\mathcal{M}^n)/\#(\mathcal{M}_6)$ goes in 6 periods from $1/4$ to 1 (see Figure 3(c)).

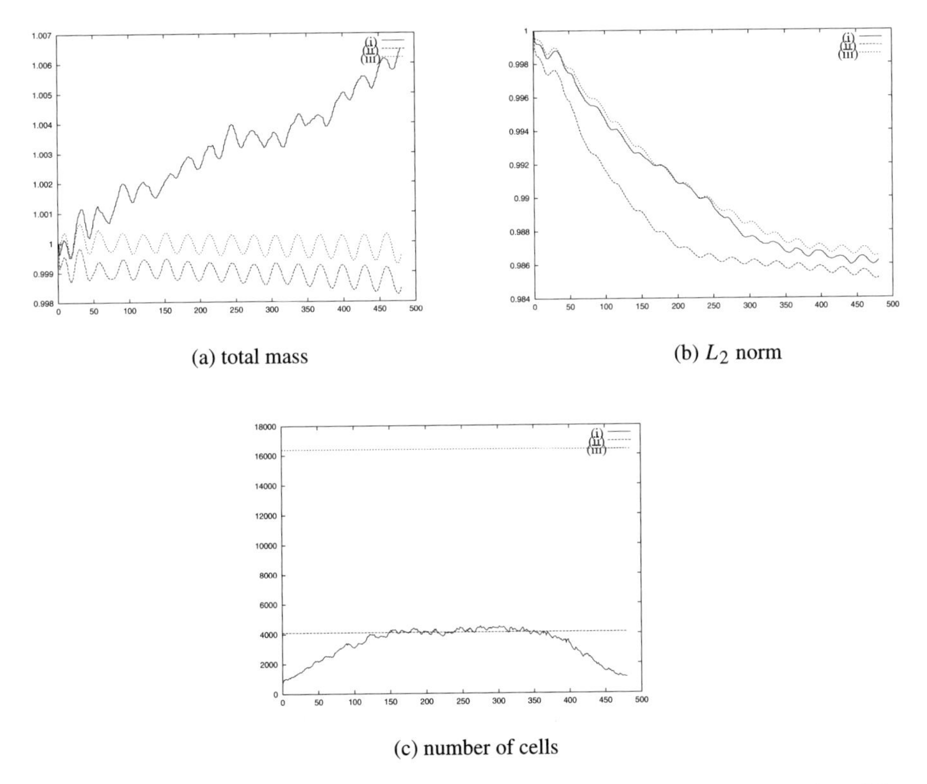

(a) total mass

(b) L_2 norm

(c) number of cells

Figure 3. Evolution of the relative *total mass*, L_2 *norm* of the distribution function f for the semi-Gaussian beam and the *number of cells* in terms of the number of iterations (32 iterations = 1 period) in the adaptive case with (i) $J = 7$ and $\varepsilon = 10^{-3}$, in the uniform case with (ii) $J = 6$ and with (iii) $J = 7$.

After 3 periods (Figure 5), the adaptive solution seems to better catch the nonlinear effects than the coarse uniform one, it remains in fact as accurate as the fine uniform solution. In Figure 6, we approach the full filamentation zone; the adaptive solution remains better than the coarse one, with always the same order of cells (see also 3(c)), but this time can not reach the accuracy of the fine uniform one. After 15 periods, the diffusion occurs and the derefining process works: we turn to the initial number of cells (see Figure 7 and 3(c)).

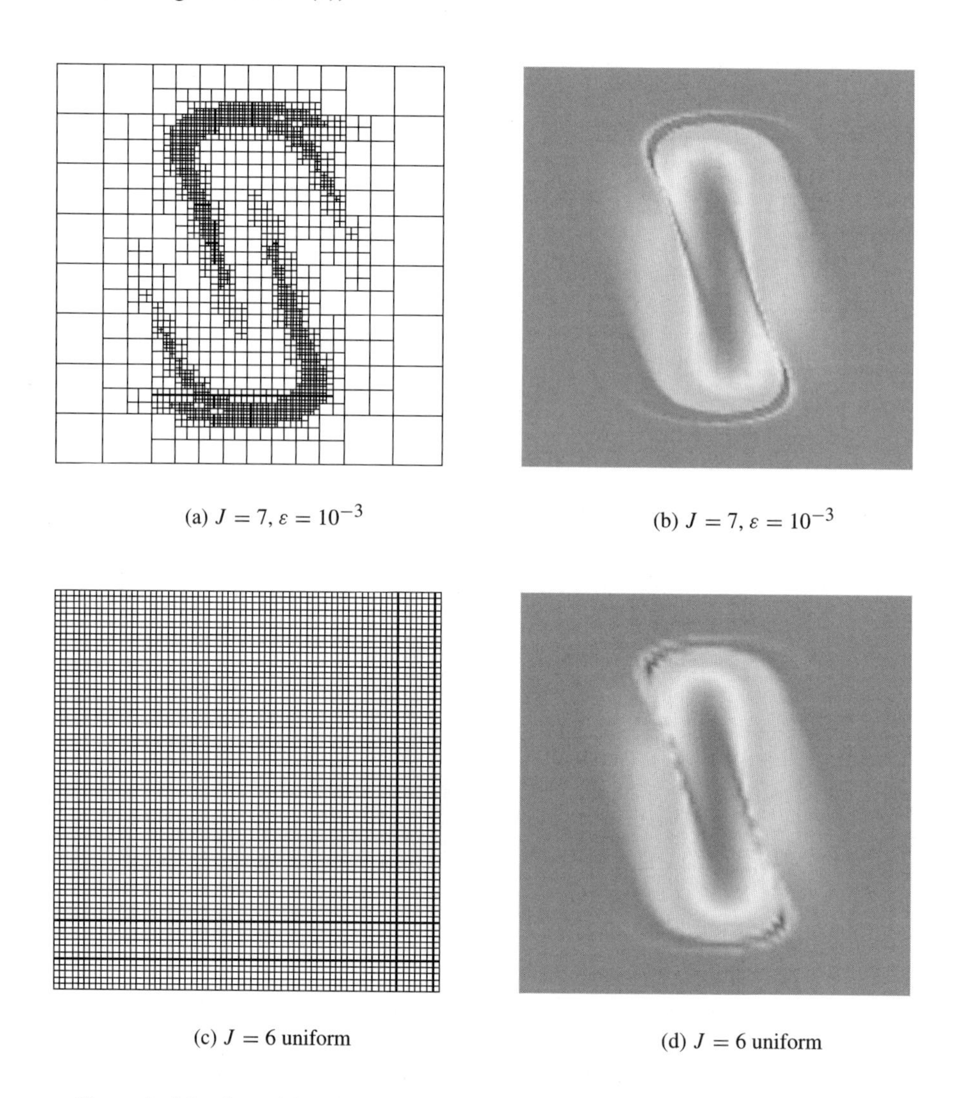

(a) $J = 7, \varepsilon = 10^{-3}$

(b) $J = 7, \varepsilon = 10^{-3}$

(c) $J = 6$ uniform

(d) $J = 6$ uniform

Figure 4. Adaptive grid and distribution function f in (v, x) phase space after 1.5 period (48 iterations).

By using biquadratic interpolation, the scheme is quite diffusive, however the adaptivity does not accelerate this phenomenon: the L_2 norm of the adaptive and fine uniform solutions are quite similar (Figure 3(b)). On the other hand, the mass is not well conserved in the adaptive case (Figure 3(a)): it tends to increase and this may come from the loss or gain of weight in the compression step and the lack of conservation of the interpolation on a non-uniform grid in the semi-Lagrangian scheme.

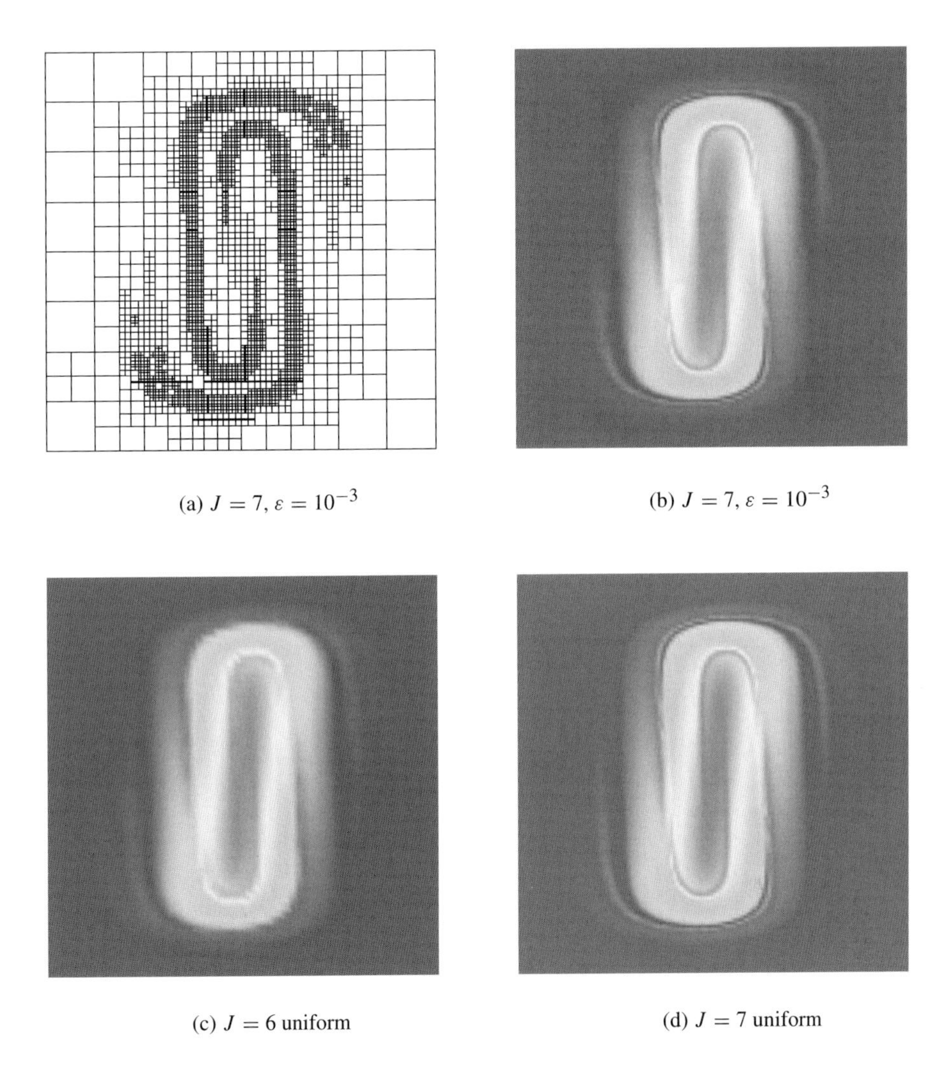

(a) $J = 7, \varepsilon = 10^{-3}$

(b) $J = 7, \varepsilon = 10^{-3}$

(c) $J = 6$ uniform

(d) $J = 7$ uniform

Figure 5. adaptive grid and distribution function f in (v, x) phase space after 3 periods (96 iterations).

Conclusion. This method seems to give the expected results. However, there is still much to do: the lack of mass conservation could be improved, and higher order polynomial interpolation should be performed in order to be more accurate. At a more theoretical point of view, we are studying the convergence and the complexity of this adaptive scheme.

Acknowledgments. The authors thank the CEMRACS organizers.

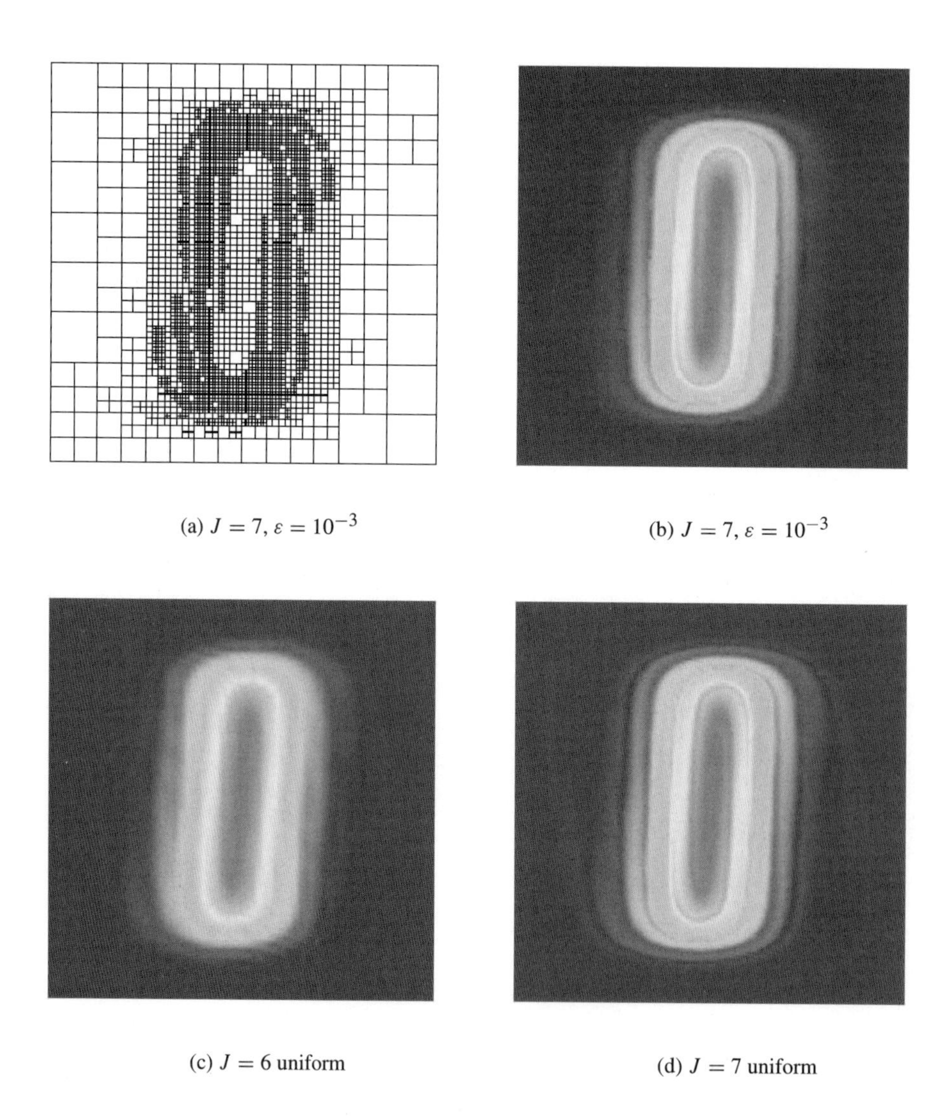

(a) $J = 7, \varepsilon = 10^{-3}$

(b) $J = 7, \varepsilon = 10^{-3}$

(c) $J = 6$ uniform

(d) $J = 7$ uniform

Figure 6. adaptive grid and distribution function f in (v, x) phase space after 6 periods (192 iterations).

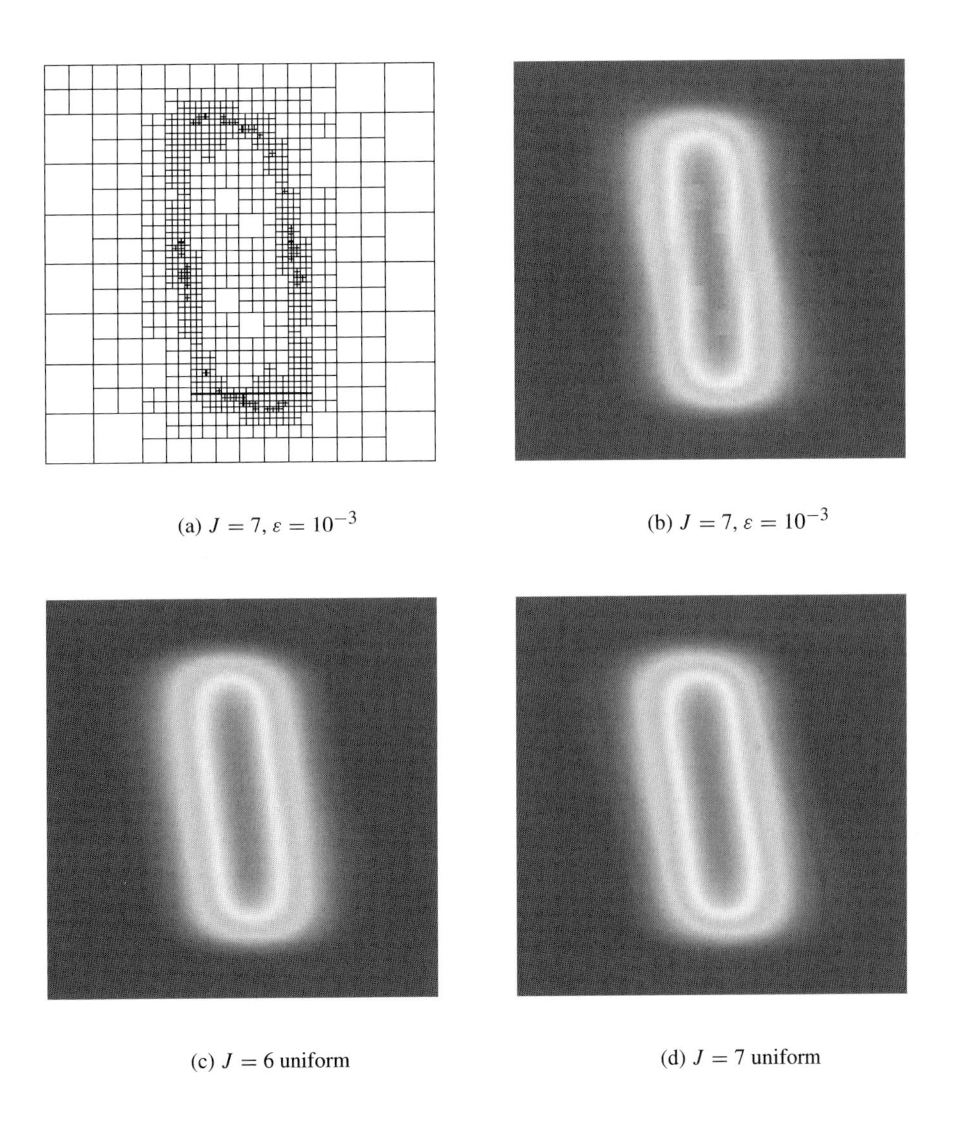

(a) $J = 7, \varepsilon = 10^{-3}$

(b) $J = 7, \varepsilon = 10^{-3}$

(c) $J = 6$ uniform

(d) $J = 7$ uniform

Figure 7. Adaptive grid and distribution function f in (v, x) phase space after 15 periods (480 iterations).

References

[1] N. Besse, Convergence of a semi-Lagrangian scheme for the one-dimensional Vlasov-Poisson system, *SIAM J. Numer. Anal.* 42 (2004), 350–382.

[2] N. Besse, F. Filbet, M. Gutnic, I. Paun, E. Sonnendrücker, An adaptive numerical method for the Vlasov equation based on a multiresolution analysis, in *Numerical Mathematics and Advanced Applications ENUMATH 2001*, ed. by F. Brezzi et al., Springer-Verlag, Berlin 2001, 437–446.

[3] N. Besse, E. Sonnendrücker, Semi-Lagrangian schemes for the Vlasov equation on an unstructured mesh of phase space, *J. Comput. Phys.* 191 (2) (2003), 341–376.

[4] C. K. Birdshall, A. B. Langdon, *Plasmaphysics via computer simulation*, McGraw-Hill, 1985.

[5] M. Campos Pinto and M. Mehrenberger, Convergence of an adaptive scheme for the one-dimensional Vlasov-Poisson system, *SIAM J. Numer. Anal.*, submitted.

[6] A. Cohen, *Numerical analysis of wavelet methods,* Stud. Math. Appl. 32, North-Holland, Amsterdam 2003.

[7] A. Cohen, S. M. Kaber, S. Müller and M. Postel, Fully adaptive multiresolution finite volume schemes for conservation laws, *Math. Comp.* 72 (241) (2003), 183–225.

[8] R. DeVore, Nonlinear approximation, in *Acta numerica* 7, ed. by A. Iserles, Cambridge University Press, Cambridge 1998, 51–150.

[9] F. Filbet, E. Sonnendrücker, Numerical Methods for the Vlasov equation, in *Numerical Mathematics and Advanced Applications, ENUMATH 2001*, ed. by F. Brezzi et al., Springer-Verlag, Berlin 2001, 459–468.

[10] F. Filbet, E. Sonnendrücker, P. Bertrand, Conservative Numerical schemes for the Vlasov equation, *J. Comput. Phys.* 172 (1) (2000), 166–187.

[11] M. Gutnic, M. Haefele, I. Paun and E. Sonnendrücker, Vlasov simulation on an adaptive phase-space grid, *Comput. Phys. Comm.* 164 (1–3) (2004), Proceedings of the 18th International Conference on the Numerical Simulation of Plasmas (September 7–10, 2003), 214–219.

[12] S. Mallat, *A wavelet tour of signal processing,* Academic Press, Diego, CA, 1998.

[13] T. Nakamura, T. Yabe, Cubic interpolated propagation scheme for solving the hyper-dimensional Vlasov-Poisson equation in phase space, *Comput. Phys. Comm.* 120 (1999), 122–154.

[14] E. Sonnendrücker, J. Roche, P. Bertrand, A. Ghizzo, The Semi-Lagrangian Method for the Numerical Resolution of Vlasov Equations, *J. Comput. Phys.* 149 (1999), 201–220.

[15] E. Sonnendrücker, F. Filbet, A. Friedman, E. Oudet and J.-L. Vay, Vlasov simulations of beams with a moving grid, *Comput. Phys. Comm.* 164 (1–3) (2004), Proceedings of the 18th International Conference on the Numerical Simulation of Plasmas (September 7–10, 2003), 390–395.

A conservative and entropic method for the Vlasov–Fokker–Planck–Landau equation

N. Crouseilles and F. Filbet

Mathématiques pour l'Industrie et la Physique, CNRS UMR 5640
Université Paul Sabatier – Toulouse 3
118 route de Narbonne, 31062 Toulouse Cedex 4, France
and
CEA-CESTA, DEV/SIS, 33114 Le Barp, France
email: `crouseilles@mip.ups-tlse.fr`

Mathématiques et Applications, Physique Mathématique d'Orléans, CNRS UMR 6628
Université d'Orléans, B.P. 6759, 45067 Orléans Cedex 2, France
email: `filbet@labomath.univ-orleans.fr`

Abstract. In this paper we compute numerical solutions to the Vlasov equation coupled with the Fokker–Planck–Landau collision operator using a phase space grid. On the one hand, the transport is treated by a flux conservative method and the distribution function is reconstructed to preserve energy and to control numerical oscillations. On the other hand, the collision operators are treated to preserve the conservation properties. Some results are presented in one dimension in space and three dimensions in velocity.

1 Introduction

The aim of this paper is to compute the evolution of a collisional plasma constituted of two species of charged particles (ions, electrons,...). The Landau or Fokker–Planck–Landau equation is a common kinetic model in plasma physics ([3], [4], [9]). It describes binary collisions between charged particles with long-range Coulomb interaction. The evolution of electrons is given by the distribution function $f(t, x, v)$ which depends on time t, position $x \in \Omega \subset \mathbb{R}^3$ and velocity $v \in \mathbb{R}^3$. This distribution function is solution to the scaled Vlasov–Fokker–Planck–Landau equation

$$\frac{\partial f}{\partial t} + v \cdot \nabla_x f + E \cdot \nabla_v f = \nu_i \, Q_i(f) + \nu_e \, Q_e(f, f), \tag{1.1}$$

where $E = E(t, x)$ is the electric field which can evolve through the Poisson equation

$$\nabla_x \cdot E(t, x) = \int_{\mathbb{R}^3} f(v) dv - \rho_i, \;\; E(t, x) = -\nabla_x \phi(t, x), \tag{1.2}$$

where ρ_i is the ion density and ϕ the electric potential. Moreover, ν_e is related to the electron-electron collision frequency and the expression of the Fokker–Planck–Landau (FPL) operator $Q_e(f, f)$, which describes collisions between two electrons, is

$$Q_e(f, f) = \nabla_v \cdot \left(\int_{\mathbb{R}^3} \Phi(v - v')(\nabla_v f(v) f(v') - \nabla_{v'} f(v') f(v)) dv' \right), \tag{1.3}$$

where $\Phi(v)$ is the 3×3 matrix

$$\Phi(v) = \frac{1}{|v|^3} \left(|v|^2 I_3 - v \otimes v \right). \tag{1.4}$$

Finally, ν_i is related to the electron-ion collision frequency and $Q_i(f)$ describes collisions between electrons and ions,

$$Q_i(f) = \rho_i \nabla_v \cdot (\Phi(v - u_i) \nabla_v f), \tag{1.5}$$

with u_i the ionic mean velocity. Some assumptions on the different species of particles leads to the simpler collision operator (1.5).

The main goal of this note is to describe a scheme in the x depending case and in the whole 3D velocity space. The first step consists in the approximation of the Vlasov equation, which is the left hand side of (1.1). We use a method which discretizes it on a phase space grid (see [5], [6], [8]). This kind of approach allows to get an accurate approximation of the distribution function in the phase space, but the non conservation of the energy due to the projection of the interpolation on the grid can be an inconvenient for the long time behaviour of the solution. In this paper, we propose a new scheme which overcomes to these inconvenient: space and velocity derivatives are approximated by a centered finite difference method and an adapted approximation of the electric field E allows us to obtain a numerical scheme that conserves the total energy. As this kind of discretization does not ensure the positivity of the unknown, we introduce slope correctors. The distribution function is reconstructed following the PFC method (see [5], [6]) of order two. When the slope correctors act, the total energy is not conserved any more; nevertheless, the variations are not very important compared to others methods such as semi-Lagrangian method.

Finally, we consider collision operators derived from (1.3). We propose conservative schemes based on the ideas of [3] to approximate this different operators. Even if these schemes give interesting properties (conservations, decay of the entropy, positivity of the distribution function), their direct implementation is very expensive in high dimensions. Thus, we adopt the multigrid method used in [1] to reduce the computational cost.

The rest of the paper is organized as follows. In the next section, we draw up the main properties of the solution to the system (1.1)–(1.2)–(1.3) and formally derive the operator intended to model collisions between different species. We then present in Section 3 a finite difference scheme for the discretization of the Vlasov–Poisson equation. Finally, some numerical results are presented in Section 4 to illustrate the efficiency of the method.

2 Description of the kinetic model

In this section we briefly recall some classical estimates on the system (1.1)–(1.5). First, the Vlasov–Poisson system (1.1)–(1.2) without collision, preserves mass, momentum and total energy. It also conserves the kinetic entropy

$$H(t) = \int_{\mathbb{R}^3\times\mathbb{R}^3} f(t)\,\log(f(t))dxdv = H(0).$$

Next, let us recall the expression of the Fokker–Planck–Landau (FPL) operator which describes collisions between two electrons

$$Q_e(f,f) = \nabla_v \cdot \left(\int_{\mathbb{R}^3} \Phi(v-v')(\nabla_v f(v) f(v') - \nabla_{v'} f(v') f(v))dv' \right), \tag{2.1}$$

where $\Phi(v)$ is the 3×3 matrix (1.4). The algebraic structure of the FPL operator leads to physical properties such as the conservations of mass, momentum and energy

$$\int_{\mathbb{R}^3} Q_e(f,f)(v) \begin{pmatrix} 1 \\ v \\ |v|^2 \end{pmatrix} dv = 0,$$

and the decay of the entropy $H(t)$. Moreover, the equilibrium states of the FPL operator, *i.e.* the distribution functions f which satisfy $Q_e(f,f) = 0$, are given by the Maxwellians whose parameters are the density, mean velocity and temperature of f (see [3], [9] for more details).

Finally, the operator Q_i describes collisions between two different species (for instance electrons and ions), and can be derived from the two species form of the full Landau operator (see [4]). If we assume that the temperature of ions T_i is negligible compared to the temperature of electrons T, we may consider that the distribution function of ions is given by a Dirac measure in velocity

$$f_i(t,x,v) = \rho_i(t,x)\,\delta_0(v - u_i(t,x)), \tag{2.2}$$

where the density ρ_i and the mean velocity u_i are given or satisfy hydrodynamic equations. The so-obtained operator then reads

$$Q_i(f) = \rho_i \nabla_v \cdot (\Phi(v-u_i)\nabla_v f), \tag{2.3}$$

where $\Phi(v)$ is the 3×3 matrix given by (1.4). The so-obtained linear collision operator (2.3) satisfies some properties which are listed in the following proposition

Proposition 2.1. *The linear collision operator $Q_i(f)$ given by* (2.3) *satisfies*

(i) *the preservation of mass and energy, i.e.*

$$\int_{\mathbb{R}^3} Q_i(f)(v)dv = 0, \quad \int_{\mathbb{R}^3} Q_i(f)(v)|v-u_i|^2 dv = 0,$$

(ii) $\text{Ker}(Q_i) = \{f(|v-u_i|^2)\}$,

(iii) *each convex function* Ψ *of* f *is an entropy for* Q_i

$$\frac{d}{dt}\int_{\mathbb{R}^3} \Psi(f)dv \le 0.$$

For the proof of this proposition, we refer the reader to [2].

3 The numerical method

In this section we introduce a new scheme for the discretization of the Vlasov–Poisson–Fokker–Planck–Landau equation (1.1)–(1.5).

3.1 Transport approximation

First, we present the approximation to the Vlasov–Poisson equation. Thanks to a discretization based on a finite difference method, joint with an adapted approximation of the electric field, we will prove that this scheme preserves the total energy. We consider a cartesian grid in one dimension in space and velocity for the sake of simplicity.

We introduce the mesh points $(x_i)_{i\in I}, (v_j)_{j\in\mathbb{Z}}$ of the computational domain $[x_{\min}, x_{\max}] \times \mathbb{R}$. We will denote by $\Delta x = x_{i+1} - x_i$ and $\Delta v = v_{j+1} - v_j$ the space and velocity steps. Finally, $t^n = n\Delta t$ is the time discretization, $n \in \mathbb{N}$.

We assume that $f_{i,j}^n$ is given by

$$f_{i,j}^n = f(t^n, x_i, v_j).$$

Then we approximate the distribution function at the time t^{n+1} by

$$f_{i,j}^{n+1} = f_{i,j}^n - \frac{\Delta t}{\Delta x}(F_{i+1/2,j}^n - F_{i-1/2,j}^n) - \frac{\Delta t}{\Delta v}(G_{i,j+1/2}^n - G_{i,j-1/2}^n),$$

where $F_{i+1/2,j}^n$ and $G_{i,j+1/2}^n$ are given by

$$F_{i+1/2,j}^n = v_j f_h(x_i + \Delta x/2, v_j), \quad G_{i,j+1/2}^n = \tilde{E}_i^n f_h(x_i, v_j + \Delta v/2),$$

where $\tilde{E}_i^n$ is an approximation at (t^n, x_i) of the electric field obtained by the Poisson equation (1.2) and f_h is an approximation of the distribution function. Assume v_j is positive, then the high order approximation $f_h(x, v)$ in the x direction is given by

$$f_h(x, v_j) = f^n_{i,j} + \frac{(x - x_i)}{\Delta x}(f^n_{i+1,j} - f^n_{i,j}).$$

As this kind of approximation can generate spurious oscillations and does not ensure the positivity of the distribution function on $[x_i - \Delta x/2, x_i + \Delta x/2]$, we introduce slope correctors

$$f_h(x, v_j) = f^n_{i,j} + \epsilon^+_{i,j}\frac{(x - x_i)}{\Delta x}(f^n_{i+1,j} - f^n_{i,j}),$$

where $\epsilon^+_{i,j}$ ensures positivity of f_h and is given by

$$\epsilon^+_{i,j} = \min\left(1, 2f^n_{i,j}/(f^n_{i+1,j} - f^n_{i,j})\right) \quad \text{if } (f^n_{i+1,j} - f^n_{i,j}) > 0. \tag{3.1}$$

For a negative velocity v_j the reconstruction on the interval $[x_i - \Delta x/2, x_i + \Delta x/2]$ writes

$$f_h(x, v_j) = f^n_{i,j} + \epsilon^-_{i,j}\frac{(x - x_i)}{\Delta x}(f^n_{i,j} - f^n_{i-1,j}), \tag{3.2}$$

with

$$\epsilon^-_{i,j} = \min\left(1, -2f^n_{i,j}/(f^n_{i,j} - f^n_{i-1,j})\right) \quad \text{if } (f^n_{i,j} - f^n_{i-1,j}) < 0. \tag{3.3}$$

Now we proceed in the same way to reconstruct f_h in the v direction depending on the sign of $\tilde{E}^n_i$.

When the slope correctors do not occur in the approximation of the velocity derivative (it is the case when the distribution function is sufficiently smooth as a function of v) we obtain a classical centered scheme. With an adapted approximation of the electric field, $\tilde{E}^n_i = (E^{n+1}_i + E^n_i)/2$, the so-obtained scheme preserves the total energy. On the other hand, the electric field at time t^n is determined through the following approximation to the Poisson equation

$$-D^\star_x E^n_i = \rho^n_i - \rho_i, \quad E^n_i = -D_x\phi^n_i, \tag{3.4}$$

where D_x is a discrete finite difference operator whereas $D^\star_x$ stands for its formal adjoint, which represents an approximation of $-\partial_x$ (for example, we can consider D_x as the usual uncentered discrete operator); hence the following equality holds:

$$\langle D^\star_x\varphi, f\rangle = \langle\varphi, D_x f\rangle \quad \text{for all sequences } \varphi, f,$$

where $\langle\, ,\rangle$ denotes an inner product as

$$\langle\varphi, f\rangle = \sum_{i\in\mathbb{Z}} \varphi_i f_i \Delta x.$$

Finally, E^{n+1}_i is a prediction of the electric field at time t^{n+1} obtained from the discretization to the Poisson equation (3.4) and to the continuity equation

$$\rho^{n+1} = \rho^n_i + \Delta t\, D^\star_x J^n_i, \tag{3.5}$$

where J_i^n and ρ_i^n are respectively an approximation of the current density $j(t,x)$ and of the charge density $\rho(t,x)$

$$J_i^n = \Delta v \sum_{j\in\mathbb{Z}} v_j f_{i,j}^n, \quad \rho_i^n = \Delta v \sum_{j\in\mathbb{Z}} f_{i,j}^n. \tag{3.6}$$

The approximation of the Vlasov–Poisson equation obtained from this algorithm satisfies the conservation of mass and total energy (see [2] for the proof).

3.2 Collision operators approximation

For the discretization of the linear collision operator (1.5), we use the log formulation and replace the ∇_v operator by its discrete version D and the $-\nabla_v\cdot$ operator by its formal adjoint. For example, for any test sequence $(\varphi_j)_{j\in\mathbb{Z}^3}$, we set $(D_\pm\varphi_j)_{j\in\mathbb{Z}^3}$ as a sequence of vectors of $\mathbb{R}^3$

$$D_+^s\varphi_j = \frac{\varphi_{j+e_s} - \varphi_j}{\Delta v}, \quad s = 1,2,3,$$

and

$$D_-^s\varphi_j = \frac{\varphi_j - \varphi_{j-e_s}}{\Delta v}, \quad s = 1,2,3,$$

where e_s denotes the s-th vector of the canonical basis of $\mathbb{R}^3$. D_+ and D_- are respectively called the forward and backward uncentered discrete operator.

Then, to preserve the energy at the discrete level, we introduce a symmetrization based on the averaging of the two uncentered discretizations

$$\begin{aligned} Q_i(f)(v_j) = -\frac{1}{2}\Bigg[& D_+^\star\left(\frac{1}{|v_j|^3} S(\tilde{v}_j^+)\, f_j\, D_+(\log f_j)\right) \\ & + D_-^\star\left(\frac{1}{|v_j|^3} S(\tilde{v}_j^-)\, f_j\, D_-(\log f_j)\right)\Bigg], \end{aligned} \tag{3.7}$$

where $S(v) = |v|^2 I_3 - v\otimes v$, and

$$\tilde{v}_j^{+,s} = \frac{1}{2}(v_j^s + v_{j+e_s}^s) \quad \text{and} \quad \tilde{v}_j^{-,s} = \frac{1}{2}(v_j^s + v_{j-e_s}^s), \quad s = 1,2,3. \tag{3.8}$$

Proposition 3.1. *The discretization* (3.7)–(3.8) *conserves mass, energy and decreases the discrete entropy. Moreover, if the sequence $(f_j)_{j\in\mathbb{Z}^3}$ is symmetric in all the directions (i.e. $f_k = f_j$, with $k_s = -j_s$, $s = 1,2,3$), then the discrete collision operator* (3.7)–(3.8) *conserves the momentum.*

For the proof of this proposition we refer the reader to [2].

Finally, $Q_e(f,f)$ is approximated in the same way as $Q_i(f)$. It is an easy matter to verify that the discrete version of $Q_e(f,f)$ preserves mass, momentum and energy, and decreases the entropy. The method was first introduced by [3]. Nevertheless, this kind of approximation remains too expensive. Hence we use the multigrid method used in [1] to treat the computational cost issue.

4 Numerical simulations

4.1 Linear Landau damping

In this section we investigate the Landau damping in the linear context. To that purpose, we apply the above approximation to discretize the full system (1.1)–(1.5) where the collision operators are given by (2.1) and (2.3) with $\rho_i = 1$ and $u_i = 0$. Besides, the collision parameters are chosen as follows: $\nu_e = \nu/Z$ and $\nu_i = \nu$, where Z is the number of charge. The initial condition is chosen as a periodic perturbation of the global equilibrium

$$f_0(x, v) = \frac{1}{(2\pi)^{3/2}} \exp(-|v|^2/2)(1 + A\cos(kx)), \quad x \in [0, 2\pi/k],\ v \in \mathbb{R}^3, \quad (4.1)$$

where A is the amplitude of the perturbation, and k denotes the wave number. In this subsection we only consider linear regimes and A is taken equal to $A = 10^{-5}$.

To capture the Landau damping, the size of the velocity domain must be chosen greater than the phase velocity $v_\phi = \omega/k$ (with $\omega^2 \approx 1 + 3k^2$). Then we set $v_{\max} = 5.75$ where the velocity grid extends from $-v_{\max}$ to $v_{\max}$. We use $N_v = 32$ grid points in each direction for the velocity grid whereas the number of spatial grid points is $N_x = 50$. The boundary conditions for the distribution function are periodic in x whereas the distribution function is truncated to zero for large velocities.

We are interested in the evolution of the square root of the electric energy approximated by

$$\mathcal{E}_h(t) = \Big(\sum_i \Delta x E_i^2(t)\Big)^{1/2}. \quad (4.2)$$

In Figure 1 we plot $\mathcal{E}_h(t)$ in log scale and its associated damping line (which is determined numerically) as a function of time. We study the influence of the collisions and consider values of ν equal to $\nu = 0$, 0.01 and 0.05 whereas $k = 0.3$ and $Z = 1$. The value of $\nu = 0$ corresponds to the Vlasov–Poisson equation. The amplitude of the electric field is damped exponentially in time behaviour and we observe the influence of the collisions on the electric energy: the damping rate becomes stronger when the collisions are more important, *i.e.* when ν is bigger.

In Figure 2 we study the influence of Z (with $\nu = 0.01$) on the damping of $\mathcal{E}_h(t)$. We observe that the damping rate is not very sensible to Z. We recover the fact that *e-i* collisions play a more important role than *e-e* ones in the damping of electronic plasma waves.

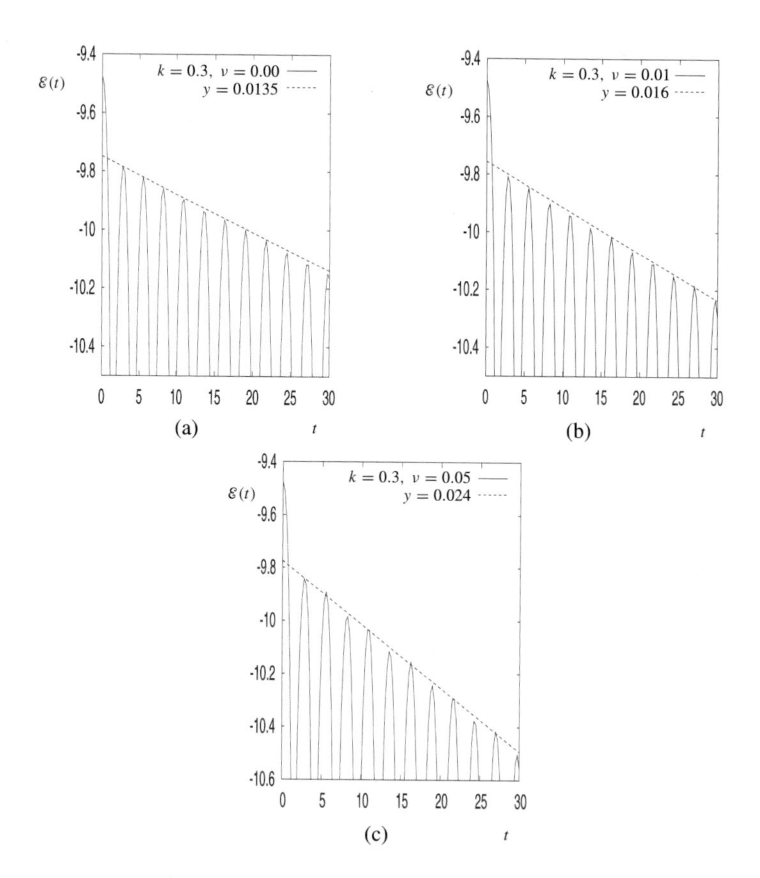

Figure 1. Study of the influence of ν: $k = 0.3$ and $\nu = 0,\ 0.01,\ 0.05$.

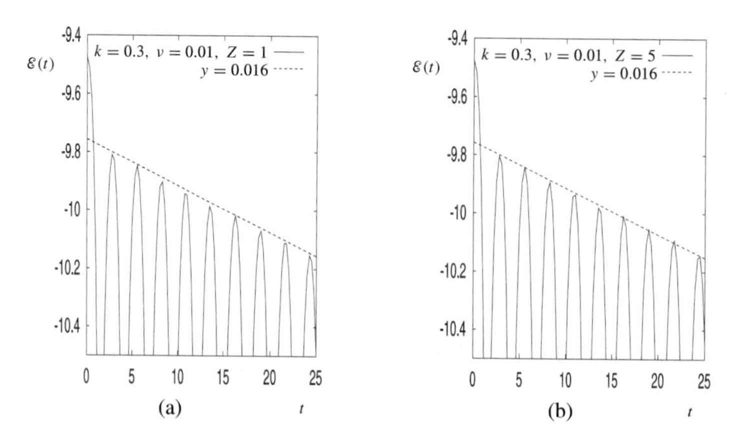

Figure 2. Study of the influence of Z: $k = 0.3$, $\nu = 0.01$ and $Z = 1,\ 5$.

The results of these tests case can be compared to estimated results. In fact, when the amplitude perturbation A is small enough, we can linearize the initial model. The solution of the so-obtained linearized model is a plasma wave the frequency and damping of which can be computed. For the damping rate, one finds $\gamma = \gamma_L + \gamma_C$, where γ_L comes from the Landau damping and γ_C from collisional effects

$$\begin{aligned} \gamma_L &= -\sqrt{\frac{\pi}{8}\frac{1}{k^3}} \exp(-1/(2k^2) - 3/2 - 3k^2), \\ \gamma_C &= -\frac{1}{3}\sqrt{\frac{2}{\pi}}\nu\left(1 + \frac{5}{2}k^2 + \frac{4\sqrt{2}}{5Z}k^2\right). \end{aligned} \tag{4.3}$$

One notes that our numerical results are in very good agreement with formulas (4.3), even for subtle phenomena such as the variation of Z.

4.2 Ion acoustic waves

The aim of this subsection is the study of the ion acoustic waves (see [4], [7], [10]). To that purpose, we only consider the Vlasov equation coupled with the Fokker–Planck–Landau operator (1.3) which models the ion-ion collisions. The initial condition is

$$f_0(x, v) = \frac{1}{(2\pi)^{3/2}} \exp(-|v|^2/2)(1 + A\cos(kx)), \quad x \in [0, 2\pi/k],\ v \in \mathbb{R}^3, \tag{4.4}$$

where A is the amplitude of the perturbation, and k denotes the wave number.

As the ionic mean free path is bigger than the electron one, we can consider a constant electron temperature; in this case, the electric field is written

$$E = -\frac{T_e}{T_i}\nabla_x(\log\rho), \tag{4.5}$$

where T_e and T_i are respectively the electron and ion temperature, and ρ the ion density (see [7]). We will take the ratio T_e/T_i equal to 4, $\nu = Z = 1$ and k varies from 0.05 to 3.

As in the previous subsection, we only consider linear regimes and take $A = 10^{-3}$. The size of the velocity domain extends from $-v_{\max}$ to $v_{\max} = 5.75$ and use $N_v = 32$ grid points in each direction for the velocity grid whereas the number of spatial grid points is $N_x = 64$. The numerical scheme presented in Section 3 is used here. However, the electric field is not obtained through the Poisson equation in this subsection. Thus, $\tilde{E}_i^n$ is an approximation of (4.5) given by

$$\tilde{E}_i^n = -\frac{T_e}{T_i} D_c(\log\rho^n)_i,$$

where D_c is the usual discrete centered operator. In this case, the total energy is not conserved any more.

We are interested in the evolution of (4.2). Its amplitude is expected to be exponentially and periodically decreasing. Then we compute numerically the damping rate γ and the frequency ω and compare our results to the results obtained in [7].

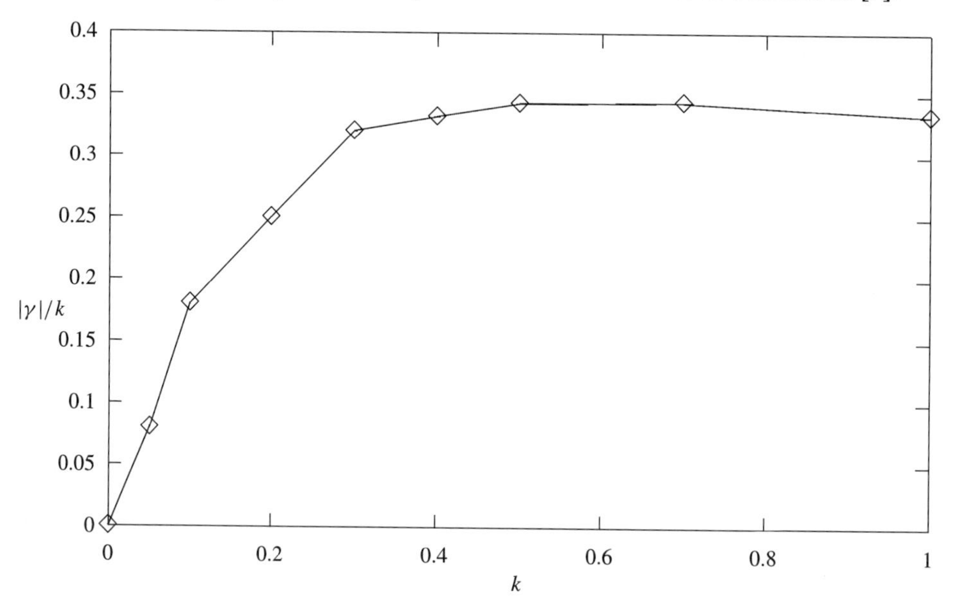

Figure 3. Study of the influence of k on the damping γ.

Table 1. Comparison of γ/k with [7].

	$k=0.05$	$k=0.1$	$k=0.2$	$k=0.3$	$k=0.4$	$k=0.5$	$k=0.7$	$k=1$
Numerical results	0.08	0.18	0.25	0.32	0.332	0.343	0.344	0.333
Results [7]	0.093	0.165	0.258	0.331	0.344	0.342	0.337	0.33
Relative difference	16.2 %	8.3 %	3.2 %	3.4 %	3.6 %	0.3 %	2 %	0.9 %

In Figures 3 and 4, the quantities $|\gamma|/k$ and ω/k are plotted as a functions of k. One should note that the behavior of our curves is similar to [7]. Indeed, for small k, $|\gamma|/k$ is proportional to k, and for large values of k, $|\gamma|/k$ becomes stabilized and decreases slowly. Concerning the frequency, we observe that ω/k slowly increases as k increases whereas in [7], ω/k becomes nearly constant for large values of k. From a quantitative point of view, our results for the damping rate are in good agreement with the values presented in [7], especially for large k (the collisionless limit ($k \to +\infty$) is well reached). For the frequency ω, the results are also in good agreement with those obtained in [7]: the relative error is only about 4.3 %.

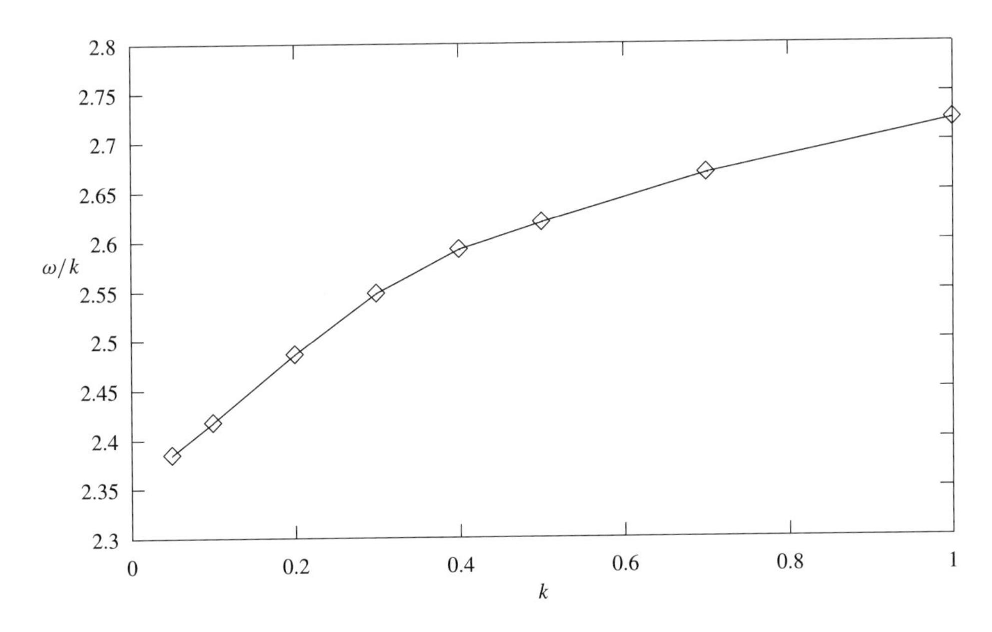

Figure 4. Study of the influence of k on the frequency ω.

Table 2. Comparison of ω/k with [7].

	$k=0.05$	$k=0.1$	$k=0.2$	$k=0.3$	$k=0.4$	$k=0.5$	$k=0.7$	$k=1$
Numerical results	2.384	2.517	2.485	2.547	2.591	2.618	2.668	2.721
Results [7]	2.38	2.39	2.42	2.59	2.66	2.73	2.76	2.78
Relative difference	0.2 %	5 %	2.6 %	1.7 %	2.7 %	4.3 %	3.4 %	2.2 %

5 Conclusion

We have developed a numerical simulation algorithm for realistic and collisional plasmas in one dimension in the physical space and in three dimensions in velocity space. This method takes into account two different species of particles (*e.g.* electrons and ions). A new discretization of the Vlasov–Poisson equation, from which we proved the important properties of conservation (mass and total energy) has been written. In the applications to Landau damping, our results are in good agreement with the estimations. On the other hand, we accurately recover the results of [7] in the applications of ion acoustic waves. We hope that this algorithm will be useful tool to help physicists in modeling and will give accurate approximates of collisional plasmas. We finally refer to [2] for a more detailed description of the method and different numerical tests for the Vlasov–Poisson system coupled with collision operators.

Acknowledgements. This work was partially performed at the Commissariat à l'énergie atomique (CEA-CESTA) and within the framework of CEMRACS 2003 whose organizers (S. Cordier, T. Goudon, M. Gutnic and E. Sonnendrücker) we wish to acknowledge. The authors are also grateful to B. Dubroca and V. Tikhonchuk for suggesting the problem and for very fruitful discussions.

References

[1] C. Buet, S. Cordier, P. Degond, M. Lemou, Fast algorithms for numerical, conservative, and entropy approximations of the Fokker-Planck-Landau equation, *J. Comput. Phys.* 133 (1997), 310–322.

[2] N. Crouseilles, F. Filbet, Numerical approximation of collisional plasmas by high order methods, *J. Comput. Phys.* 201 (2004), 546–572.

[3] P. Degond, B. Lucquin-Desreux, An entropy scheme for the Fokker-Planck collision operator of plasma kinetic theory, *Numer. Math.* 68 (1994), 239–262.

[4] J. P. Delcroix, A. Bers, *Physique des plasmas*, Savoirs Actuels, InterEditions, CNRS Editions, 1994.

[5] F. Filbet, L. Pareschi, Numerical method for the accurate solution of the Fokker-Planck-Landau equation in the non homogeneous case, *J. Comput. Phys.* 179 (2002), 1–26.

[6] F. Filbet, E. Sonnendrücker, P. Bertrand, Conservative Numerical schemes for the Vlasov equation, *J. Comput. Phys.* 172 (2001), 166–187.

[7] M. Ono, R. M. Kulsrud, Frequency and damping of ion acoustic waves, *The Physics of Fluids* 10 (1975), 1287–1293.

[8] M. Shoucri, G. Knorr, Numerical integration of the Vlasov equation, *J. Comput. Phys.* 14 (1974), 84–92.

[9] C. Villani, A review of mathematical topics in collisional kinetic theory, in *Handbook of mathematical fluid dynamics*, Vol. 1, ed. by S. Frielander and D. Serre, North-Holland, Amsterdam 2002, 71–305.

[10] J. Zheng, C. X. Yu, Ion-collisional effects on ion-acoustic waves: an eigenvalue technique via moment expansion, *Plasma Phys. Control. Fusion* 42 (2000), 435–441.

Numerical studies for nonlinear Schrödinger equations: the Schrödinger–Poisson-Xα model and Davey–Stewartson systems

C. Besse, N. J. Mauser and H. P. Stimming

Laboratoire de Mathématiques pour l'Industrie et la Physique, CNRS UMR 5640
Université Paul Sabatier, 118 Route de Narbonne, 31062 Toulouse Cedex 4, France
email: besse@mip.ups-tlse.fr

Wolfgang Pauli Institute, c/o Institut für Mathematik, Universität Wien, Strudlhofgasse 4, 1090 Wien, Austria
email: norbert.mauser@univie.ac.at
hans.peter.stimming@univie.ac.at

Abstract. We study the numerical approximation and analysis of two different time dependent nonlinear Schrödinger equations (NLS) : the "Schrödinger–Poisson-Xα" equation, which plays a role in the modeling of quantum particle dynamics, and the Davey–Stewartson system, which is a 2-d equation modeling unidirectional water surface waves. For both equations we use a Time Splitting Spectral scheme, which had previously shown to be a good tool for numerical simulation of cubic NLS. This scheme is particularly useful for calculations in the "semi-classical regime", where the scaled Planck constant is taken to be small. Extensive numerical results of position density and Wigner measures in 1d, 2d and 3d for the S-P-Xα model with/without an external potential are presented. These results give an insight to understand the interplay between the nonlocal ("weak") and the local ("strong") nonlinearity.

For the Davey–Stewartson system, we give a convergence analysis for the semi-discrete version of the scheme. Numerical results are presented for various blowup phenomena of the equation, including blowup of defocusing, elliptic-elliptic Davey–Stewartson systems and simultaneous blowup at multiple locations in the focusing elliptic-elliptic system. Also the modeling of exact soliton type solutions for the hyperbolic-elliptic (DS2) system is studied.

1 Introduction

In the recent mathematics literature the name "Nonlinear Schrödinger equation" (NLS) mostly refers to the following class of Schrödinger equations with a particular "local nonlinearity":

$$i\varepsilon\partial_t u + \frac{\varepsilon^2}{2}\Delta u = \pm\,\varepsilon^{\alpha}|u|^p u, \quad x \in \mathbb{R}^d,\ t \in \mathbb{R}. \tag{1.1}$$

Here u is the (complex valued) "wave function" and ε is the scaled Planck constant. The equation is called "focusing NLS" for the minus sign in front of the nonlinear term and "defocusing NLS" for the plus sign. Existence and uniqueness results on the initial value problem (IVP) have first been obtained by Ginibre and Vélo [19]. Much work has been done on this equation since, for an extensive coverage of the subject we refer to [14] and [37]. For short times, existence and uniqueness of a solution in the energy space H^1 holds if $p < 4/(d-2)$ if $d \geq 3$, and for all $p < \infty$ for $d = 1,\ 2$. The long term behaviour depends on whether the nonlinearity has focusing or defocusing sign. For defocusing NLS, the short time solution extends to all times. For the focusing NLS, this is true only if the power p is below the critical value $4/d$; for p at or above this value, finite time blowup can occur.

A quite difficult and mostly open question is the "(semi-)classical" limit of nonlinear Schrödinger equations, i.e. the passage to the limit when the "scaled Planck constant" tends to zero ($\varepsilon \to 0$). For the NLS (1.1), some results on particular cases are known. For for one dimensional defocusing cubic NLS ($p = 2$), Levermore, D. McLaughlin and Jin obtained that limit by using inverse scattering theory [22], which works for integrable systems only. Kamvissis, K. McLaughlin and Miller obtained the limit for the focusing cubic NLS by the same method [23]. Carles used a geometric argument to obtain the limit in cases where the scale exponent α is large enough and a particular geometry is given e.g. by a harmonic potential [15]. For other cases, the semiclassical limit is an open problem.

The first model we investigate is the "Schrödinger–Poisson-Xα" (S-P-Xα) equation, which is proposed as a "local one particle approximation" of the time dependent Hartree–Fock equations. It describes the time evolution of electrons in a quantum model respecting the Pauli principle in an approximate fashion [28]. It reads

$$i\varepsilon\partial_t\psi = -\frac{\varepsilon^2}{2}\Delta\psi + CV_{\text{Hartree}}\psi - \alpha|\psi|^{2/d}\psi, \quad x \in \mathbb{R}^d,\ t \in \mathbb{R}, \tag{1.2}$$

$$\Delta V_{\text{Hartree}} = -|\psi|^2 \tag{1.3}$$

It is an NLS with two nonlinear terms of different nature: the nonlocal Hartree potential and a local power term, with focusing sign and an exponent that is subcritical for finite time blowup. Such "time dependent density functional theory" models yield approximations of the time-dependent Hartree–Fock (TDHF) equations that are much easier to solve numerically: firstly, for a large number N of particles the system of coupled NLS of the TDHF system becomes too large, and secondly, the exchange terms are very costly to calculate. We will discuss the derivation of this equation in Section 1.1.

The "semiclassical limit" for equation (1.2) is mostly an open problem, rigorous results have recently been given by Carles, Mauser, Stimming for a model with quadratic confinement potential only in the case of a scaling where α is $o(\varepsilon)$ [16].

Our numerical approximations yield interesting insight in the behaviour of the solution in the limit, in particular they indicate the critical relative scaling of the 2 nonlinearities.

In the regime of "small" ε, it would be desirable that a numerical scheme permits a choice of discretization steps that is independent of the scale ε. Studies on finite difference discretizations of linear Schrödinger equations [29], [30] show that in this case this is not possible, on the contrary the scale ε has to be over-resolved, which means discretization steps have to be of order $o(\varepsilon)$ to guarantee a correct approximation. On the other hand, for a scheme like ours, the work of Bao et al. [13] shows that it is possible to choose the time-step independent of ε while the space discretization step can be $O(\varepsilon)$. This scheme has proven to be a quite efficient method and hence is the method of choice for the problem. We have developed a parallel version of the scheme that also allows for 3-d semiclassical simulations of time dependent NLS.

The second model equation treated here is the Davey–Stewartson (DS) system, which is a 2-dimensional nonlinear Schrödinger equation coupled to a potential equation. It reads, in dimensionless form,

$$
\begin{aligned}
i\partial_t u + \lambda\partial_x^2 u + \partial_y^2 u &= \nu|u|^2 u + u\,\partial_x\phi\,, \\
\alpha\partial_x^2\phi + \partial_y^2\phi &= \chi\,\partial_x(|u|^2), \qquad (x,y)\in\mathbb{R}^2;\ t\in\mathbb{R}.
\end{aligned}
\tag{1.4}
$$

Here u is the (complex) amplitude of the wave and ϕ the (real) potential which is generated by the mean velocity. The parameters α and λ can have both signs, according to which the system can be classified as elliptic-elliptic (E-E), elliptic-hyperbolic (E-H), hyperbolic-elliptic (H-E) and hyperbolic-hyperbolic (H-H). This system is a model for a surface wave of a fluid over a flat ground which is propagating mostly in one direction. It is valid in a situation where both gravity and the surface tension of the fluid influence the motion.

The equations can be viewed as a generalization of the cubic NLS (1.1) ($p = 2$ in (1.1)). Like the 1-d cubic NLS, the 2-d Davey–Stewartson system is (for specific values of the parameters) integrable by inverse scattering methods ([1], [2]). With respect to finite time blowup, the focusing cubic NLS is the critical case in 2 space dimensions, and hence an interesting question is given by the blowup behaviour of (focusing) DS equations and its relations to the NLS case.

We use the same time-splitting spectral scheme as before. In the present work we show numerical results for the blowup of focusing E-E equations and for exact soliton type solutions of the H-E system. We add a study of multi-focusing of the E-E system and an investigation of blowup in the H-E system, which are treated for the first time in this work (to our knowledge).

1.1 The Schrödinger–Poisson-Xα equation: an NLS for one-particle quantum dynamics of electrons

The (nonrelativistic) quantum dynamics of a system of N electrons is given by the linear N particle Schrödinger equation with Coulomb interaction. However, even for moderate N a numerical solution of this equation is out of question. A successful method to rigorously derive "mean field approximations" and other nonlocal "one particle" Schrödinger equations from a (linear) many-particle equation is "weak interaction limits" (see e.g. [4], [6]), i.e. a limit where the number N of particles tends to infinity and the interaction potential among the particles is rescaled with $1/N$. Depending on the "ansatz" for the (initial) N particle wave function, different asymptotic limits of the linear N particle Schrödinger equation are obtained. For the Hartree ansatz, the case of a bounded interaction potentials has been solved as well as the Coulomb interaction case which leads to the Schrödinger–Poisson equation [4], [6].

By the Hartree–Fock ansatz, in which the antisymmetrized wave function of N fermions is taken as a "Slater determinant", a minimization of total energy yields the Hartree–Fock (HF) equations for the time-independent case. For a rigorous analysis of this stationary Hartree–Fock system and references see e.g. [27].

For the time-dependent case, the rigorous derivation of the HF equations, by means of "mean field limits", is given in [5] for the bounded potential case.

The HF exchange potential, which is the key part of this equation (system) presents a problem in numerical simulations since it is very costly to calculate, especially if the number of particles in the model becomes large. An approximation of the exchange potential is due to Slater [36] who replaces it by the local density taken to the power $1/3$. This expression was first given implicitly by Dirac in the context of the exchange energy as a correction in the Thomas–Fermi model. It is also named after Gaspar and Kohn–Sham [25] where it appears with a difference of $2/3$ in the factor in front. By the name "$X\alpha$ method" [17], [35], these expressions are summarized in the sense that the value of the factor is named α and taken as parameter tunable in a certain range.

Despite the successful use of this kind of local approximations of the HF exchange potential, rigorous derivations are still missing. In a particular setting of a high density limit on the torus a rigorous version of Slater's heuristic arguments to derive the $X\alpha$ exchange potential was given in [10], [11].

In order to take into account exchange effects in a time-dependent one-particle approximation, we can most simply take the more or less rigorously derived expression of the stationary case and hence add the "$X\alpha$" term, with t as additional variable, to the effective potential in the Schrödinger–Poisson model. This corresponds to "time dependent density functional theory", where the energy is expressed in terms of the local density $n(x,t)$. This yields an NLS with a "weak" nonlocal nonlinearity, and a "strong" local nonlinearity with a potential that is a power of the local density.

In a model with d space dimensions, the approximated exchange term is proportional to $n^{1/d}$, according to the derivation in [10]. In 1d, i.e. $d = 1$ in (1.5), the $X\alpha$ term is exactly what is called the "focusing cubic NLS", i.e. $-\alpha\, |\psi|^2\psi$.

We call the equations (1.5)–(1.7) the "Schrödinger–Poisson-$X\alpha$" (S-P-Xα) model:

$$i\varepsilon\partial_t\psi = -\frac{\varepsilon^2}{2}\Delta\psi + C\, V_{\text{Hartree}}\psi - \alpha|\psi|^{2/d}\psi + V_{\text{ext}}\psi, \quad \boldsymbol{x} \in \mathbb{R}^d,\ t \in \mathbb{R}, \tag{1.5}$$

$$\Delta V_{\text{Hartree}} = -|\psi|^2, \tag{1.6}$$

$$\psi(\boldsymbol{x}, t = 0) = \psi_I(\boldsymbol{x}), \quad \boldsymbol{x} \in \mathbb{R}^d; \tag{1.7}$$

where $C > 0$ is a fixed constant, $\alpha > 0$ a parameter and V_{ext} is a given external potential, for example a confining potential.

Since $C > 0$ and $\alpha > 0$, we have a repulsive Hartree interaction and a focusing local nonlinearity. The wave function ψ is used to compute the physical observables, e.g. the position density

$$n(\boldsymbol{x}, t) = |\psi(\boldsymbol{x}, t)|^2, \quad \boldsymbol{x} \in \mathbb{R}^d, \ t \geq 0. \tag{1.8}$$

At ε fixed, the analysis of equation (1.5)–(1.7) in 3-d can be done by a straightforward application of standard results on NLS [14], [19], [24].

For the 1-d case, the existence and uniqueness analysis of the S-P-Xα equation is given in [39]. For the case of 2 space dimensions, the analysis of the S-P-Xα equation is open.

1.2 Davey–Stewartson systems: NLS for surface waves

Davey–Stewartson equations model free surface waves subject to the effects of both gravity and capillarity ("Gravity–Capillary waves"). In dimensionless form, they read

$$\begin{aligned} i\partial_t u + \lambda\partial_x^2 u + \partial_y^2 u &= \nu|u|^2 u + u\,\partial_x\phi, \\ \alpha\partial_x^2\phi + \partial_y^2\phi &= \chi\,\partial_x(|u|^2), \quad (x, y) \in \mathbb{R}^2;\ t \in \mathbb{R}. \end{aligned} \tag{1.9}$$

u is the (complex) amplitude of the wave and ϕ the (real) potential which is generated by the mean velocity. The parameters α and λ can have both signs, according to which the system can be classified as elliptic-elliptic (E-E), elliptic-hyperbolic (E-H), hyperbolic-elliptic (H-E) and hyperbolic-hyperbolic (H-H).

The equations can be viewed as a generalization of the cubic nonlinear Schrödinger equation (NLS)

$$i\partial_t u + \Delta u = \nu|u|^2 u. \tag{1.10}$$

The 2-d Davey–Stewartson (DS) system is (for specific values of the parameters) integrable by inverse scattering methods ([2]), as is the cubic NLS in 1-d. Therefore it generalizes the inverse Scattering theory in 2 dimensions [1]. With respect to finite time blowup, the focusing cubic NLS is the critical case in 2 space dimensions, and an interesting question is the blowup behaviour of (focusing) DS equations and its relations to the NLS case.

For the following invariance properties to hold we need to assume that the solution (u, ϕ) of (1.9) is sufficiently smooth and decaying at infinity. This restricts the result to the case $\alpha > 0$ where the equation for ϕ is of elliptic type. If this assumption is satisfied, then the following functionals are independent of time:

$$N(u) = \int_{\mathbb{R}^2} |u|^2\, dxdy \tag{1.11}$$

$$J_x(u) = \int_{\mathbb{R}^2} (u\overline{\partial_x u} - \overline{u}\partial_x u)\, dxdy \tag{1.12}$$

$$J_y(u) = \int_{\mathbb{R}^2} (u\overline{\partial_y u} - \overline{u}\partial_y u)\, dxdy \tag{1.13}$$

$$H(u) = \int_{\mathbb{R}^2} \mu|\partial_x u|^2 + \lambda|\partial_y u|^2 + \frac{1}{2}(\nu|u|^4 - \nu_1\beta(\partial_x\phi^2 + \alpha\partial_y\phi^2))\, dxdy \tag{1.14}$$

The (straightforward) proof is omitted here and can be found for example in [37].

For the derivation from a physical model and an overview of existence results by Ghidaglia *et al.* and Hayashi *et al.* we refer to [9], [18], [20].

The paper is organized as follows: in Section 2, we will discuss the time splitting scheme and give a convergence proof for the DS equations for the Strang splitting method. In Section 1.1, we consider the S-P-Xα model and present numerical results of position density and Wigner measures with/without an external potential. Section 1.2 deals with the DS system, where we present numerical results on H-E DS and focusing and defocusing E-E DS.

2 Time splitting spectral methods for nonlinear Schrödinger equations

In order to approximate (1.2) and (1.4) numerically, we adapt the time-splitting spectral code of [13] ("Fourier split-step method" [40]). The method goes back to the 1970s ([21]) and was recently used and studied for NLS in the semi-classical regime. It showed much better spatial and temporal resolution than finite difference methods [29], [30]. It is hence also the method of choice for the equations treated here. White and Weideman [41] applied it for the first time to DS systems in the E-H and H-E cases.

The semi-discrete version of the splitting method was recently proved convergent for nonlinear Schrödinger equations with local Lipschitz nonlinearities [7]. In [34], Papanicolaou *et al.* studied the self-focusing of E-E DS equations by a numeric resolution of the blowup. Besse and Bruneau [8] have successfully applied a finite difference method with time relaxation to the general DS equations.

The idea behind the time-splitting method is to decouple the nonlinear system into a linear PDE with constant coefficients, which can be discretized in space by a spectral method or by a classical finite difference method, and a nonlinear equation which can be solved exactly.

Let us consider a very general form of the NLS, which allows to summarize both models we consider, i.e. (1.5) and (1.9):

$$\left.\begin{aligned} &i\varepsilon\partial_t u + \varepsilon^2 L_\lambda u = F(u), && \boldsymbol{x} \in \mathbb{R}^d,\ t > 0, \\ &u(\boldsymbol{x}, 0) = u_0(\boldsymbol{x}), && \boldsymbol{x} \in \mathbb{R}^d, \end{aligned}\right\} \tag{2.1}$$

Here $L_\lambda = \Delta$ in case of (1.2) or $L_\lambda = \lambda\partial_x^2 + \partial_y^2$, with $\lambda = \pm 1$, for (1.4). ε is set to one in (1.4), and $F(u)$ denotes the nonlinear terms in the respective equations. The split-step method is based on a decomposition of the flow of (2.1).

Indeed, let us define the flow X^t of the linear Schrödinger equation

$$\left.\begin{aligned} &i\varepsilon\partial_t v + \varepsilon^2 L_\lambda v = 0, \quad \boldsymbol{x} \in \mathbb{R}^2,\ t > 0,\\ &v(\boldsymbol{x}, 0) = v_0(\boldsymbol{x}), \qquad\qquad \boldsymbol{x} \in \mathbb{R}^2, \end{aligned}\right\} \tag{2.2}$$

and Y^t as the flow of the nonlinear differential equation

$$\left.\begin{aligned} &i\varepsilon\partial_t w = F(w), \qquad \boldsymbol{x} \in \mathbb{R}^2,\ t > 0,\\ &w(\boldsymbol{x}, 0) = w_0(\boldsymbol{x}), \qquad \boldsymbol{x} \in \mathbb{R}^2. \end{aligned}\right\} \tag{2.3}$$

Then the splitting method consists of combining the two flows X^t and Y^t. The best known methods are the following: the most simple "Lie formula" given by $Z^t_{L_1} = X^t Y^t$ (or $Z^t_{L_2} = Y^t X^t$) and the "Strang splitting" $Z^t_S = Y^{t/2} X^t Y^{t/2}$ (or $X^{t/2} Y^t X^{t/2}$), which usually yields a higher order of convergence.

2.1 Convergence of the split-step method for DS systems

In this section, we are concerned with equation (1.9) (so ε is set to 1) in the case of a subsonic wave packet, that is the E-E or H-E versions of the DS system. This system can then be written as a single equation for u (see [18]). We set $\phi_x = E(|u|^2)$, where the singular integral operator E is defined in Fourier variables by

$$\widehat{E(f)}(\xi_1, \xi_2) = \frac{\xi_1^2}{\xi_1^2 + \xi_2^2} \hat{f}(\xi_1, \xi_2). \tag{2.4}$$

Therefore, the E-E or H-E DS system reduces to a nonlinear, non-local Schrödinger equation

$$i\partial_t u + L_\lambda u = \nu|u|^2 u + \nu_1 E(|u|^2)u = F(u) \tag{2.5}$$

where $L_\lambda = \lambda\partial_x^2 + \partial_y^2$ with $\lambda = \pm 1$.

We use the first Strang formulation $Z^t_S = Y^{t/2} X^t Y^{t/2}$. For a general nonlinear Schrödinger equation, the convergence proof for this method has been done in [7]. Note that there Besse *et al.* actually prove the convergence result only for initial data belonging to H^4, which does not work for E-E or H-E versions of DS. Hence the proof must be adapted for initial data belonging to H^2.

To state the convergence result, we define a new flow S^t which gives the solution of (2.5) as

$$u(\boldsymbol{x}, t) = S^t u_0(\boldsymbol{x}) = X^t u_0(\boldsymbol{x}) - i\int_0^t X^{t-s} F(u(\boldsymbol{x}, s))\, ds. \tag{2.6}$$

The result, proven in [9], is:

Theorem 2.1. *For all u_0 in H^2 and for all $0 < T < T^*$, there exists C and h_0 such that for all $h \in (0, h_0]$ and for all n such that $nh \leq T$,*

$$\left\| \left(Z_S^h\right)^n u_0 - S^{nh} u_0 \right\| \leq C(\|u_0\|_{H^2}) h \|u_0\|_{H^2}. \tag{2.7}$$

Here $\|\cdot\|$ denotes the L^2 norm. We will give a sketch of the proof which is done in [9].

The key of the proof is to prove that the nonlinearity $F(u)$ is Lipschitz, and an application of the following estimates of the separated flows:

2.1.1 Estimates on the flows X^t and Y^t. From the definition of the Schrödinger flow, we first have

$$\dot{X}^t = i L_\lambda X^t = i X^t L_\lambda. \tag{2.8}$$

As the flow X^t is a unitary group on H^2, we have

$$\|X^t u_0\|_{H^2} = \|u_0\|_{H^2}. \tag{2.9}$$

For the nonlinear flow Y^t, by a simple calculation ([13]), it follows from (2.3) and the explicit form of F that

$$|Y^t w| = |w|,$$

and with this result, an alternative way to define it is

$$Y^t w = e^{-i(\nu|w|^2 + \nu_1 E(|w|^2))t} w. \tag{2.10}$$

We immediately get for all $t \in [0, T]$

$$\|Y^t w\|_{H^2} \leq \|w\|_{H^2}. \tag{2.11}$$

By a Gronwall argument and the above results, the error for a single step of length t of the Lie method L_ν can be bounded by a $O(t^2)$ term. From this the same can be deduced for the Strang splitting.

Next it is shown that the Strang method Z_S^t is Lipschitz, with a Lipschitz constant $1+C_0 t$. By an expansion via the triangle inequality, the error term in (2.7) is expressed by the single step error which gives the result.

2.2 The time splitting spectral scheme for DS Systems

The crucial advantages of the time-splitting spectral method are that it is fully explicit, unconditionally stable, time reversible, and time-transverse invariant [13].

We use the Strang splitting method, $Z_S^t = Y^{t/2} X^t Y^{t/2}$. The linear flow X^t is solved by a spectral Fourier method and the nonlinear flow Y^t by exact integration via formula (2.4).

We choose a square domain $[a, b]^2$ and the same grid size in x- and y-direction, $\Delta x = \Delta y = (b - a)/M$ for an even, positive integer M. The time step is Δt and the grid points are $(x_j, y_k) = (a + j\Delta x, a + k\Delta x)$, $t_n = n\Delta t$. The approximation of $u(x_j, y_k, t_n)$ is denoted u^n_{jk}, and u^n is the solution matrix at time t_n with components u^n_{jk}. We impose periodic boundary conditions for a sufficiently fast decaying u in order to apply the spectral method.

In detail, the method is as follows:

$$u^*_{jk} = \frac{1}{M^2} \sum_{l=-M/2}^{M/2-1} \sum_{m=-M/2}^{M/2-1} e^{-i(\alpha\mu_l^2+\mu_m^2)\,\Delta t/2}\, \widehat{(u^n)}_{lm}\, e^{i(\mu_l(x_j-a)+\mu_m(x_k-a))},$$
$$j, k = 0, 1, 2, \ldots, M-1,$$

$$E^*_{jk} = \frac{1}{M^2} \sum_{-M/2\le l,m\le M/2-1,(l,m)\neq(0,0)} \frac{\beta\nu_1\mu_l^2}{\mu_l^2+\mu_m^2} \widehat{(|u^*|^2)}_{lm}\, e^{i(\mu_l(x_j-a)+\mu_m(x_k-a))},$$
$$j, k = 0, 1, 2, \ldots, M-1,$$

$$u^{**}_{jk} = e^{-i(E^*_{jk}+\nu|u^*_{jk}|^2)\,\Delta t}\, u^*_{jk}, \qquad j, k = 0, 1, 2, \ldots, M-1,$$

$$u^{n+1}_{jk} = \frac{1}{M^2} \sum_{l=-M/2}^{M/2-1} \sum_{m=-M/2}^{M/2-1} e^{-i(\alpha\mu_l^2+\mu_m^2)\,\Delta t/2}\, \widehat{(u^{**})}_{lm}\, e^{i(\mu_l(x_j-a)+\mu_m(x_k-a))},$$
$$j, k = 0, 1, 2, \ldots, M-1.$$

Here $\widehat{U}_{l,m}$ $(l, m = -M/2, \ldots, M/2-1)$ is the 2-D discrete Fourier transform of a (periodic) matrix U:

$$\widehat{U}_{l,m} = \sum_{j=0}^{M-1} \sum_{k=0}^{M-1} U_{jk}\, e^{-i(\mu_l(x_j-a)+\mu_m(x_k-a))}, \quad \mu_l = \frac{2\pi}{b-a} l,$$
$$l, m = -\frac{M}{2}, \ldots, \frac{M}{2} - 1.$$

2.3 Time-splitting spectral scheme for the Schrödinger–Poisson-Xα equation

As for the previous equation, we impose periodic boundary conditions for convenience to use the spectral method. By choosing a sufficiently large domain of computation we can avoid spurious effects for the time regime we regard.

For simplicity of notation we introduce the method in one space dimension ($d = 1$). Generalizations to $d > 1$ are immediate for tensor product grids.

We choose the spatial mesh size $h = \Delta x = (b-a)/M > 0$ with M an even positive integer, the time step $k = \Delta t > 0$, and denote $x_j = a + jh$ $(j = 0, \ldots, M)$,

$t_n = nk$ $(n = 0, 1, \ldots)$ the grid points. Let u_j^n be the approximation of $u(x_j, t_n)$ and u^n be the solution vector with components u_j^n.

The linear flow X^t will be discretized in space by the Fourier spectral method due to the periodic boundary conditions and integrated in time exactly. For $t \in [t_n, t_{n+1}]$, the nonlinear flow Y^t leaves $|u|$ invariant in t [13] and therefore becomes

$$\partial_{xx} V_{\text{Hartree}} = -|u(x, t_n)|^2, \tag{2.12}$$

$$i\varepsilon\partial_t u(x,t) = C V_{\text{Hartree}}\, u(x,t) - \alpha|u(x,t_n)|^{2/d} u(x,t) + V_{\text{ext}} u(x,t). \tag{2.13}$$

Equation (2.12) will be discretized by the Fourier spectral method when $|u|$ is given and (2.13) can be integrated exactly. From time $t = t_n$ to $t = t_{n+1}$, we use the Strang splitting formula:

$$u_j^* = \frac{1}{M} \sum_{l=-M/2}^{M/2-1} e^{-i\varepsilon\mu_l^2 k/4}\, \widehat{(u^n)}_l\, e^{i\mu_l(x_j-a)}, \qquad j = 0, 1, 2, \ldots, M-1,$$

$$(V_{\text{Hartree}})_j^* = \frac{1}{M} \sum_{-M/2 \le l \le M/2-1,\, l \ne 0} \widehat{(|u^*|^2)}_l/\mu_l^2\; e^{i\mu_l(x_j-a)}, \quad j = 0, 1, 2, \ldots, M-1,$$

$$u_j^{**} = e^{-i[C\,(V_{\text{Hartree}})_j^* + V_{\text{ext}}(x_j) - \alpha|u_j^*|^{2/d}]k/\varepsilon}\, u_j^*, \qquad j = 0, 1, 2, \ldots, M-1,$$

$$u_j^{n+1} = \frac{1}{M} \sum_{l=-M/2}^{M/2-1} e^{-i\varepsilon\mu_l^2 k/4}\, \widehat{(u^{**})}_l\, e^{i\mu_l(x_j-a)}, \qquad j = 0, 1, 2, \ldots, M-1,$$

where $\widehat{U}_l$ $(l = -M/2, \ldots, M/2-1)$, the Fourier coefficients of a vector $U = (U_0, U_1, \ldots, U_M)^T$ with $U_0 = U_M$, are defined as

$$\mu_l = \frac{2\pi l}{b-a}, \quad \widehat{U}_l = \sum_{j=0}^{M-1} U_j\, e^{-i\mu_l(x_j-a)}, \qquad l = -\frac{M}{2}, \ldots, \frac{M}{2} - 1. \tag{2.14}$$

2.4 Realization on a parallel machine

For simulations in 3 space dimensions with a satisfactory space resolution a large amount of memory is needed, exceeding the limitations of standard single processor computers. For such simulations parallel computers are very appropriate. This requires of course an adaptation of the code which can not be done automatically.

We are using the cluster "Schrödinger 2" of the University of Vienna, which currently features 192 Pentium IV processors, with about 1 GB memory each, linked by a switched gigabit network. The machine's LINPACK performance ranges among the fastest available.

As the Table 1 shows, 64 nodes of this parallel machine are sufficient for 3-d simulations with a resolution with about 1000 points in each space dimension.

Table 1. Memory requirement for 3-d calculation.

Resolution	Memory requirement
256 points per dimension	$8 \cdot 10^8$ byte = 768 MB
512 points per dimension	$6.4 \cdot 10^9$ byte = 6.4 GB
1024 points per dimension	$5.1 \cdot 10^{10}$ byte = 48 GB
16 byte numbers (type 'double complex'), 3 instances needed	

We use the MPI parallelization interface to adapt the code for distributed memory parallelization and compile it with "ifc", a Fortran compiler made by Intel. Hence both the compiler and the processor hardware come from the same manufacturer, which made it relatively easy to generate code optimized specifically for the used processor type and to obtain a quite good performance.

The main workload of the scheme, and the only part of the actual algorithm which needs to be adapted for parallelization, consists of Fast Fourier transforms. We implemented a parallel version FFT code on the cluster "Schrödinger 2" and test the performance of the parallelization which is listed in Table 2.

We see from Table 2 that the code has a good degree of parallelization in the sense that most of the work seems to be evenly distributed among the nodes and calculation time decreases linearly with the number of nodes used. (The final version of the machine is actually faster than in Table 2, reducing the required CPU time once more.)

Table 2. Performance.

	256^3 grid points, walltime	
Number of nodes	50 time steps	500 time steps
2	1527 s	14760 s
4	812 s	7860 s
8	432 s	4020 s

In all our 3d simulations, we use the parallel code of the time-splitting spectral method (TSSP) to compute our numerical results. We also used the parallel version of the scheme for the study of blowup in E-E DS systems.

3 Numerical results

3.1 Numerical results for S-P-Xα

In this section, we will present 1-d and 3-d numerical results of the S-P-Xα model (1.2) by using the time-splitting spectral discretization.

In our computations, the initial condition (1.7) for (1.5) is always chosen in WKB form:

$$\psi(\boldsymbol{x}, t=0) = \psi_I(\boldsymbol{x}) = A_I(\boldsymbol{x}) e^{i\, S_I(\boldsymbol{x})/\varepsilon}, \quad \boldsymbol{x} \in \mathbb{R}^d, \tag{3.1}$$

with A_I and S_I real valued, regular and with A_I decaying to zero sufficiently fast as $|\boldsymbol{x}| \to \infty$. We always compute on a domain which is large enough (as controlled by the initial data and how long in time to compute) such that the periodic boundary conditions do not introduce a significant aliasing error relative to the whole space problem. To visualize our numerical results, we consistently present the position density $n(\boldsymbol{x}, t)$ which is defined as

$$n(\boldsymbol{x}, t) = |\psi(\boldsymbol{x}, t)|^2, \quad \boldsymbol{x} \in \mathbb{R}^d.$$

Example 1 (1-d S-P-Xα model). We choose $d = 1$, $V_{\text{ext}} \equiv 0$, $C = 1$ in (1.5). Note that the local interaction term in (1.5) is the "focusing cubic NLS interaction" in the case $d = 1$. The initial condition (3.1) is hence taken the same as in the simulations of [32]:

$$A_I = e^{-x^2}, \quad \frac{d}{dx} S_I(x) = -\tanh(x), \quad -\infty < x < \infty. \tag{3.2}$$

Note that S_I is such that the initial phase is "compressive". This means that even the linear evolution develops caustics in finite time. We solve this problem either on the interval $x \in [-4, 4]$ or on $x \in [-8, 8]$ depending on the time for which the solution is calculated.

We present numerical results for four different regimes of α.

Case I. $\alpha = 0$, i.e. Schrödinger–Poisson regime;

Case II. $\alpha = \varepsilon$, i.e. Schrödinger–Poisson equation with $O(\varepsilon)$ cubic nonlinearity;

Case III. $\alpha = \sqrt{\varepsilon}$, i.e. Schrödinger–Poisson equation with $O(\sqrt{\varepsilon})$ cubic nonlinearity;

Case IV. $\alpha = 1$, i.e. Schrödinger–Poisson equation with $O(1)$ cubic nonlinearity.

Figure 1 displays comparisons of the position density $n(x, t)$ at fixed time for the above four different parameter regimes with different ε. Figure 2 plots the evolution of the position density and the Wigner transforms of the wave function for $\alpha = 1$ and $\varepsilon = 0.025$. Figure 3 shows the analogous results for attractive Hartree interaction, i.e. $C = -1$, $\alpha = 0.5$ and $\varepsilon = 0.025$.

From Figure 1, we can see that before the break (part a) and b)), the result is essentially independent of ε. After the break the behavior of the position density

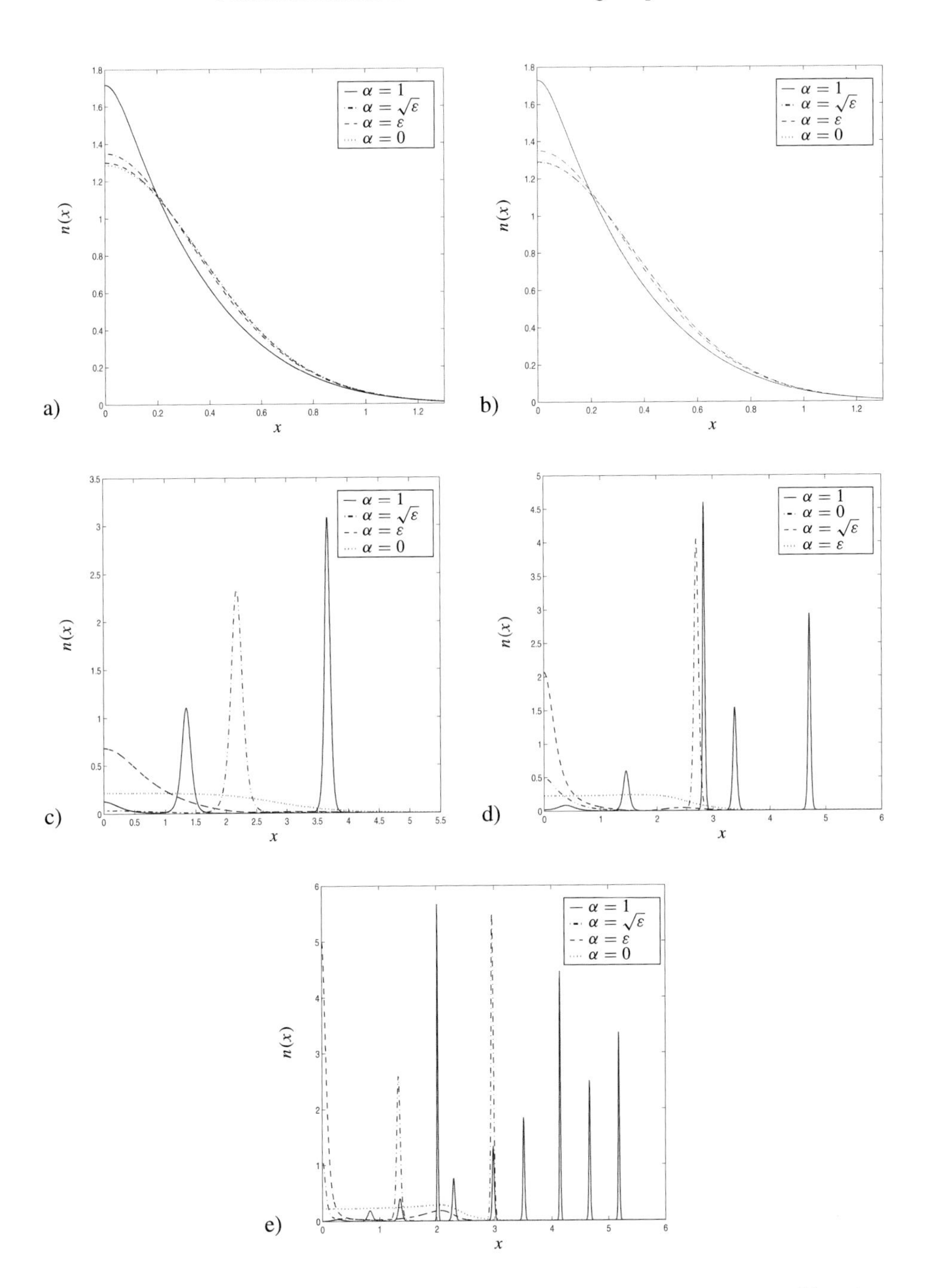

Figure 1. Numerical results for different scales of the Xα term, i.e. $\alpha = 1, \sqrt{\varepsilon}, \varepsilon, 0$.
a) and b): small time $t = 0.25$, pre-break, a) for $\varepsilon = 0.05$, b) for $\varepsilon = 0.0125$;
c)–e): large time, $t = 4.0$, post-break, c) for $\varepsilon = 0.1$, d) for $\varepsilon = 0.05$, e) for $\varepsilon = 0.025$.

changes substantially with respect to the different regimes of α. For $\alpha = 0$ the solution stays smooth. For $\alpha = \varepsilon$ it stays also smooth, but it concentrates at the origin. For $\alpha = \sqrt{\varepsilon}$ a pronounced structure of peaks develop, that look like the soliton structure typical for the NLS [23]. The number of peaks is doubled when ε is halved. For $\alpha = 1$, the number of peaks increases again and they occur at different locations than for $\alpha = \sqrt{\varepsilon}$.

We can see that the scaling $\alpha = O(\sqrt{\varepsilon})$ is critical in the sense that the solution has a substantially different behavior than for the smaller scales of α. Beyond this scaling, the semiclassical limit cannot be obtained by naive numerics.

Figure 2 b) shows how the Wigner function completely changes its qualitative behaviour after the "break time" ("caustic") and develops a rich structure of oscillations.

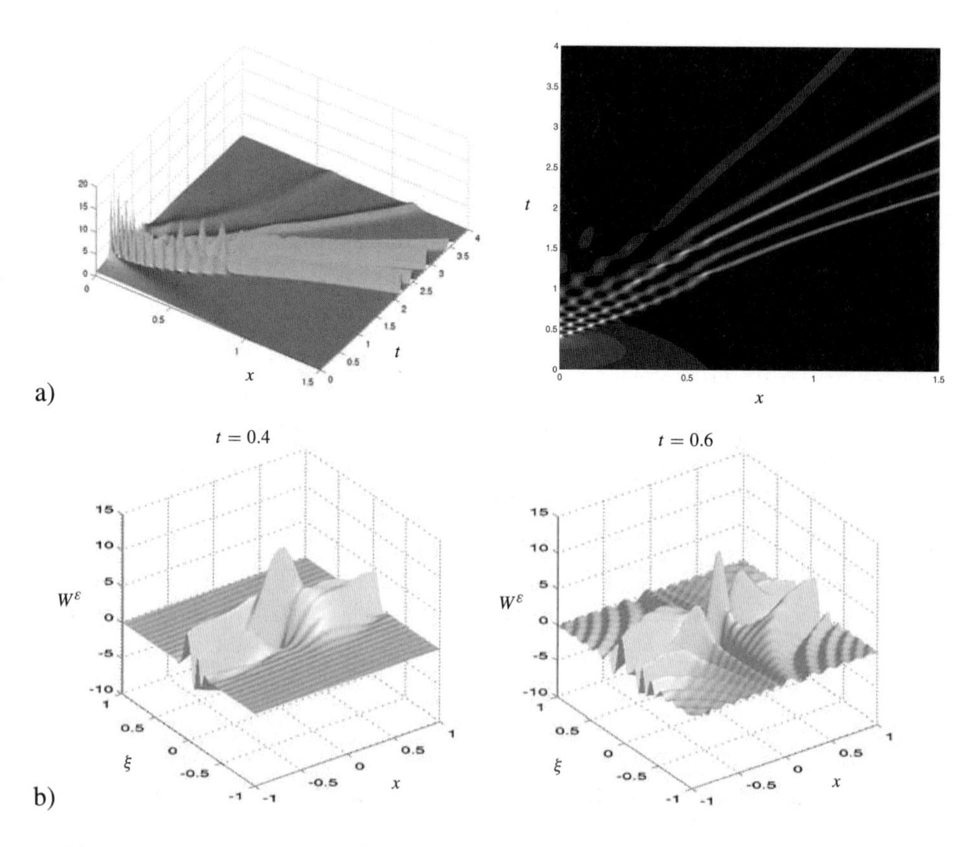

Figure 2. Numerical results for Xα term at $O(1)$, i.e. $\alpha = 0.5$, with $\varepsilon = 0.025, h = 1/512$ and $k = 0.0005$. a) Time evolution of the position density $n(x,t) = |\psi(x,t)|^2$: Left: surface plot; Right: pseudocolor plot. b) Wigner function $W[\psi(x,t)]$ at different times.

Figure 3 is a test to see what happens if the Hartree potential is attractive instead of repulsive, with all other parameters kept the same, i.e. Figure 2 a) and Figure 3 differ

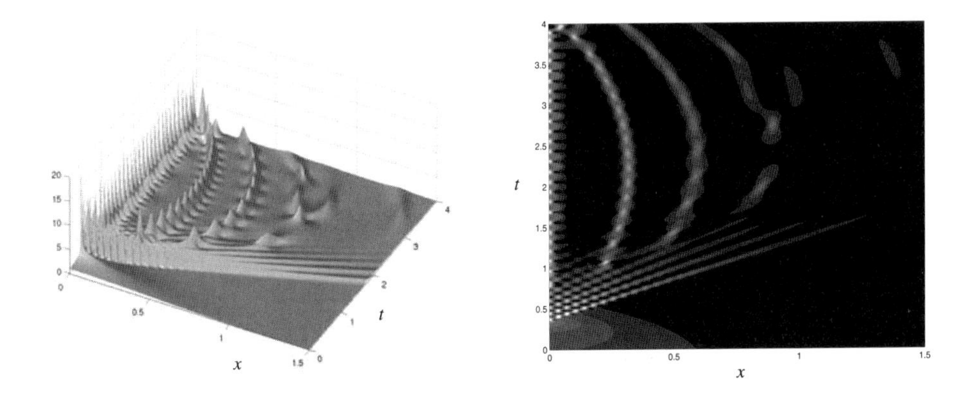

Figure 3. Time evolution of the position density for attractive Hartree interaction, i.e. $C = -1$, $\alpha = 0.5$, $\varepsilon = 0.025$, $\Delta t = 0.00015$. Left: surface plot; right: pseudocolor plot.

only by the sign of C. The resulting effect corresponds to the physical intuition that the pattern of caustics that is typical for the focusing NLS would be enhanced and focused in physical space by an additional attractive force.

Example 2 (3-d S-P-Xα model). We choose $d = 3$, $V_{\mathrm{ext}} \equiv 0$, $C = 1$ in (1.5). We consider the following initial data with nonzero phase:

$$A_I(x, y, z) = e^{-(x^2+y^2+z^2)}, \quad S_I(x, y, z) = -\ln\cosh\left(\sqrt{x^2 + y^2 + z^2}\right). \tag{3.3}$$

We solve this problem on the box $[-8, 8]^3$. We present numerical results for four different regimes of α.

Case I. $\alpha = 0$, i.e. Schrödinger–Poisson regime;

Case II. $\alpha = \varepsilon$, i.e. Schrödinger–Poisson equation with Xα nonlinearity at $O(\varepsilon)$;

Case III. $\alpha = \sqrt{\varepsilon}$, i.e. Schrödinger–Poisson equation with Xα nonlinearity at $O(\sqrt{\varepsilon})$;

Case IV. $\alpha = 1$, i.e. Schrödinger–Poisson equation with Xα nonlinearity at $O(1)$.

Figure 4 displays comparisons of the position density $n(x, y, z = 0, t = 4)$ and evolution of the position density $n(x, 0, 0, t)$ for the 4 different parameter regimes case I to IV, with $\varepsilon = 0.1$.

The simulations of Figure 4 show again that the critical scaling is at $\alpha = O(\sqrt{\varepsilon})$ – a careful examination of figure b) shows also a less regular behaviour than for c) and d). For $\alpha = O(\varepsilon^2)$ the figure is virtually the same as for $\alpha = 0$, so we skipped that plot. Hence in some physical situations, we can conclude that in case that the "local exchange term" occurs only at $O(\varepsilon^2)$ the effect of the Pauli principle can be neglected and the Schrödinger–Poisson model is sufficiently precise.

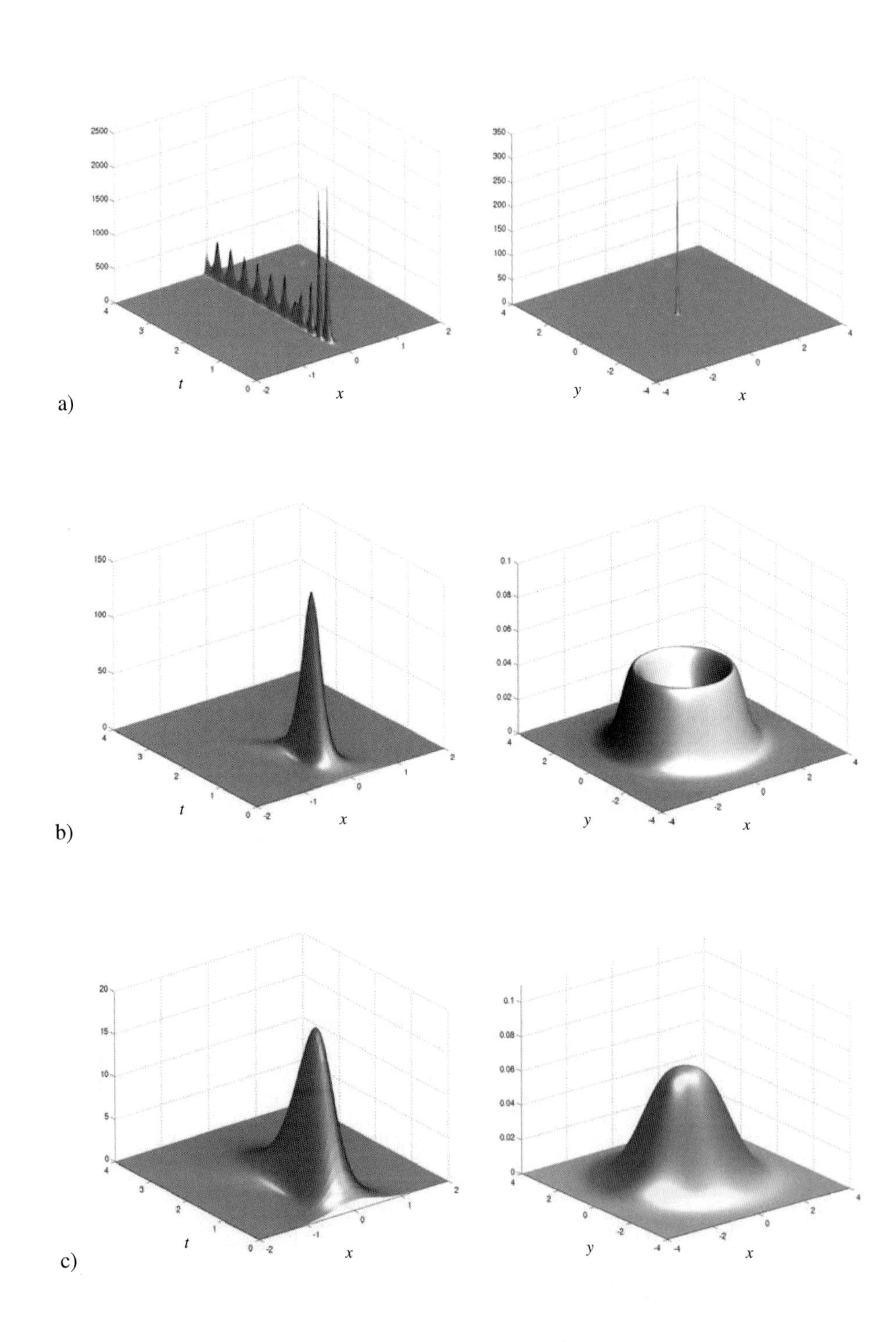

2500
2000
1500
1000
500
0
t
x
350
300
250
200
150
100
50
y
x
a)
150
100
50
t
x
0.1
0.08
0.06
0.04
0.02
y
x
b)
20
15
10
5
t
x
y
x
c)

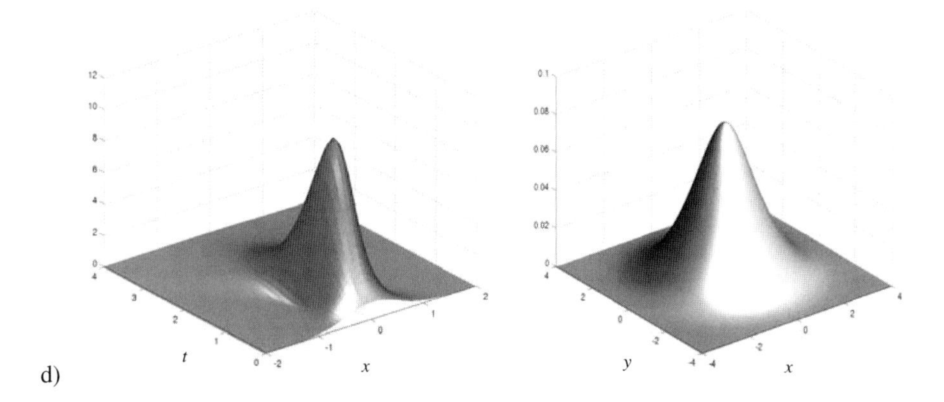

d)

Figure 4. Numerical results for 3d S-P-Xα model in Example 2 with $C = 1$, $\varepsilon = 0.1$, $V_{\text{ext}} \equiv 0$ for different regime of the parameter α. Left: Evolution of position density on $y = z = 0$, i.e. $n(x, 0, 0, t)$; Right: Position density at time $t = 4$ on $z = 0$, i.e. $n(x, y, 0, t = 4)$. a) $\alpha = 1$; b) $\alpha = \sqrt{\varepsilon}$; c) $\alpha = \varepsilon$; d) $\alpha = 0$.

3.2 Numerical Results for Davey–Stewartson systems

3.2.1 Hyperbolic-Elliptic DS System.

Exact solutions. In the long wavelength limit, the system (1.9) becomes

$$i\partial_t\psi - \sigma_1\partial_\xi^2\psi + \partial_\eta^2\psi = \sigma_1|\psi|^2\psi + \psi\partial_\xi\phi, \tag{3.1}$$

$$\sigma_1\partial_\xi^2\phi + \partial_\eta^2\phi = \partial_\xi(|\psi|^2), \quad (\xi, \eta) \in \mathbb{R}^2;\ t \in \mathbb{R}. \tag{3.2}$$

where $\sigma_1 = \pm 1$. This system has the remarkable property of being integrable by inverse scattering and is called "DS1" or "DS2" system according to the value of σ_1.

The Hyperbolic-Elliptic system (3.1)–(3.2) with $\sigma_1 = 1$, is called "DS2". Arkadiev *et al.* [2] proved, by inverse scattering methods, the existence of a class of exact solutions. These show localized structures getting displaced without dispersion and for this reason are said to be of "Soliton type". To formulate these solutions we first rewrite the equation as

$$\begin{aligned} &iu_t + \partial_x^2 u - \partial_y^2 u = -2\chi|u|^2 u - u\phi_x \\ &\Delta\phi = -4\chi(|u|^2)_x. \end{aligned}$$

Then the exact solution is

$$u(x, y, t) = \frac{2\bar{\nu}\exp(x(\lambda - \bar{\lambda}) + iy(\lambda + \bar{\lambda}) + 2it(\lambda^2 + \bar{\lambda}^2))}{|x + iy + \mu - 2i\lambda t|^2 - \chi|\nu|^2}.$$

where $\nu, \mu, \lambda \in \mathbb{C}$ are parameters. We choose $\chi = -1$ and $\nu = \mu = \lambda = 1$. With this choice the expression of the solution is

$$u(x, y, t) = \frac{4\exp(iy + 2it)}{(x+1)^2 + (y-2t)^2 + 1}.$$

Figure 5 shows the initial data at $t_0 = -3.5$, it is a hump centered at $x = -1$, $y = -7 = 2 \cdot t_0$. This solution has only geometric decay towards infinity, which

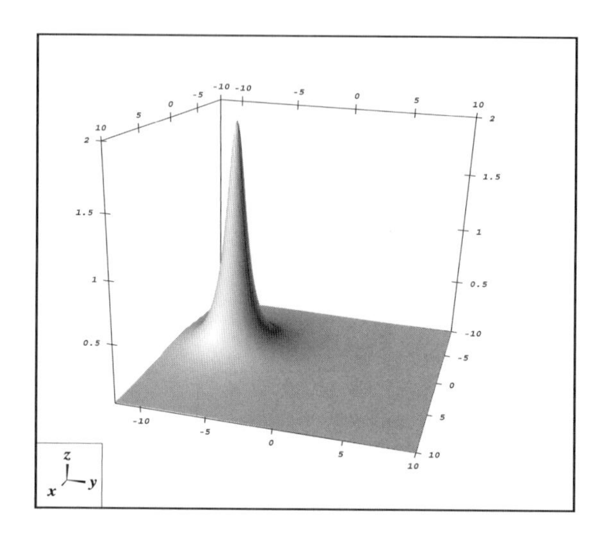

Figure 5. Exact solution at time $t_0 = -3.5$.

poses a problem for our scheme which needs to rely on a rapid decay of solutions to prevent errors from the artificial periodic boundary conditions. To prevent such errors we need to take a rather large domain, we choose $\Omega = [-16, 16]^2$. Figure 6 shows contour plots of $|u|$ at different times during the evolution. We can see that the "soliton structure" is traveling in y-direction at speed 2 as required. Also the shape of the solution is preserved accurately.

Blowup of H-E System. The next test case for the DS2 equation is the case of a finite time blowup studied by Ozawa [33], where an exact solution is constructed which blows up at a given time. The initial data are $u_0(x, y) = \frac{\exp(i(x^2-y^2))}{1+x^2+y^2}$, which is a localized lump with algebraic decay (as in the previous case). By the construction we know that the blowup time is $t_* = 0.25$ and $\|u(t)\|_{L_\infty} = 1/(1-4t)$ holds for he exact solution.

In this test we found the code to be very sensitive to cutoff errors from the artificial boundary conditions. We need to calculate on the domain $[-40, 40]^2$ to get reasonable results. Figure 7 shows the blowup for three different choices of the resolutions (upper line $\Delta x = 0.01953$, $\Delta t = 10^{-4}$, middle line $\Delta x = 0.039$, $\Delta t = 2.5 \cdot 10^{-4}$, lower

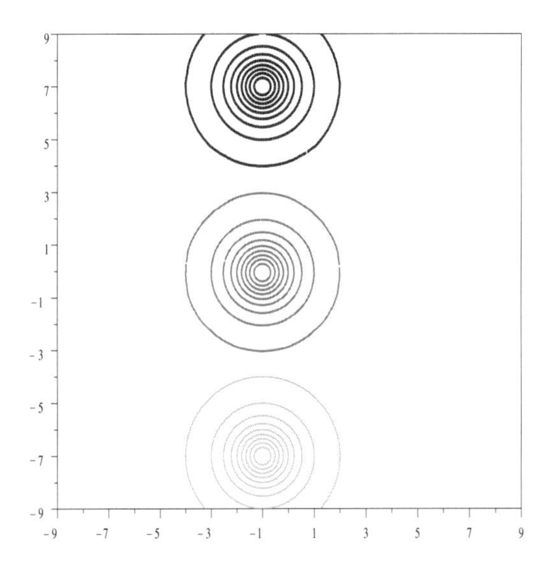

Figure 6. Traveling "soliton" at times $t = -3.5, 0, 3.5$.

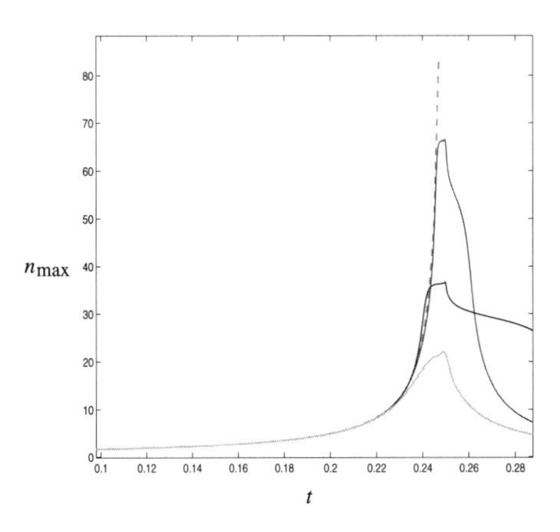

Figure 7. L^{∞}-norm of blowup solution for increasingly refined discretizations. Dashed line is analytic blowup rate.

line $\Delta x = 0.0781$, $\Delta t = 5 \cdot 10^{-4}$), and the exact rate stated above as a dashed line. We can see that, if the discretization is chosen fine enough, we can recover the correct blowup rate. Figure 8 shows the position density close to the blowup time.

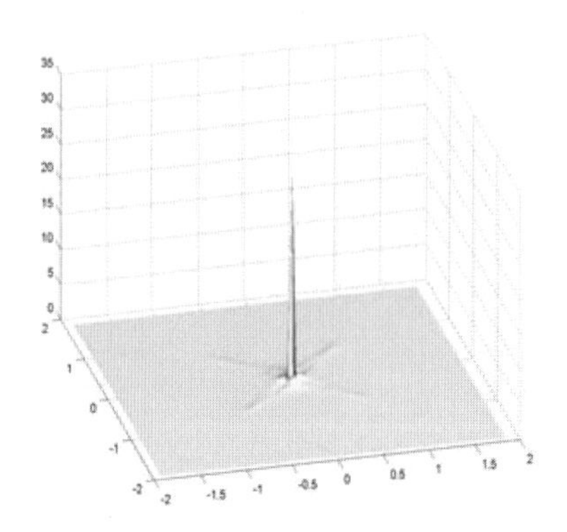

Figure 8. Position density close to blowup.

3.2.2 Elliptic–elliptic DS system

Finite time blowup. In this section we treat the elliptic–elliptic DS system

$$\begin{aligned} iu_t + \Delta u &= \chi |u|^2 u + u\phi_x \\ \Delta \phi &= -\gamma(|u|^2)_x. \end{aligned} \tag{3.3}$$

Finite time blowup of this system was studied by Ghidaglia and Saut [18]. (3.3) always leads to blowup either if $\chi = -1$, for any γ, or if $\chi = 1$ and $\gamma > \chi$. We choose the initial data profile

$$u_I(x, y) = 4 \exp\left(\frac{-x^2 - y^2}{4}\right).$$

First we investigate the case $\chi = -1$ (focusing nonlinearity) and set $\chi = -1$, $\gamma = 1$. This case is analogous to the focusing NLS, since for $\gamma = 0$, (3.3) reduces to the focusing cubic NLS, which in two space dimensions is the critical case for finite-time blowup. We expect the blowup mechanism to be similar to the one of that equation. We find that the solution blows up at time $t = 0.1311$. Figure 9 shows the L^∞-norm of u. Papanicolaou *et al.* ([34]) analytically found that the solution blows up at the rate

$$L(t) \approx (t_* - t)^{\frac{1}{2}} \left(\ln \ln \frac{1}{t_* - t}\right)^{-\frac{1}{2}}.$$

This rate is the same as the one for the critical NLS (Landman *et al.* [26], see also [37]). In Figure 9, we added as a dashed line the analytic blowup by the above formula,

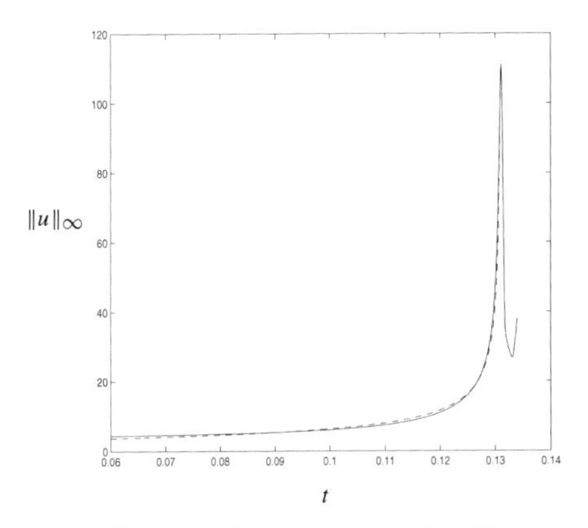

Figure 9. L^∞ norm of $|\Psi|$ and analytical blowup rate.

with t_* at the value we found for our computation. It can be observed that the rate of blowup is recovered quite accurately.

Figure 10 shows contours of $|u|^2$ near the focus. A distinct anisotropy can be observed.

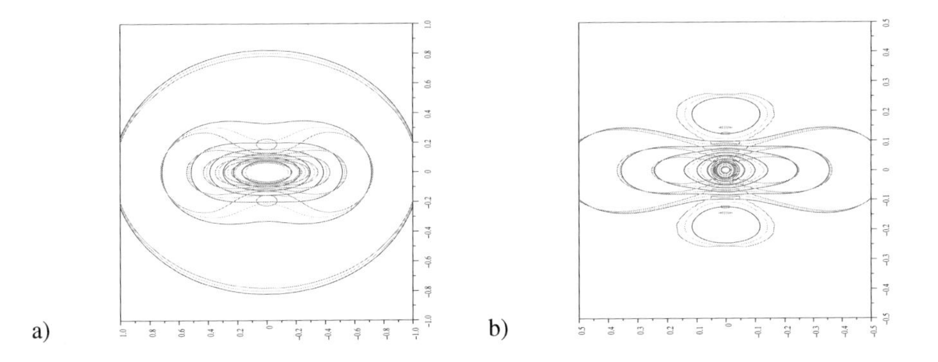

Figure 10. Contour plot of position densities: a) before blowup, at $t = 0.12$, $t = 0.124$, $t = 0.128$, b) close to blowup, at $t = 0.1298$, $t = 0.1304$, $t = 0.1308$.

Next we investigate the case $\chi = 1$ (defocusing nonlinearity) and choose $\chi = 1$, $\gamma = 5$. In this case the second nonlinear term is causing the blowup. The corresponding L^∞-norm of u is shown in Figure 11. In this case the blowup is more complicated, there is a first concentration at $t = 0.1$ before the blowup at $t = 0.1115$. A much finer time and space resolution is needed to capture this blowup correctly. Figure 12 shows the maximum values of $|u|^2$ around the blowup time for several resolutions.

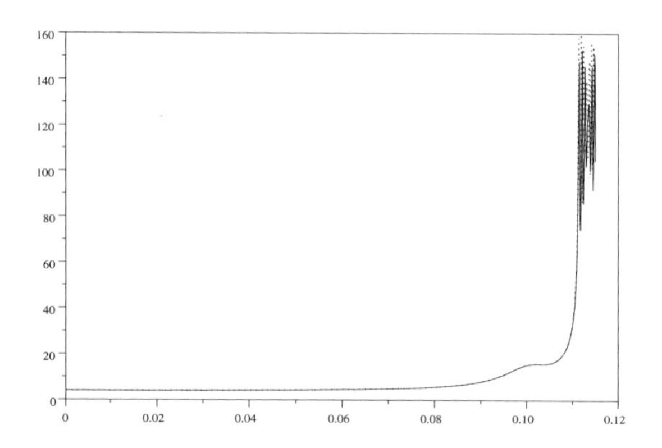

Figure 11. Blowup of defocusing ell.-ell. system ($\nu = 1, \beta = 5$). L^∞ norm of $|\Psi|$ for different time discretization steps. Solid line: $\Delta t = 5{\cdot}10^{-5}$; dotted line: $\Delta t = 2.5{\cdot}10^{-5}$.

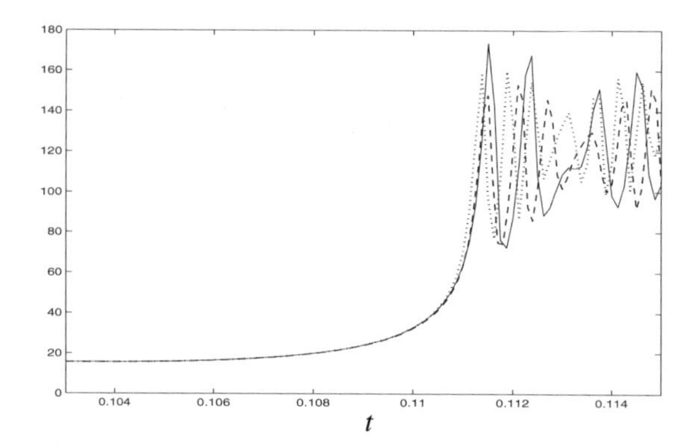

Figure 12. Blowup of defocusing ell.-ell. system. L^∞ norm of $|\Psi|$ for various space and time discretization steps, zoom. Dashed line: $\Delta x = 0.0156$, $\Delta t = 5 \cdot 10^{-5}$; dotted line: $\Delta x = 0.0156$, $\Delta t = 2.5 \cdot 10^{-5}$, solid line: $\Delta x = 0.00781$, $\Delta t = 2.5 \cdot 10^{-5}$.

Figure 13 shows contours of $|u|^2$ close to the blowup, Figure 14 shows surface plots. We can see that there is more anisotropy in the results for this case than in the previous, NLS-dominated case.

Multi-focusing. For the focusing critical NLS, according to a result of Merle [31] it is possible to construct a solution which blows up exactly at a prescribed set of points. We will state his result. Let, in $\mathbb{R}^d$ with a general dimension d, $R_0, R_1, \ldots$ be the infinite sequence of radial solutions to the equation

$$\Delta R - R + R^{\frac{4}{d}+1} = 0$$

such that R_k has exactly k nodes as a function of r and decreases exponentially at infinity.

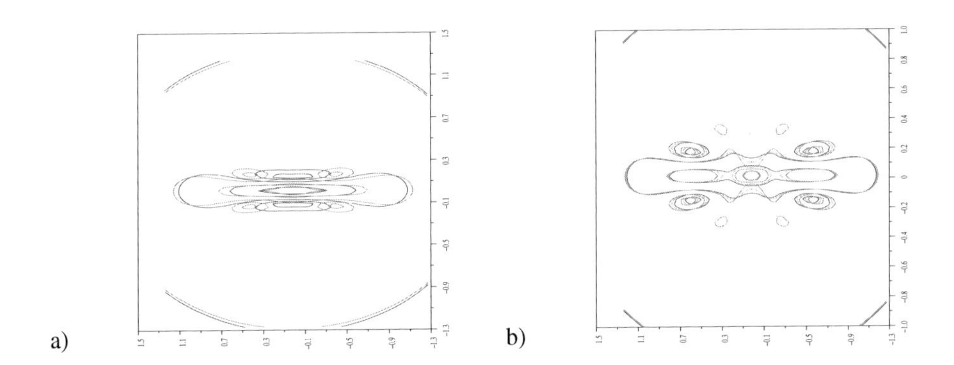

Figure 13. a) Contour plot of position densities before blowup. $t = 0.099, t = 0.1045$. b) Contour plot of position densities close to blowup. $t = 0.1091, t = 0.1101, t = 0.111$.

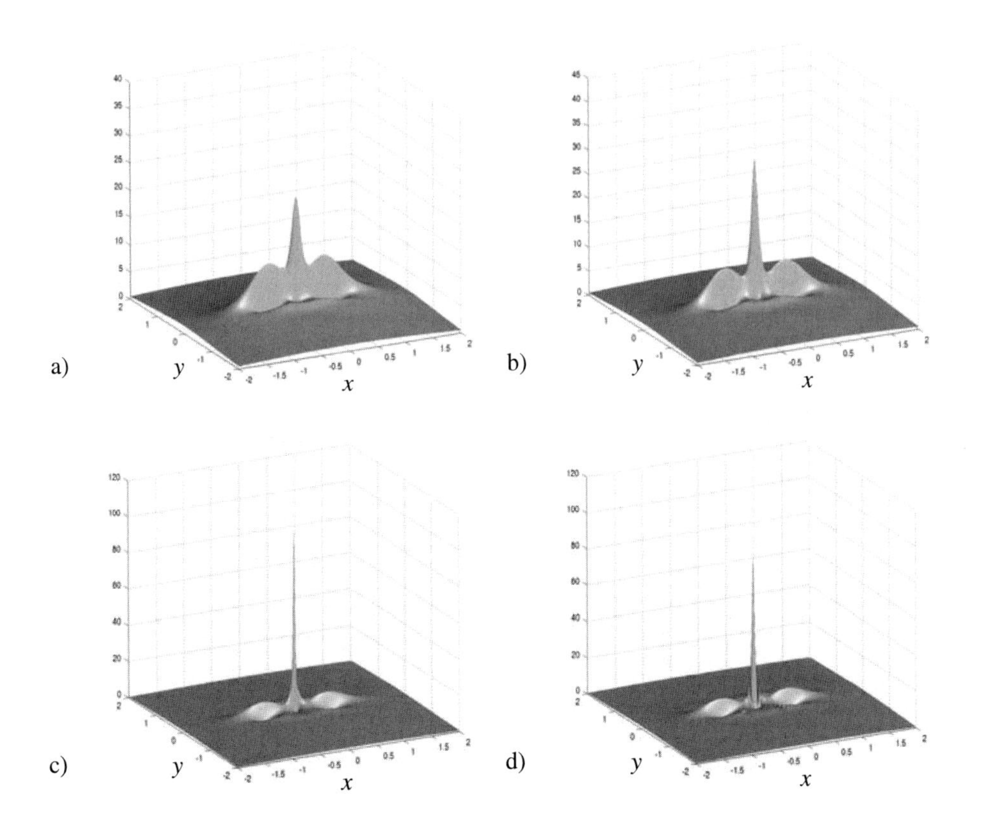

Figure 14. Position density at several times close to blowup. a) $t = 0.1109$, b) $t = 0.111$, c) $t = 0.1111$, d) $t = 0.1112$.

Theorem 3.1 ([31]). *Let $\boldsymbol{x}_1, \ldots, \boldsymbol{x}_k$ be given in $\mathbb{R}^d$. There is a constant ω_0 such that for any constants $\omega_1, \ldots, \omega_k$ all strictly larger than ω_0, there exists a solution ψ of the critical (focusing) NLS that blows up in a finite time t_* such that:*

1. *The set of blowup points in $L^{2+4/d}$ and H^1 is $\{\boldsymbol{x}_1, \ldots, \boldsymbol{x}_k\}$.*
2. *For $i = 1, \ldots, k$ and all A such that the balls $B_i = B(\boldsymbol{x}_i, A)$ are disjoint, $\lim_{t\to t_*} \|\psi(t)\|_{L^2(B_i)} = \|R_i\|_{L^2}$.*
3. *$\lim_{t\to t_*} \|\psi(t)\|_{L^2(\overline{B})} = 0$, where $\overline{B} = \mathbb{R}^d \setminus \cup_{i=1,\ldots,k} B_i$.*

In addition, there is a constant $\gamma > 0$ such that on $[0, t_*)$,

$$\left\| \psi(t) - \sum_{i=1}^{k} \frac{1}{|(t_* - t)\omega_i|^{d/2}} e^{\frac{-i}{(t_*-t)\omega_i^2} + \frac{i|x|^2}{4(t_*-t)}} R_i\left(\frac{\boldsymbol{x} - \boldsymbol{x}_i}{\omega_i(t_* - t)}\right) \right\|_{L^{2+4/d}} \leq e^{-\frac{\gamma}{t_*-t}}.$$

Papanicolaou *et al.* [34] proved that there is a ground state solution to the focusing E-E DS system (3.3). Since there is only one ground state, we decide instead of calculating with the exact ground-state profile to choose a test profile $Q(\boldsymbol{x}) = \exp(-|\boldsymbol{x}|^2)$, which is radial and decaying sufficiently fast. In order to realize the asymptotics described by the above result, we take the initial data as

$$u_I = \sum_{j=1}^{k} \frac{1}{\tilde{t}\omega} e^{-\frac{i}{\tilde{t}\omega^2} + i\frac{|x|^2}{4\tilde{t}}} Q\left(\frac{\boldsymbol{x} - \boldsymbol{x}_j}{\tilde{t}\omega}\right)$$

where $\tilde{t}$ is an estimate of the blowup time and ω a constant (and $\boldsymbol{x} = (x, y) \in \mathbb{R}^2$). We solve the equation on the domain $[-10, 10]^2$, and choose the focus points $\boldsymbol{x}_1 = (4, 4)$, $\boldsymbol{x}_2 = (-4, 4)$, $\boldsymbol{x}_3 = (-4, -4)$, $\boldsymbol{x}_4 = (4, -4)$ and the constants $\tilde{t} = 0.03$, $\omega = 13.3$. We find that blowup occurs at $t_* = 0.032$. Figure 15 shows the position densities at initial time and at the blowup. We can see that the solution blows up exactly at the four chosen points.

3.3 Conclusion

We show that the "time splitting spectral method" is a very appropriate numerical method for a large class of time dependent nonlinear Schrödinger equations and that precise numerical simulations, even in 3 space dimensions and in the semiclassical regime are possible using modern parallel machines. The range of models we consider goes from the standard "cubic" NLS with its strong local nonlinearity to weakly nonlinear NLS of the Schrödinger–Poisson type and mixed versions of these two types, as well as generalizations like the different types of the Davey Stewartson system were the NLS structure as such is modified.

Key properties of this numerical method are that it is explicit, unconditionally stable, time reversible and time transverse invariant. We use the code written by Bao

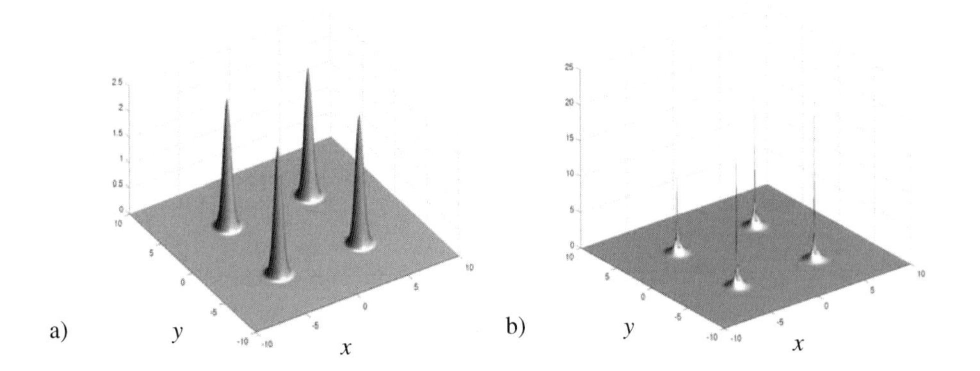

Figure 15. Multi-focusing solution. a) Initial data, b) result shortly after blowup time.

et al. that we have adapted for parallel computation. The adaption of this numerical code to the different types of NLS is relatively straightforward.

We present a first numerical study of the Schrödinger–Poisson-Xα (S-P-Xα) equation as a "local effective one particle approximation" of the time dependent Hartree–Fock equations. This particular NLS is the simplest model for quantum dynamics of electrons that respects the Pauli principle.

Thanks to 'good' ε-resolution of the numerical scheme we can study the S-P-Xα model in the semi-classical regime in 3-d.

Extensive numerical results of position density and Wigner measures in 1 and 3 space dimensions with/without an external potential are presented.

The interplay of the smoothing nonlocal nonlinearity, i.e. the repulsive "direct Coulomb interaction" ("Hartree potential"), with the strong local nonlinearity, i.e. the local approximation of the exchange interaction ("Xα potential"), is systematically studied by varying the scaling of the 2 nonlinearities between the Schrödinger–Poisson equation and an "exchange only" model. A critical scaling occurs when the Hartree term is $O(1)$ and the $X\alpha$ term is $O(\varepsilon)$.

In all simulations a critical time, the "break time", can be clearly distinguished and its "semiclassical limit" can be numerically estimated by comparing simulations with decreasing ε. Such simulations require a fine discretization which could be achieved also in 3-d by implementing the numerical code on a parallel machine.

The results in 1-d show a similarity to simulations of the "focusing cubic" NLS, with the smoothing effect of the additional Poisson equation nonlinearity. The well known "soliton" structure of the NLS (see e.g. [32]) is preserved in the S-P-Xα model when the cubic nonlinearity is dominant.

In 3-d, our simulations show a similarity with simulations of Bose–Einstein-Condensates modeled by the time dependent Gross–Pitaevski equation, where the third root of the density of the S-P-Xα model is replaced by the density itself [12], i.e.

a cubic NLS. In both cases, the local nonlinearity has the "focusing sign". However, the additional smoothing effect of the Hartree potential and the lower exponent of the local nonlinearity show a somewhat "smoother" structure of the solution unless the $X\alpha$ term and the Hartree term are of the same order of magnitude.

We have shown that the S-P-Xα model allows for simulations of quantum dynamics in 3 dimensions, also in the semiclassical regime. The inclusion of the exchange interaction to the widely used Schrödinger–Poisson model leads to qualitative changes in the transient behaviour of the solution that become very pronounced beyond a relative scaling of $O(\varepsilon)$ and change the behaviour completely towards the typical "soliton-like structures" of the focusing cubic NLS beyond a relative scaling of $O(\sqrt{\varepsilon})$.

The second model we study numerically by the "time splitting spectral method" is the Davey–Stewartson (DS) system. The method was first used for the hyperbolic-elliptic (H-E) DS equations by White, Weideman [41] ten years ago. The new code we use has been developed independently of this previous work.

We not only study H-E DS systems, like White/Weideman, but also E-E DS systems. Exact soliton type solutions of H-E DS can be recovered accurately, eliminating numerical dispersion effects which appear in earlier results of [41] and [8]. For the finite time blowup solution of H-E DS, the blowup rate can be recovered.

For the E-E DS systems, we study finite time blowup of the focusing and defocusing equations. In the focusing case the analytic blowup rate is recovered accurately and blowup profiles are presented, in full agreement to the results in [8]. The blowup mechanism in this case is very similar to the one of critical focusing NLS. In the defocusing case, however, a resolution fine enough to approximate the blowup is new. High numerical resolution in the results admits a detailed representation of blowup profiles in both cases.

For the first time, we investigate the phenomenon of simultaneous blowup at a predefined, exact number of points. This is known for focusing NLS; we find that focusing E-E DS can show the same behaviour.

Continuation of the solution after the blowup time is a very demanding challenge for numerical simulations. Further work in this direction based on our code will be performed.

References

[1] M. J. Ablowitz, P. A. Clarkson, *Solitons, nonlinear evolution equations and inverse scattering*, London Math. Soc. Lecture Note Ser. 149, Cambridge University Press, Cambridge 1991

[2] V. A. Arkadiev, A. K. Pogrebkov and M. C. Polivanov, Inverse scattering transform method and soliton solutions for the Davey-Stewartson II equation, *Physica D* 36 (1989), 189–196.

[3] W. Bao, N. J. Mauser and H. P. Stimming, Effective one particle quantum dynamics of electrons: a numerical study of the Schrödinger–Poisson-Xα model, *Commun. Math. Sci.* 1 (4) (2003), 809–831.

[4] C. Bardos, F. Golse and N. J. Mauser, Weak coupling limit of the N-particle Schrödinger equation, *Math. Anal. Appl.* 7 (2) (2000), 275–293.

[5] C. Bardos, F. Golse, A. Gottlieb and N. J. Mauser, Mean field dynamics of fermions and the time-dependent Hartree-Fock equation, *J. Math. Pures Appl.* (9) 82 (6) (2003), 665–683.

[6] C. Bardos, L. Erdös, F. Golse, N. J. Mauser and H.-T. Yau, Derivation of the Schrödinger-Poisson equation from the quantum N-particle Coulomb problem, *C. R. Math. Acad. Sci. Paris* 334 (6) (2002), 515–520.

[7] C. Besse, B. Bidégaray, S. Descombes, Order estimates in time of the splitting methods for the nonlinear Schrödinger equation, *SIAM J. Numer. Anal.* 40 (1) (2002), 26–40.

[8] C. Besse, C. H. Bruneau, Numerical study of elliptic-hyperbolic Davey-Stewartson system: dromions simulation and blow-up, *Math. Models Methods Appl. Sci.* 8 (8) (1998), 1363–1386.

[9] C. Besse, N. J. Mauser, H. P. Stimming, Numerical study of the Davey-Stewartson system, *Math. Mod. Num. Anal.* 38 (6) (2004), 1035–1054.

[10] O. Bokanowski and N. J. Mauser, Local approximation for the Hartree-Fock exchange potential: a deformation approach, *Math. Models Methods Appl. Sci.* 9 (6) (1999), 941–961

[11] O. Bokanowski, B. Grébert and N. J. Mauser, Local density approximation for the Energy of a Periodic Coulomb Model, *Math. Models Methods Appl. Sci.* 13 (8) (2003), 1185–1217.

[12] W. Bao, D. Jaksch and P. A. Markowich, Numerical solution of the Gross-Pitaevskii Equation for Bose-Einstein condensation, *J. Comput. Phys.* 187 (1) (2003), 318–342.

[13] W. Bao, S. Jin, P. A. Markowich, Time-splitting spectral approximations for the Schrödinger equation in the semiclassical regime, *J. Comput. Phys.* 175 (2) (2002), 487–524.

[14] T. Cazenave, *Introduction to nonlinear Schrödinger equations*, Textos de Métodos Matemáticos 26, Rio de Janeiro, Instituto de Matemática - UFRJ, 1996.

[15] R. Carles, Semi-classical Schrödinger equations with harmonic potential and nonlinear perturbation, *Ann. Inst. H. Poincaré Anal. Non Linéaire* 20 (3) (2003), 501–542.

[16] R. Carles, N. J. Mauser, H. P. Stimming, Semiclassical Hartree equation with harmonic potential, submitted.

[17] J. W. D. Conolly, The $X\alpha$ method, in *Semi-empirical methods of electronic structure calculations*, ed. by G. A. Segal, Plenum Press, 1977.

[18] J. M. Ghidaglia and J. C. Saut, On the initial value problem for the Davey-Stewartson systems, *Nonlinearity* 3 (1990), 475–506.

[19] J. Ginibre and G. Velo, On a class of nonlinear Schrödinger equations. I. The Cauchy Problam, general case, *J. Funct. Anal.* 32 (1979), 1–32.

[20] N. Hayashi and H. Hirata, Global existence and asymptotic behaviour of small solutions to the elliptic-hyperbolic Davey-Stewartson system, *Nonlinearity* 9 (1996), 1387–1409.

[21] R. H. Hardin, F. D. Tappert, Applications of the split-step Fourier method to the numerical solution of nonlinear and variable coefficient wave equations, *SIAM Review, Chronicle* 15 (1973), 423.

[22] S. Jin, C. D. Levermore and D. McLaughlin, The Semiclassical Limit of the Defocusing NLS Hierarchy, *Comm. Pure Appl. Math.* 52 (5) (1999), 613–654.

[23] S. Kamvissis, K. T.-R. McLaughlin, P. D. Miller, *Semiclassical Soliton Ensembles for the Focusing Nonlinear Schrödinger Equation*, Ann. of Math. Stud. 154, Princeton University Press, Princeton 2003.

[24] C. Kenig, G. Ponce and L. Vega, On the IVP for the nonlinear Schrödinger equations, in *Harmonic analysis and operator theory* (Caracas, 1994), ed. by S. A. M.Marcantognini et al., Contemp. Math. 189 (1995), Amer. Math. Soc., Providence, RI, 1995, 353–367.

[25] W. Kohn and L. J. Sham, Self-consistent equations including exchange and correlation effects, *Phys. Rev. A* 140 (1965), 1133.

[26] M. J. Landman, G. C. Papanicolaou, C. Sulem, P.-L. Sulem, Rate of blowup for solutions of the Nonlinear Schrödinger equation at critical dimension, *Phys. Rev. A* 38 (1988), 3837–3843.

[27] P. L. Lions, Solution of Hartree-Fock equations for Coulomb systems, *Comm. Math. Phys.* 109 (1987), 33–97.

[28] N. J. Mauser, Quantum steady states: The Bloch-Poisson model, in *Proc. "Numsim 91"*, ed. by P. Deuflhard, Techn. Rep. 91-8, ZIB, Berlin 1991, 62–67.

[29] P. A. Markowich, P. Pietra and C. Pohl, A Wigner measure approach to the analysis of difference methods for the Schrödinger equation, in *ENUMATH 97* (Heidelberg), World Sci. Publishing, River Edge, N.J., 1998, 453–460.

[30] P. A. Markowich, P. Pietra, C. Pohl, H.-P. Stimming, A Wigner-measure analysis of the Dufort-Frankel scheme for the Schrödinger equation, *SIAM J. Numer. Anal.* 40 (4) (2002), 1281–1310.

[31] F. Merle, Construction of solutions with exactly k blowup points for the Schrödinger equation with critical nonlinearity, *Comm. Math. Phys.* 129 (1990), 223–240.

[32] P. Miller and S. Kamvissis, On the semiclassical limit of the focusing nonlinear Schrödinger equation, *Phys. Lett. A* 247 (1–2) (1998), 75–86.

[33] T. Ozawa, Exact blow-up solutions to the Cauchy problem for the Davey-Stewartson systems, *Proc. R. Soc. A* 436 (1992), 345–349.

[34] G. C. Papanicolaou, C. Sulem, P.-L. Sulem, X. P. Wang, The focusing singularity of the Davey-Stewartson equations for gravity-capillary surface waves, *Physica D* 72 (1994), 61–86.

[35] R. G. Parr and W. Yang, *Density functional theory of Atoms and Molecules*, Oxford University Press, Oxford 1989.

[36] J. C. Slater, A simplification of the Hartree-Fock method, *Phys. Rev.* 81 (3) (1951), 385–390.

[37] C. Sulem, P.-L. Sulem, *The Nonlinear Schrödinger Equation: Self-Focusing and Wave Collapse*, Appl. Math. Sci. 139, Springer-Verlag, New York 1999.

[38] E. Skovsen, H. Stapelfeldt, S. Juhl and K. Molmer, Quantum state tomography of dissociating molecules, *Phys. Rev. Lett.* 91 (9) (2003), 406–410.

[39] H. P. Stimming, The IVP for the Schrödinger-Poisson-Xα equation in one dimension, *Math. Mod. Meth. Appl. Sci.* (2005), to appear.

[40] J. A. C. Weideman and B. M. Herbst, Split-step methods for the solution of the nonlinear Schrödinger equation, *SIAM J. Numer. Anal.* 23 (3) (1986), 485–507.

[41] P. W. White, J. A. C. Weideman, Numerical simulation of solitons and dromions in the Davey-Stewartson system, *Math. Comput. Simulation* 37 (4–5) (1994), 469–479.

Ionospheric plasmas: model derivation, stability analysis and numerical simulations *

C. Besse[1], *J. Claudel*[2], *P. Degond*[1], *F. Deluzet*[1], *G. Gallice*[2] *and C. Tessieras*[2]

[1]*Laboratoire de Mathématiques pour l'Industrie et la Physique, CNRS UMR 5640 Université Paul Sabatier, 118 Route de Narbonne, 31062 Toulouse Cedex 4, France*
email: besse@mip.ups-tlse.fr, degond@mip.ups-tlse.fr deluzet@mip.ups-tlse.fr

[2]*CEA–Centre d'Etudes Scientifiques et Techniques d'Aquitaine, BP2 33114 Le Barp, France*
email: jean.claudel@cea.fr, Gerard.Gallice@cea.fr Christian.Tessieras@cea.fr

Abstract. This paper is concerned with a review of models of ionospheric plasma. Two hierarchies of models are presented, both of them leading to the so-called striation model. A linear stability analysis is performed and shows that the striation model can developed plasma instabilities. A dissipation process has disappeared during the limiting process. In order to add a dissipation mechanism to the striation model, we propose to use a statistical approach to turbulence and derive a so-called 'turbulent striation model'. Numerical simulations are proposed to show the gain of stability.

Acknowledgments. Support of the 'Centre d'Etudes Scientifiques d'Aquitaine' of the 'Commissariat à l'Energie Atomique' is acknowledged. Support by the European network HYKE, funded by the EC as contract HPRN-CT-2002-00282, is also acknowledged.

1 Introduction

The earth atmosphere is continuously subject to cosmic radiations and in particular is under the influence of solar wind. These particles of high energy interact with the high altitude atmosphere constituted of plasma, the so-called ionosphere. It creates major disturbances, the more famous of which are auroras. They can perturb significantly the communications between earth and satellites and even long range communica-

*This work was supported by the 'Centre d'Etudes Scientifiques d'Aquitaine' of the 'Commisariat à l'Energie Atomique'and by the European network HYKE, funded by the EC as contract HPRN-CT-2002-00282.

tions. There are numerous ionospheric instabilities going from equatorial electrojet to 'Spread F' (see [29], [19], [8], [2], [12], [13] and [28] for a review of ionospheric physics and the generation of ionospheric instabilities).

In this article, we focus on instabilities which take place at altitudes ranging between a few hundred and a thousand kilometers (F region) and at middle latitudes. They may be natural or created by possible artificial causes (such as e.g. thermonuclear explosions [24], [35], [17]). In the lower layers of the ionosphere, the neutral wind blows and carries the ionospheric plasma with collisions. As a result, a net electrical current flows across the magnetic field lines. An induced electric field E is created, orthogonal both to the earth magnetic field B and to the direction of plasma propagation. As the electron conductivity is extremely high along the earth magnetic field lines, the induced electric field is immediately transposed in the higher layers of the ionosphere. The gradient drift instability (the so-called $E \times B$ instability) appears as a response of the external electric field. This is the so-called ionospheric dynamo effect [1]. In the presence of a gradient in the plasma density, the neutral wind can trigger the $E \times B$ drift instability, which bears strong analogies with the Rayleigh–Taylor instability in fluid mechanics [9]. This instability produces strong inhomogeneities (the ionospheric striations) which soon propagate over hundreds of kilometers along the magnetic field lines (see Figure 1).

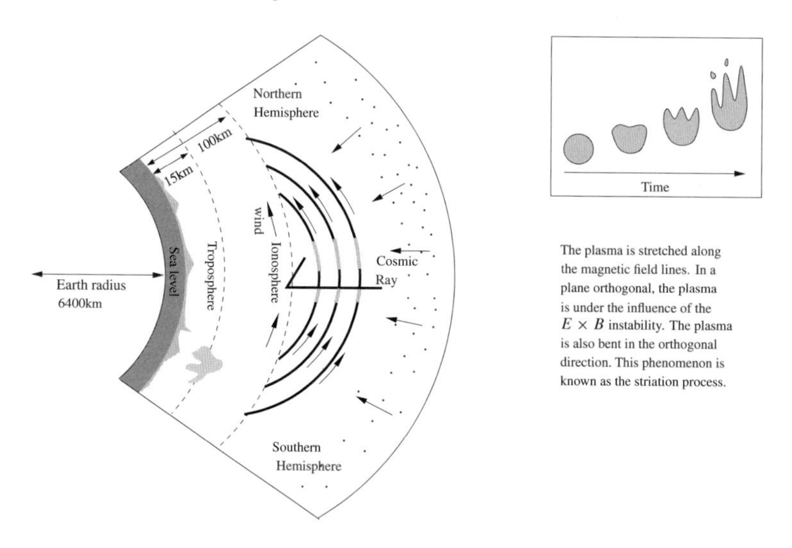

Figure 1. Instabilities evolution.

Striations as well as related instability phenomena of the ionospheric plasma have been the subject of a wide literature (see e.g. discussions of the 'Spread F' in [36], [21], [26], of the equatorial electrojet in [7], [33], [30], [31] and of Barium releases experiments in [11]). The well-accepted mathematical model for these phenomena is the 'Dynamo' model [36], [11], which consists of mass and momentum balance equations for the plasma species. A simpler model, the 'striation' model, is obtained when the field-aligned mobilities are supposed infinite. Both models need a strong assumption on the external magnetic field B. A new model, the Massless-MHD model, is developed in [3] to take into account possible variations of B.

The derivation of this hierarchy of models for the modeling of striation process is reviewed in [3]. We briefly recall their derivation in Section 2 and present the striation model in a uniform and non-uniform magnetic field.

Under influence of the $E \times B$ drift term, the behavior of the ionospheric plasma becomes chaotic. We perform in Section 3 a linear stability analysis of the striation model. We restrict the analysis to an exponential density profile which allows to make explicit computations via Fourier analysis. In a companion work [4], we extend the linear stability analysis for discontinuous density profiles which are more realistic. In both cases the striation model is proved to be unstable. The instability is linked to the neutral wind orientation. During the limit process of the Dynamo model to the striation model a dissipation mechanism was erased. We also perform a linear stability analysis on the Dynamo model, which reveals that there exist some dissipation processes. The stability analysis of the Massless-MHD model will be developed in a future work.

A dimensional analysis shows that the involved parameters are too weak to allow a stabilization process during a numerical simulation. In fluid mechanics, it is well known that turbulence may enhance dissipation. We therefore investigate the possible influence of fluid turbulence and develop an analysis similar to the $k - \varepsilon$ model for fluids. This method involves a statistical approach. We identify an averaged striation model named turbulent striation model. Its linear stability analysis proves that there exists some stability region.

Finally, Section 7 is devoted to numerical simulations. They show that the turbulent striation model, contrary to the classical striation model, produces persistent structures. We also make numerical experiments in the case of a non-uniform magnetic field.

2 A hierarchy of models

The ionospheric plasma irregularities in middle latitude regions at altitudes contained between 250 and 500 km can be modeled by a system of coupled Euler equations for the two charged species ions and electrons and the Maxwell equation for the evolution of the electric and magnetic fields. To simplify the modeling, the ions are supposed singly charged. In the bottom of the ionosphere, the charged particles are assumed so dilute that they have no influence on the dynamics of the neutrals. Therefore, the velocity of the neutrals $u_n(x, t)$ (also called the neutral wind) is supposed known. We denote by $n_e(x, t)$ and $n_i(x, t)$ the electron and ion densities, by $u_e(x, t)$ and $u_i(x, t)$ their velocities, by p_e and p_i their pressures, by $E(x, t)$ and $B(x, t)$ the electric and magnetic fields, and finally by $\rho(x, t) = e(n_i - n_e)$ and $j(x, t) = e(n_i u_i - n_e u_e)$ the charge density and current. These are functions of a three-dimensional position vector $x \in \mathbb{R}^3$ and of the time $t > 0$. Additionally, we introduce m_e, m_i, the electron and ion masses, e, the elementary charge, ε_0 and μ_0, the vacuum permittivity and permeability, $c = (\varepsilon_0 \mu_0)^{-1/2}$ the speed of light, which are given physical constants. We shall also introduce the electron-neutral and ion-neutral collision frequencies ν_e, ν_i. Collisions between electrons and ions will be modeled by a rate constant K. The data ν_e, ν_i and K may be functions of (x, t), in a way which will not be further detailed. The 2-fluids isothermal or isentropic Euler–Maxwell system is written

$$\frac{\partial n_{e,i}}{\partial t} + \nabla \cdot (n_{e,i} u_{e,i}) = 0, \tag{2.1}$$

$$\frac{\partial}{\partial t}(m_{e,i} n_{e,i} u_{e,i}) + \nabla \cdot (m_{e,i} n_{e,i} u_{e,i} u_{e,i}) + \nabla p_{e,i}(n_{e,i}) \tag{2.2}$$

$$= \pm e n_{e,i}(E + u_{e,i} \times B) - \nu_{e,i} m_{e,i} n_{e,i}(u_{e,i} - u_n) - K n_e n_i (u_{e,i} - u_{i,e}),$$

$$\frac{1}{c^2} \frac{\partial E}{\partial t} - \nabla \times B = -\mu_0 j, \tag{2.3}$$

$$\frac{\partial B}{\partial t} + \nabla \times E = 0, \tag{2.4}$$

$$\nabla \cdot E = \frac{1}{\varepsilon_0} \rho, \tag{2.5}$$

$$\nabla \cdot B = 0\,. \tag{2.6}$$

We replace the energy equations by the assumptions that the electron and ion pressures are local functions of the electron and ion densities. Actually, we assume through an adiabatic assumption that $p_{e,i} = n_{e,i} k_B T_{e,i}$, where k_B refers to the Boltzmann constant.

This model describes the sharp interactions between the charged and neutrals particles and the electric and magnetic field. However, the Euler–Maxwell model is too expensive to perform realistic numerical simulations in desired time and length scales. Indeed, we are looking for evolutions of plasmas blobs over regions of typical size of a hundred of kilometers, over time scales between ten minutes to an hour. In order to reduce the computational cost, we study various limiting regimes of this system and introduce some scaling units to do so. Let us define dimensionless variables and unknowns according to Table 1 below. After scaling the equations, there remain seven dimensionless parameters, which can be combined in a variety of ways. We choose to express them in the way described in Table 2 below.

Our first limiting process leads to the Hall-MHD system, which is obtained by passing to the limit in the electron to ion mass ratio ε and the ratio of ion sound speed to the speed of light α, while keeping the other parameters finite. The system is quasi neutral.

Among the dimensionless parameters appearing in the Hall-MHD system, κ, the ratio of the collision frequency against neutrals to the cyclotron frequency in the earth

Table 1. Scaling units.

Quantity	Scaling unit	Value
Time	$\bar{t}$	10^3 s
Length	$\bar{x}$	10^5 m
Velocity	$\bar{u} = \bar{x}/\bar{t}$	10^2 ms^{-1}
Density	$\bar{n}$	10^{12} m^{-3}
Temperature	$\bar{T}$	10^3 K
Magnetic field	$\bar{B}$	10^{-5} T
Electric field	$\bar{E} = \bar{u}\bar{B}$	10^{-3} Vm^{-1}
e-n collision frequency	$\bar{\nu}_e$	10^2 s^{-1}
i-n collision frequency	$\bar{\nu}_i = \frac{m_e}{m_i}\bar{\nu}_e$	10^{-2} s^{-1}
Electron-ion collision	$\bar{K}$	10^{-39} kg m^3 s^{-1}
Charge density	$\bar{\rho} = e\bar{n}$	10^{-7} Cm^{-3}
Current density	$\bar{j} = e\bar{n}\bar{u}\kappa$	10^{-9} Am^{-2}

Table 2. Dimensionless parameters.

Dimensionless parameter	Meaning	Value
$\varepsilon = \dfrac{m_e}{m_i}$	Electron to ion mass ratio	10^{-4}
$\tau = \dfrac{1}{\bar{\nu}_i\bar{t}} = \dfrac{m_i}{m_e}\dfrac{1}{\bar{\nu}_e\bar{t}}$	Mean-time between i-n collisions (dimensionless)	10^{-1}
$\eta = \dfrac{k_B\bar{T}}{m_i\bar{n}\bar{u}^2}\dfrac{1}{\bar{\nu}_i\bar{t}}$	Measure of the thermal energy	10^{1}
$\kappa = \dfrac{m_e\bar{\nu}_e}{e\bar{B}} = \dfrac{m_i\bar{\nu}_i}{e\bar{B}}$	Number of e-n (or i-n) collisions per rotation period in $\bar{B}$-field	10^{-4}
$\kappa_{ei} = \dfrac{\bar{K}}{\mu_0\bar{x}e^2\bar{u}}$	Measure of the e-i collision frequency	10^{-2}
$\alpha = \dfrac{\bar{u}^2}{c^2}$	squared reciprocal of light speed (dimensionless)	10^{-12}
$\beta = \dfrac{m_i\bar{n}\bar{u}^2}{\bar{B}^2/\mu_0}\bar{\nu}_i\bar{t}$	Drift energy relative to magnetic energy	10^{-5}

magnetic field, and β, the strength of the self-consistent magnetic field perturbation induced by the dynamics of the plasma relative to earth magnetic field, tend to zero faster than the other ones (see Table 2). A third one, τ, plays also an important role, and is small when the characteristic time scales of interest are large compared with the mean collision time against neutrals.

Two strategies can be devised and are summarized in Figure 2.

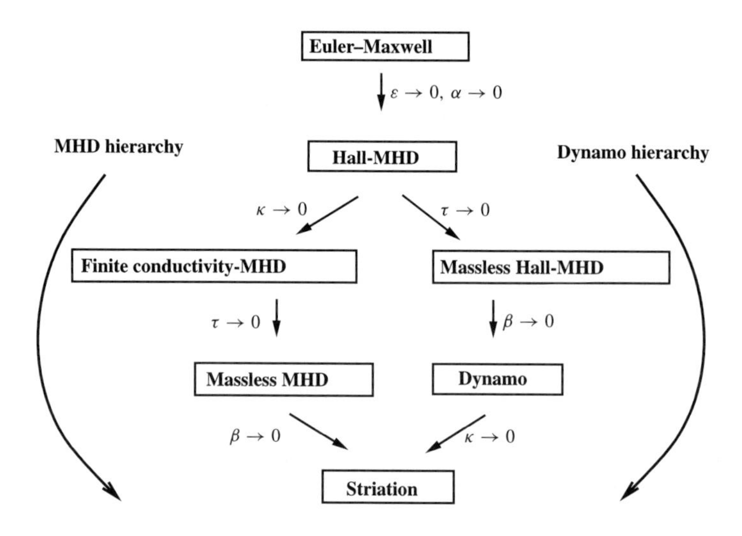

Figure 2. Hierarchy of ionospheric models.

The Dynamo hierarchy is obtained when the limit $\beta \to 0$ is taken first while the MHD hierarchy develops when the limit $\kappa \to 0$ is taken first. However, the limit $\beta \to 0$ leads to simpler models if it is preceded by the limit $\tau \to 0$. This is why the Dynamo hierarchy will consist of the successive limits $\tau \to 0$, $\beta \to 0$ and $\kappa \to 0$, while the MHD hierarchy is obtained through $\kappa \to 0$, $\tau \to 0$ and $\beta \to 0$.

2.1 The Dynamo hierarchy

In the Dynamo hierarchy the most widely used model is the Dynamo model (obtained after $\tau \to 0$ and $\beta \to 0$). In the limiting process the magnetic field B converges to a solution of the system

$$\nabla \times B = 0, \quad \nabla \cdot B = 0, \tag{2.7}$$

i.e. B is a given external magnetic field, which will be supposed independent of t for simplicity. In the context of ionospheric modeling, B is the earth magnetic field and is actually independent of t. The electric field becomes irrotational. The resulting electric potential is the solution of an elliptic equation deduced from the current conservation equation. It is coupled with the usual density and (massless) momentum equations. The Dynamo model is written as follows:

$$\frac{\partial n}{\partial t} + \nabla \cdot (n u_i) = 0, \tag{2.8}$$

$$E + u_i \times B = \kappa \left[\nu_i (u_i - u_n) + \eta T \nabla \log n \right], \tag{2.9}$$

$$E + u_e \times B = -\kappa \left[\nu_e (u_e - u_n) + \eta T \nabla \log n \right], \tag{2.10}$$

$$\nabla \cdot j = 0, \tag{2.11}$$

$$\nabla \times E = 0, \tag{2.12}$$

$$\kappa j = n(u_i - u_e), \tag{2.13}$$

where $n = \lim n_e = \lim n_i$, and T is the common electron and ion temperature.

The Dynamo model can be recast in a more appropriate form. In a local reference frame in which the last basis vector is aligned with the magnetic field, the ion and electron mobility matrices $\mathbb{M}_e$ and $\mathbb{M}_i$ are given by

$$\mathbb{M}_e = \begin{pmatrix} \mu_e^P & -\mu_e^H & 0 \\ \mu_e^H & \mu_e^P & 0 \\ 0 & 0 & \mu_e^{\parallel} \end{pmatrix}, \quad \mathbb{M}_i = \begin{pmatrix} \mu_i^P & \mu_i^H & 0 \\ -\mu_i^H & \mu_i^P & 0 \\ 0 & 0 & \mu_i^{\parallel} \end{pmatrix},$$

where the electron and ion Pedersen, Hall and field-aligned mobilities are respectively defined by

$$\mu_{e,i}^P = \frac{\kappa \nu_{e,i}}{(\kappa \nu_{e,i})^2 + |B|^2}, \quad \mu_{e,i}^H = \frac{|B|}{(\kappa \nu_{e,i})^2 + |B|^2}, \quad \mu_{e,i}^{\parallel} = \frac{1}{\kappa \nu_{e,i}}.$$

Since the electric field is irrotational, we can write $E = -\nabla \phi$, where ϕ is the resulting electric potential.

Due to the mobility matrices the equations (2.9), (2.10) and (2.11) may be rewritten as

$$u_i = \mathbb{M}_i \left(E + \kappa (\nu_i u_n - \eta T \nabla \log n) \right), \tag{2.14}$$

$$u_e = \mathbb{M}_e \left(-E + \kappa (\nu_e u_n - \eta T \nabla \log n) \right), \tag{2.15}$$

$$\begin{aligned} &-\nabla \cdot (n(\mathbb{M}_i + \mathbb{M}_e) \nabla \phi)) \\ &\quad = -\kappa \nabla \cdot (n[\mathbb{M}_i(\nu_i u_n - \eta T \nabla \log n) - \mathbb{M}_e(\nu_e u_n - \eta T \nabla \log n)]). \end{aligned} \tag{2.16}$$

It is clear that the conductivity matrix $n(\mathbb{M}_i + \mathbb{M}_e)$ is positive definite (provided that ν_i or ν_e is positive and finite). Therefore, (2.16) is a three-dimensional elliptic equation for ϕ.

2.2 The MHD hierarchy

In the MHD hierarchy a new model is found (after letting $\kappa \to 0$ and $\tau \to 0$) in which the electron inertia is neglected. It is referred to as the Massless-MHD model. The major difference with respect to the Dynamo model is that the magnetic field B is not constraint to be constant. The dynamics of the plasma may influence the earth magnetic field. The Massless-MHD model is written as

$$\frac{\partial n}{\partial t} + \nabla \cdot (nu) = 0, \tag{2.17}$$

$$\eta T \nabla n = j \times B - \nu n (u - u_n), \tag{2.18}$$

$$E + u \times B - \beta \kappa_{ei} K j = 0, \tag{2.19}$$

$$\nabla \times B = \beta j, \tag{2.20}$$

$$\frac{\partial B}{\partial t} + \nabla \times E = 0, \tag{2.21}$$

$$\nabla \cdot B = 0, \tag{2.22}$$

where $n = \lim n_e = \lim n_i$, and T is the common electron and ion temperature.

This model may be reduced to a coupled system of convection-diffusion for the plasma density and the magnetic field. Indeed, from (2.18) and (2.20) we deduce

$$\begin{aligned} u &= u_n + \frac{1}{\nu n}(j \times B - \eta T \nabla n) \\ &= u_n - \frac{\eta T}{\nu} \nabla n + \frac{1}{\beta \nu n} (\nabla \times B) \times B, \end{aligned} \tag{2.23}$$

and from (2.19) we obtain

$$\begin{aligned} E &= -u \times B + \beta \kappa_{ei} K j \\ &= -u_n \times B + \frac{\eta T}{\nu} \nabla n \times B + \frac{1}{\beta \nu n} (|B|^2 \mathrm{Id} - BB)(\nabla \times B) \\ &\quad + \kappa_{ei} K \nabla \times B. \end{aligned} \tag{2.24}$$

Then inserting (2.23) into (2.17) and (2.24) into (2.21) leads to

$$\frac{\partial n}{\partial t} + \nabla \cdot (n u_n) + \frac{1}{\beta} \nabla \cdot \left(\frac{1}{\nu} (\nabla \times B) \times B \right) - \eta \nabla \cdot \left(\frac{T}{\nu} \nabla n \right) = 0, \tag{2.25}$$

$$\begin{aligned} \frac{\partial B}{\partial t} + \frac{1}{\beta} \nabla \times \left(\frac{1}{n\nu} (|B|^2 \mathrm{Id} - BB)(\nabla \times B) \right) + \eta \nabla \times \left(\frac{T}{\nu} \nabla n \times B \right) \\ - \nabla \times (u_n \times B) + \kappa_{ei} \nabla \times (K \nabla \times B) = 0. \end{aligned} \tag{2.26}$$

The investigation and application to ionospheric plasma modeling of the Massless-MHD model will be developed in future work.

2.3 The 'striation' model

The terminal models of the two previous hierarchies are the same, the so-called Striation model (see Figure 2). This model is a reduction of the Dynamo model in which the electric field is forced to be orthogonal to the magnetic field, thus giving rise to a 2-dimensional elliptic equation. This equation is still coupled with the 3D density and (massless) momentum equations. The Striation model has been extensively used in the physics literature because the reduction to a 2-dimensional elliptic equation generates a considerable increase of computational efficiency. We present only in this section the reduction of the Dynamo model to the striation model. The construction of the striation model as a limit of the Massless-MHD model is performed in [3].

2.3.1 The 'striation' model in a uniform magnetic field. In a first approximation, the magnetic field can be considered constant and uniform. We assume that it is directed in the z direction (see Figure 3). Therefore, the magnetic field B is defined with $B = |B|\hat{z}$, where $\hat{z}$ is a unit vector. We denote by $(\hat{x}, \hat{y}, \hat{z})$ the orthonormal coordinates basis, by $x_\perp = (x, y)$ the position vector in the 2-dimensional plane orthogonal to B and by $\nabla_\perp = (\partial_x, \partial_y)^t$ the 2-dimensional gradient. For any vector $a = (a_x, a_y, a_z)$, we define $a_\perp = (a_x, a_y)^t$ its projection onto this plane.

When the electron and ion Pedersen mobilities go to infinity ($\kappa \to 0$), the Dynamo model reduces to the so-called striation model [3], which reads as follows:

$$\frac{\partial n}{\partial t} + \nabla \cdot (nu) = 0, \tag{2.27}$$

$$u = \frac{E \times B}{|B|^2} + \left(\left(u_n - \eta T \frac{\nabla \log n}{\nu} \right) \cdot \frac{B}{|B|} \right) \frac{B}{|B|}. \tag{2.28}$$

$$E = -\nabla_\perp \phi. \tag{2.29}$$

$$\nabla_\perp \cdot J_\perp = 0, \tag{2.30}$$

$$J_\perp = \frac{1}{|B|^2} (-\sigma(x_\perp) \nabla_\perp \phi + U_n \times B - 2\eta T \nabla_\perp \log n_\perp \times B), \tag{2.31}$$

with $\phi = \phi(x_\perp), \sigma(x_\perp) = \int n\nu dz, U_n = \int n\nu u_n dz, \log n_\perp = \int \log n dz, \nu = \nu_i + \nu_e$ and $u = u_i = u_e$.

The striation model involves a three-dimensional convection diffusion equation (2.27)–(2.28) for the density n with a 2-dimensional elliptic equation for the electric potential ϕ. The coefficients of the elliptic equation (2.30) are integrals of the density n over the third position coordinate, i.e. along the magnetic field lines. The infinite conductivity of the plasma along the magnetic field lines constrains the electric potential to be constant along these lines, i.e. to depend only on the 2-dimensional coordinate $x_\perp$. The coefficient $\sigma/|B|^2$ is referred to in the physics literature as the 'field-integrated Pedersen conductivity' (because $n\nu/|B|^2$ itself is the local Pedersen conductivity). Due to the integration of quantities along magnetic field lines, the com-

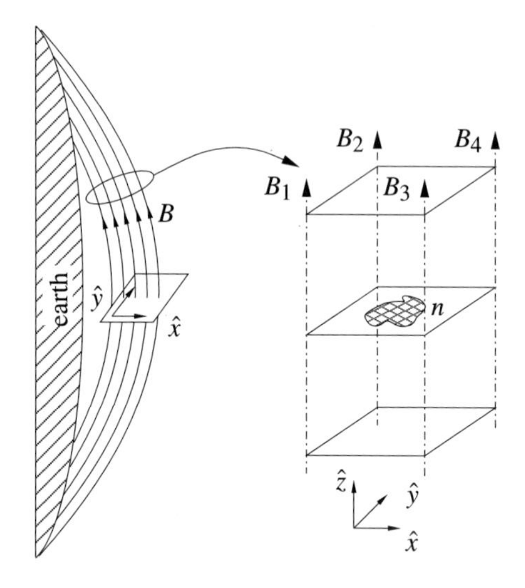

Figure 3. Geometry of the earth environment.

putational domain is a field tube, which is divided in a finite number of layers in the z coordinate(see Figure 3). This leads to a model which is often referred to as the Multi-Layer striation model. The special case of a single layer is obtained when u_n is orthogonal to B and all data and unknowns are independent of z. In this case the mono layer striation model is written

$$\frac{\partial n}{\partial t} + \nabla \cdot (nu) = 0, \tag{2.32}$$

$$u = \frac{E \times B}{|B|^2}, \tag{2.33}$$

$$E = -\nabla \phi, \tag{2.34}$$

$$\nabla \cdot (nh) = 0, \tag{2.35}$$

$$h = \frac{\nu}{|B|^2} \left(-\nabla \phi + u_n \times B - 2\eta T \frac{\nabla \log n}{\nu} \times B \right), \tag{2.36}$$

where now all variables and vectors are 2-dimensional (except B itself) and the underlying of 2-dimensional vectors has been omitted. The quantity h represents the electron-ion relative velocity. The fact that this model is 2-dimensional implies considerable time savings for numerical simulations. Since $E = -\nabla \phi$, we remark that $\nabla \cdot u = 0$. Therefore, we can write the relation (2.32)

$$\frac{\partial n}{\partial t} + (u \cdot \nabla) n = 0. \tag{2.37}$$

As we shall next see, the temperature does not play a role in the stability analysis of the striation model. In order to simplify the computations, we assume that $T = 0$. Therefore, the relation (2.36) becomes

$$h = \frac{\nu}{|B|^2}(-\nabla\phi + u_n \times B). \tag{2.38}$$

We now refer to the striation model as the combination of the equations (2.32)–(2.35) and (2.38), which we write

$$\frac{\partial n}{\partial t} + \nabla \cdot (nu) = 0, \tag{2.39}$$

$$u = \frac{-\nabla\phi \times B}{|B|^2}, \tag{2.40}$$

$$\nabla \cdot (nh) = 0, \tag{2.41}$$

$$h = \frac{\nu}{|B|^2}(-\nabla\phi + u_n \times B). \tag{2.42}$$

2.3.2 The 'striation' model in a non-uniform magnetic field. We present in this subsection an extension of the striation model to a non-uniform magnetic field [3]. We restrict to the case of an axisymmetric field. More precisely, we let (r, θ, φ) a spherical coordinate system and suppose that $B = (B_r(r, \varphi), 0, B_\varphi(r, \varphi))$ in this coordinate system. As B is divergence free and irrotational, there exist two functions $\beta(r, \varphi)$ and $\gamma(r, \varphi)$ such that $\partial_r \beta = -r \sin(\varphi) B_\varphi$, $\partial_\varphi \beta = r^2 \sin(\varphi) B_r$, $\partial_r \gamma = -B_r$ and $\partial_\varphi \gamma = -r B_\varphi$. The variable $\beta/(r \sin\varphi)$ is the θ-component of the vector potential of the magnetic field while γ is its scalar potential. As functions of x, the triple (θ, β, γ) forms an orthogonal curvilinear coordinate system. We denote by $(\hat{\theta} = \nabla\theta/|\nabla\theta|$, $\hat{\beta} = \nabla\beta/|\nabla\beta|$, $\hat{\gamma} = \nabla\gamma/|\nabla\gamma|)$, the direct orthonormal basis so constructed. In this orthogonal system, the metric is given by

$$ds^2 = \frac{1}{|B|^2}\left(r^2 \sin^2\varphi |B|^2 d\theta^2 + \frac{d\beta^2}{r^2 \sin^2\varphi} + d\gamma^2\right),$$

and the expressions for the gradient and divergence operators are

$$\nabla f = \left(\frac{1}{r \sin\varphi}\frac{\partial f}{\partial \theta}, r \sin\varphi |B| \frac{\partial f}{\partial \beta}, |B| \frac{\partial f}{\partial \gamma}\right),$$

$$\nabla \cdot A = |B|^2 \left(\frac{\partial}{\partial \theta}\left(\frac{A_\theta}{r \sin\varphi |B|^2}\right) + \frac{\partial}{\partial \beta}\left(\frac{r \sin\varphi}{|B|} A_\beta\right) + \frac{\partial}{\partial \gamma}\left(\frac{1}{|B|} A_\gamma\right)\right),$$

where the components of the gradient are written in the basis $(\hat{\theta}, \hat{\beta}, \hat{\gamma})$ and $(A_\theta, A_\beta, A_\gamma)$ denote the components of the vector A in this basis.

Using the previous relations of differential geometry, it is easy to extend the striation model to the non-uniform B case. The striation system (2.27)–(2.31) is written

$$\frac{\partial n}{\partial t} + \nabla \cdot (nu) = 0, \tag{2.43}$$

$$u_\theta = r \sin\varphi \frac{\partial V}{\partial \beta}, \quad u_\beta = -\frac{1}{r\sin\varphi |B|}\frac{\partial V}{\partial \theta}, \quad u_\gamma = u_{n\gamma} - \eta \frac{T|B|}{\nu}\frac{\partial \log n}{\partial \gamma}, \tag{2.44}$$

$$\frac{\partial J_\theta}{\partial \theta} + \frac{\partial J_\beta}{\partial \beta} = 0, \tag{2.45}$$

$$J_\theta = -A_\theta \frac{\partial V}{\partial \theta} - (U_{n\beta} - \eta \Pi_\beta), \quad J_\beta = -A_\beta \frac{\partial V}{\partial \beta} + (U_{n\theta} - \eta \Pi_\theta), \tag{2.46}$$

$$A_\theta = \int n\nu \frac{d\gamma}{r^2 \sin^2\varphi |B|^4}, \quad A_\beta = \int n\nu \frac{r^2 \sin^2 \varphi d\gamma}{|B|^2}, \tag{2.47}$$

$$U_{n\theta} = \int n\nu u_{n\theta} \frac{r\sin\varphi d\gamma}{|B|^2}, \quad U_{n\beta} = \int n\nu u_{n\beta} \frac{d\gamma}{r\sin\varphi |B|^3}, \tag{2.48}$$

$$\Pi_{n\theta} = \int \frac{\partial p}{\partial \theta}\frac{d\gamma}{|B|^2}, \quad \Pi_{n\beta} = \int \frac{\partial p}{\partial \beta}\frac{d\gamma}{|B|^2}. \tag{2.49}$$

In this case the striation model has the same structure as in the uniform magnetic field case: it couples a 3-dimensional convection-diffusion equation for n (eqs. (2.43), (2.44)) with a 2-dimensional elliptic equation for V (eqs. (2.45), (2.46)). The coefficients of the elliptic system are integrals over the field-line coordinate γ of various expressions involving the density n (eqs. (2.47)–(2.49)). The only but important difference is that vector field $u_\perp$ (where the index $\perp$ denotes the projection onto the (θ, β) plane) is no more divergence free.

If we define B as a dipole magnetic field in the coordinate system (r, θ, φ), it is possible to compute an explicit relation for β and γ. Actually, they are given by

$$\beta = -C\frac{\sin^2\varphi}{r}, \quad \gamma = -C\frac{\cos\varphi}{r^2}. \tag{2.50}$$

where the constant C is linked to the dipole moment $\mathcal{M}$ by $C = (\mu_0/4\pi)\mathcal{M}$. In this situation, the earth is surrounded by a dipole magnetic field with local coordinates (θ, β, γ). The field tube associated with this coordinate system is represented in Figure 4.

3 Stability analysis of the striation model for a uniform magnetic field

The striation model (2.39)–(2.42) in the case of a uniform magnetic field is 2-dimensional. The numerical computations are therefore easier than for the full 3-dimensional Dynamo model. However, we must guarantee that this model is well posed in the

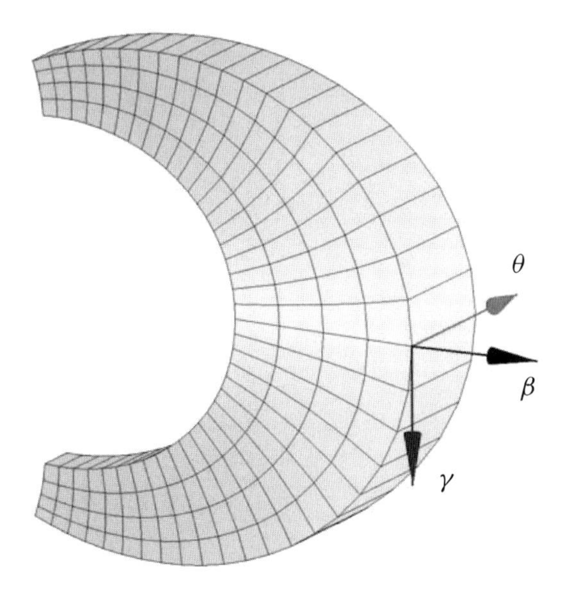

Figure 4. Geometry of a field tube for a non-uniform magnetic field.

Hadarmard sense. In a recent work [6], local-in-time existence and uniqueness of solutions for this model has been proved. The other step is to perform a stability analysis. The linear stability analysis often provides a first idea of the behavior of the solutions. Before proceeding rigorously to this analysis, we briefly give a phenomenological view. We consider a steady state consisting of a discontinuous density

$$n(x) = \begin{cases} \underline{n}, & y < 0 \\ \overline{n} > \underline{n}, & y > 0, \end{cases}$$

with $\nabla\phi = 0$ and $u_n = (0, U)$.

We slightly perturb the steady state with a perturbation n_1 of the surface of discontinuity. The interface is now represented by the graph of the function

$$(n + \varepsilon n_1)(x) = \begin{cases} \underline{n}, & y < \varepsilon \sin(\xi_1 x) \\ \overline{n} > \underline{n}, & y > \varepsilon \sin(\xi_1 x). \end{cases}$$

The parameter ε represents the scale of the perturbation ($\varepsilon \ll 1$) and ξ_1 the spatial frequency of the oscillations.

The term $u_n \times B$ of the electron-ion relative velocity creates a charge accumulation along the discontinuity. The sign of this charge accumulation changes with the monotony of the sine wave. If $U > 0$, $u_n \times B$ points at the $x > 0$ direction. When the sine wave increases, the charge accumulation is positive and negative in the decreasing parts. A polarization electric field E appears, which is negative in the convexity part of the wave and positive elsewhere. The velocity of the charged particles $u = E \times B/|B|^2$ points at $y > 0$ direction when the wave is convex and at $y < 0$ elsewhere. In this configuration, the equilibrium is stable (see Figure 5).

If we change the direction of u_n, all the signs are changed and u points at the $y < 0$ direction when the wave is convex and at the $y > 0$ direction elsewhere. In this case, the velocity u increases the instability (see Figure 6).

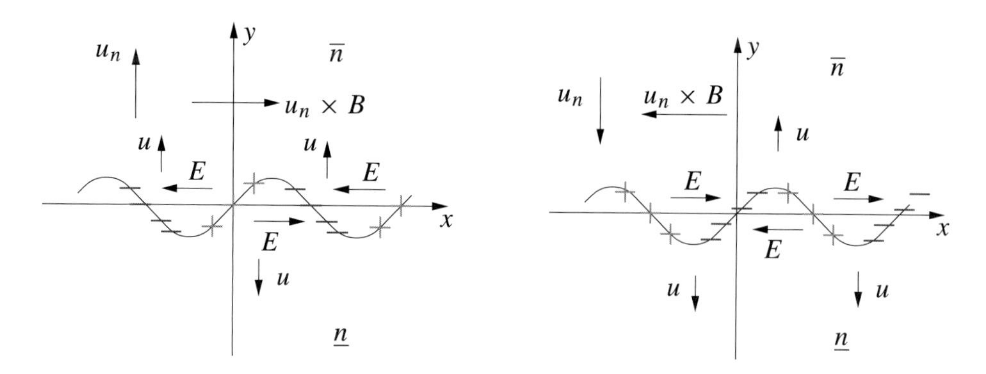

Figure 5. Stable configuration.

Figure 6. Unstable configuration.

The instability is due to the drift term $E \times B$. It is quite similar to the type II instabilities of the spread F ([13], [28]), which play a prominent role in the comprehension of the plasma bubbles upwelling in the ionosphere F layer ([36], [26], [21]). Nevertheless, the striations instabilities can not be assimilated to these type II instabilities which are actually involved by the gravity. Indeed, the major phenomena leading to the striations instabilities is the neutral wind.

This behavior can be recovered by the linear stability analysis. We proceed to this analysis for an exponential increasing density in one direction. It is possible to extend this analysis to discontinuous density profiles (see [4]).

3.1 Linear stability analysis

The plasma inhomogeneity is represented with an unperturbed state consisting in an exponential profile of density in the y-direction

$$n^0 = Ne^{y/\lambda}, \quad \lambda > 0, \tag{3.1}$$

where λ is the gradient length. The neutral wind has the general expression $u_n = (V, U)^t$. In order to have a steady state, the velocity of the charge particles u^0 should be equal to $u^0 = (V, 0)^t$ and the electron-ion relative velocity equal to $h^0 = (\nu U/|B|, 0)^t$. Finally, we should take $E^0 = -\nabla\phi^0 = (0, V|B|)$.

Then we introduce the perturbation

$$\begin{aligned} n &= n^0(1 + \varepsilon n^1 + O(\varepsilon^2)), \qquad & u &= u^0 + \varepsilon u^1 + O(\varepsilon^2), \\ \phi &= \phi^0 + \varepsilon \phi^1 + O(\varepsilon^2), & h &= h^0 + \varepsilon h^1 + O(\varepsilon^2), \end{aligned} \tag{3.2}$$

with $\varepsilon \ll 1$ in the striation model. We leave out all the terms less than ε. The neutral wind is supposed to not contain pertubative terms. We note that $\nabla n^0 = (0, n^0/\lambda)$. An easy computation gives the linearized system governing the first-order perturbed state:

$$\frac{\partial n^1}{\partial t} + \frac{1}{\lambda |B|} \frac{\partial \phi^1}{\partial x} + V \frac{\partial n^1}{\partial x} = 0, \tag{3.3}$$

$$-\frac{1}{\lambda} \frac{\partial \phi^1}{\partial y} - \Delta \phi^1 + U|B| \frac{\partial n^1}{\partial x} = 0. \tag{3.4}$$

Remark 3.1. If the temperature is retained in the striation model, the steady state is only modified through the expression of h^0. The electron-ion relative velocity should be taken as $h^0 = (\nu U/|B| - 2T/(\lambda |B|), 0)^t$. However, the equations generated with the introduction of the perturbation leads to the same expressions (3.3) and (3.4). At a first sight, the expression of h^0 might indicate that the temperature plays the same role as the neutral wind. In fact, an algebraic compensation in (3.4) takes away a participation of the temperature in the stability process of the striation model.

We develop the solution into plane waves (dropping the superscripts '1' for clarity)

$$n = \bar{n} \exp i \left(\xi_1 \frac{x}{\lambda} + \xi_2 \frac{y}{\lambda} - \omega t \frac{|U|}{\lambda} \right), \tag{3.5}$$

$$\phi = \bar{\phi} \lambda |U| \exp i \left(\xi_1 \frac{x}{\lambda} + \xi_2 \frac{y}{\lambda} - \omega t \frac{|U|}{\lambda} \right), \tag{3.6}$$

where $\xi = (\xi_1, \xi_2)$ is the (normalized) wave-vector of the perturbation and ω its frequency.

Introducing the expressions (3.5) and (3.6) in (3.3), (3.4), we get

$$-\omega \bar{n} + \xi_1 \bar{\phi} = 0, \tag{3.7}$$

$$i \sigma \xi_1 \bar{n} + (\xi_1^2 + \xi_2^2 - i \xi_2) \bar{\phi} = 0, \tag{3.8}$$

with $\sigma = \text{sign}(U) \in \{-1, 1\}$. This system has a non trivial solution iff its determinant is non vanishing. This condition yields the dispersion relation

$$\omega = \frac{-i \sigma \xi_1^2}{(\xi_1^2 + \xi_2^2)^2 + \xi_2^2} (\xi_1^2 + \xi_2^2 + i \xi_2). \tag{3.9}$$

We now recall the following standard

Definition 3.2. The perturbation is stable if n and ϕ stay bounded for all times $t \geq 0$ and unstable in the converse situation. Therefore, a perturbation is stable iff

$\mathrm{Im}(\omega) \leq 0$. On the contrary, a perturbation is unstable iff $\mathrm{Im}(\omega) > 0$. A stationary state is called stable if all the perturbations are stable for all wave vectors ξ. It is unstable as soon as it exists a wave vector ξ giving rise to an unstable perturbation.

Thanks to (3.9), we have

$$\mathrm{Im}(\omega) = \frac{-\sigma \xi_1^2(\xi_1^2 + \xi_2^2)}{(\xi_1^2 + \xi_2^2)^2 + \xi_2^2}. \tag{3.10}$$

The conclusion is then easy and is summarized in the following proposition.

Proposition 3.3. *The steady state configuration with an exponential density profile of the striation model is stable if and only if $U \geq 0$. In the case $U < 0$, all the wave vectors $\xi \neq 0$ are unstable and for $\xi_2 = 0$, the growing rate $\dfrac{|U|}{\lambda}$ is independent of ξ_1.*

3.2 Nonlinear stability analysis

The linear instability of the striation model does not guarantee its nonlinear instability. Indeed, the nonlinearity can control the unstable linear modes and avoid their growth. So, we must verify that an unstable linear mode is actually a nonlinear one. This study is far more tricky than the classical linear stability analysis. Some recent works have presented solutions [16], [18], [10]. The most recent contribution was done by [20] for the Rayleigh–Taylor nonlinear instabilities for the incompressible Euler equations.

Their methodology may be adapted to the study of the striation model instabilities. The first key point is the study of the Cauchy problem. The next point is to find unstable linear modes and a global mode to control their behaviors. With these hypotheses, it is possible to show the nonlinear instability. The rigorous mathematical result for the striation model is developed in [6]. We just give here the two main results.

Theorem 3.4. *Let $n_0 \in H^k(\mathbb{R}^2)$, $n_0 \geq \kappa > 0$. Then the Cauchy problem for the striation model is locally well posed, i.e. there exists T and a solution $(n, u) \in C([0, T], H^k(\mathbb{R}^2)) \cap \mathrm{Lip}([0, T], H^{k-1}(\mathbb{R}^2))$.*

Theorem 3.5. *Let $k \geq 3$ and let us consider the steady state $(\bar{n}(x), 0)$ with the neutral wind $u_n = (0, U(x))$. Then the striation model is linearly unstable if there exists x_0 such that $-(U\frac{\partial}{\partial x}\bar{n})|_{x_0} > 0$. Moreover, there is ε_0 such that for all small parameters $\delta > 0$, there exists a solution (n^δ, ϕ^δ) and a time T^δ such that*

$$|n^\delta(0) - \bar{n}|_{H^k(\mathbb{R}^2)} \leq \delta$$

and

$$|n^\delta(T^\delta) - \bar{n}|_{L^2(\mathbb{R}^2)} + |\nabla\phi^\delta(T^\delta)|_{L^2(\mathbb{R}^2)} \geq \varepsilon_0.$$

Thus, smooth stationary profiles which are linearly unstable are actually nonlinearly unstable. Moreover, we also show that the instability growth time is previous to the existence time of the solution.

The striation model is at once linearly and nonlinearly unstable when the neutral wind $u_n = (V, U)$ is chosen such that $U < 0$. The $E \times B$ drift instability contributes to the development of smaller and smaller structures in the plasma. Moreover, the small scale disturbances (small wavelength induces big wave number) grow faster than the big scale ones. Quickly, the plasma becomes chaotic.

The model lacks a diffusive process capable to stabilize the small scales. Adding a stabilization term should counterbalance the role of the instabilities and soften the big wave numbers. However, the diffusion vanished from the striation model into the limit $\kappa \to 0$ of the Dynamo model. This limit comes down to consider an infinite electronic or ionic Pedersen mobility. The Hall mobility only remains to stabilize the model, but is too small to be able to play a role. In order to reproduce the effect of the Pedersen mobility, two cures can be devised. It is obviously possible to come back to the full Dynamo model, but the computation cost may be really expensive. We choose an other alternative and provide a statistical model which gives the mean evolution of the involved quantities. In this approach, the idea is that the instabilities generate some turbulence. We will process in a way similar to the $k - \varepsilon$ modeling of the fluid mechanics turbulence. We examine these possibilities in the two following sections.

4 Influence of finite ionic conductivity and finite temperature

The stability analysis of the striation model reveals a lack of diffusion. The diffusion modes disappear in the infinite Pedersen mobility limit. We should come back to the full 3-dimensional Dynamo model to recover these processes. We reproduce the linear stability analysis for the exponential density profile. This analysis can be found, in parts, in [12], [13]. The analysis for a discontinuous density profile can not be performed anymore. On the one hand, the diffusion will spread the discontinuity and the profile will be broken. On the other hand, the theoretical analysis of the nonlinear instability is not made.

To simplify the analysis, we chose to suppose that the quantities do not depend on the z variable and that the neutral wind is orthogonal to the magnetic field B.

The steady state is given by

$$n^0 = N \exp(y/\lambda), \tag{4.1}$$

$$u_n = (V, U, 0)^T, \tag{4.2}$$

$$u_i^0 = \left(\frac{\kappa \nu_e}{|B|}(U - \frac{2\eta T}{(\nu_i + \nu_e)\lambda}) + V \right) \hat{x}, \tag{4.3}$$

$$u_e^0 = \left(-\frac{\kappa \nu_i}{|B|} (U - \frac{2\eta T}{(\nu_i + \nu_e)\lambda}) + V \right) \hat{x}, \quad (4.4)$$

$$\nabla \phi^0 = -E^0 = \begin{pmatrix} -\frac{\kappa^2 \nu_i \nu_e}{|B|} (U - \frac{2\eta T}{(\nu_i + \nu_e)\lambda}) \\ -\kappa(\nu_e - \nu_i)(U - \frac{\eta T}{(\nu_i + \nu_e)\lambda}) - |B| V \\ 0 \end{pmatrix}, \quad (4.5)$$

$$h^0 = \frac{1}{B} \left((\nu_i + \nu_e) U - \frac{2\eta T}{\lambda} \right) \hat{x}. \quad (4.6)$$

We proceed to the linear stability analysis as in Section 3.1. We introduce the perturbation $n = n^0(1 + \varepsilon n^1) + O(\varepsilon^2)$, $u_{i,e} = u_{i,e}^0 + \varepsilon u_{i,e}^1 + O(\varepsilon^2)$ and $\phi = \phi^0 + \varepsilon \phi^1 + O(\varepsilon^2)$ in the Dynamo model (2.8)–(2.12). We keep only ε-order terms and develop the solution as a plane wave (3.5)–(3.6). Let us define

$$\mu_-^H = \mu_i^H - \mu_e^H, \quad \mu_+^H = \mu_i^H + \mu_e^H, \quad \mu_-^P = \mu_i^P - \mu_e^P, \quad \mu_+^P = \mu_i^P + \mu_e^P,$$

$$X = \xi_1 \mu_i^H - \xi_2 \mu_i^P, \quad Y = \xi_1 \mu_-^H - \xi_2 \mu_+^P, \quad Z = \xi_1 \mu_+^H - \xi_2 \mu_-^P,$$

$$A_X = \mu_i^P |\xi|^2 + iX, \quad A_Y = \mu_+^P |\xi|^2 + iY, \quad A_Z = \mu_-^P |\xi|^2 + iZ.$$

Then we get the dispersion relation

$$\omega = \frac{\overline{A_Y}}{|U|^2 |A_Y|^2} \Big(\xi_1 u_{ix}^0 - i \frac{\kappa \eta T}{\lambda} A_X |U| A_Y - \xi_1 (h_x^0) |U| A_X + i \frac{\kappa \eta T}{\lambda} A_X |U| A_Z \Big). \quad (4.7)$$

The expression of $\mathrm{Im}(\omega)$ may be simplified to

$$\mathrm{Im}(\omega) = \frac{\mu_i^P \mu_e^P}{|A_Y|^2} \Bigg\{ -\frac{\sigma (\nu_i + \nu_e)^2}{\nu_i \nu_e} \xi_1^2 |\xi|^2 - \frac{2\kappa \eta T}{\lambda |U| |A_Y|^2} \Big[\mu_+^P |\xi|^6 + \mu_+^P |\xi|^2 (\xi_2^2 - \xi_1^2) - 2\mu_-^H |\xi|^2 \xi_1 \xi_2 \Big] \Bigg\}, \quad (4.8)$$

where $\sigma = \frac{U}{|U|}$ refers to the sign of U.

Letting $\alpha = \kappa^2 \nu_i^2 + |B|^2$ and $\beta = \kappa^2 \nu_e^2 + |B|^2$, we can express $\frac{\mu_i^P \mu_e^P}{|A_Y|^2}$ as

$$\frac{\mu_i^P \mu_e^P}{|A_Y|^2} \quad (4.9)$$

$$= \frac{\nu_i \nu_e}{(\nu_i + \nu_e)^2} \frac{\alpha \beta}{\big((\nu_i \nu_e \kappa^2 + |B|^2)^2 |\xi|^4 + (|B| \kappa (\nu_i - \nu_e) \xi_1 - (\nu_i \nu_e \kappa^2 + |B|^2) \xi_2)^2 \big)}$$

$$= \frac{\nu_i \nu_e}{(\nu_i + \nu_e)^2} \Theta_0,$$

with $\Theta_0 > 0$.

If $\kappa \to 0$, the Dynamo model reduces to the striation one. We should recover this property in the expression of $\mathrm{Im}(\omega)$. The formal limit leads to

$$\mathrm{Im}(\omega) = -\frac{\sigma \xi_1^2 |\xi|^2}{|\xi|^4 + \xi_2^2}, \tag{4.10}$$

which is the growth rate of the instability (3.10).

In the null temperature limit $T = 0$, the growth rate is given by

$$\mathrm{Im}(\omega) = -\Theta_0 \sigma \xi_1^2 |\xi|^2. \tag{4.11}$$

Thus the result is equivalent to the previous case. As we had mentioned in Section 3.1, the temperature alone can not stabilize the model.

Therefore, for the model to exhibit a stable range of wave-vectors, we need at the same time a finite conductivity and a finite temperature. We now show that this is indeed the case.

Proposition 4.1. *Suppose that* $\eta > 0$ *and* $\kappa > 0$. *Then there exists* $R_0(\eta, \kappa) \geq 0$ *such that the Dynamo model* (2.8)–(2.13) *linearized about the above-defined steady-states is stable for all wave-vectors* ξ *with* $|\xi| \geq R_0(\eta, \kappa)$.

Proof. We introduce polar coordinates $\xi_1 = r \cos\theta$ and $\xi_2 = r \sin\theta$. The study of $\mathrm{Im}(\omega)$ reduces to the study of the function

$$f(\xi) = -br^4(r^2 + \Sigma(\theta)), \quad \Sigma(\theta) := ab^{-1}\cos^2\theta - \cos 2\theta - b^{-1}(c/2)\sin 2\theta \tag{4.12}$$

where

$$a = -\sigma\Theta_0, \quad b = \frac{\nu_i \nu_e}{(\nu_i + \nu_e)^2}\Theta_0 \frac{2\kappa\eta}{\lambda |U||A_Y|^2}\mu_+^P, \quad c = 2\mu_-^H.$$

We remark that Σ is a bounded function of θ. Let $\Sigma_0 = \min(0, \min_\theta \Sigma)$ and $R_0 = \Sigma_0^{1/2}$. Then, for wave-vectors ξ such that $r = |\xi| \geq R_0$, the Dynamo model is stable. □

The fact that the model is stable apart from a bounded region of wave-vectors can be seen as a favorable feature. Indeed, in such a case small wave-vector (i.e. long wave-length) perturbations first grow exponentially due to the instability, but also undergo a mode cascade towards higher wave-numbers due to nonlinearity. Once the wave-vectors are large enough to reach the stability region, they are damped by the dissipation. We therefore expect that only structures of typical size R_0^{-1} will remain for long times.

However, the physical parameters in the Dynamo model are too small to ensure a stability process. Indeed, we see in Table 2 that $\kappa = 10^{-4}$. It only remains the parameter η, which models the strength of the temperature effect. Now, we just saw that the temperature alone can not stabilize the model. On the other hand, the computational cost for the simulation of the Dynamo model is very important compared to the striation one. A new approach is necessary.

5 Derivation of turbulent Dynamo model

The numerical simulation does not allow to compute scales less than mesh size. Thus, the numerical methods cut spontaneously the smaller scales. Actually, we are interested only in the evolution of the big scales and this process seems not really annoying. However, in the striation model case, the small scales density oscillations may contribute to a modification of the electric potential, and by induction to the plasma dynamics. Therefore, we must compute a new model, which describes the small scale effects on the big scale plasma dynamics. This new model should not behave as the striation model at least for the small scales.

To produce this new model we follow the $k-\varepsilon$ model, which models the effects of an homogeneous and isotropic turbulence for the incompressible Navier–Stokes equations ([25]). Precisely, we use a result of Kesten and Papanicolaou [22]. Their result has been extensively used and theoretically generalized in [27]. This is a statistical approach. We consider the small scales as random disturbances of the big scales and we model the evolution of the mean quantities. Other modelings exist with spectral methods, see e.g. [31].

We suppose that the quantities (n, u, ϕ, h) can be factorized in mean values denoted by $(\bar{n}, \bar{u}, \bar{\phi}, \bar{h})$ and random fluctuations denoted by (n', u', ϕ', h'). Since the randomness concerns the realization of the flow, the mean value operator commutes with the spatial and time derivatives. Then, for a quantity a, we have:

$$a = \bar{a} + a', \quad \bar{a} = \text{mean value}, \quad a' = \text{fluctuation}, \quad \overline{(\bar{a})} = \bar{a}, \quad \overline{a'} = 0,$$

$$\nabla a = \nabla \bar{a} + \nabla a', \quad \frac{\partial a}{\partial t} = \frac{\partial \bar{a}}{\partial t} + \frac{\partial a'}{\partial t}.$$

If b is a non-fluctuating quantity, we have $\overline{ba} = b\bar{a}$. For two random quantities a and b, we do not have $\overline{ab} = \bar{a}\bar{b}$ unless these two quantities are independent. On the other hand, we have $\overline{a\bar{b}} = \bar{a}\bar{b}$.

Since the neutral wind is a non-fluctuating quantity, the striation model (2.39)–(2.42) becomes

$$\begin{aligned}
&\frac{\partial \bar{n}}{\partial t} + \nabla \cdot (\overline{nu}) = 0, \\
&\bar{u} = -\frac{\nabla \bar{\phi} \times B}{|B|^2}, \quad u' = -\frac{\nabla \phi' \times B}{|B|^2}, \\
&\nabla \cdot (\overline{nh}) = 0, \\
&\bar{h} = -\nabla \bar{\phi} - u_n \times B, \quad h' = -\nabla \bar{\phi}'.
\end{aligned}$$

Thanks to the calculus rules given above, we have

$$\overline{nu} = \overline{(\bar{n} + n')(\bar{u} + u')} = \bar{n}\bar{u} + \bar{n}\overline{u'} + \overline{n'}\bar{u} + \overline{n'u'} = \bar{n}\bar{u} + \overline{n'u'}.$$

Since n' and u' are not independent random variables, the correlation $\overline{n'u'}$ of the density and the velocity is not null. In the same way, we have for nh that

$$\overline{nh} = \bar{n}\bar{h} + \overline{n'h'}.$$

Therefore, the striation model reads

$$\frac{\partial \bar{n}}{\partial t} + \nabla \cdot (\bar{n}\bar{u}) + \nabla \cdot (\overline{n'u'}) = 0, \tag{5.1}$$

$$\bar{u} = -\frac{\nabla \bar{\phi} \times B}{|B|^2}, \tag{5.2}$$

$$\nabla \cdot (\bar{n}\bar{h}) + \nabla \cdot (\overline{n'h'}) = 0. \tag{5.3}$$

To close the model, it remains to compute the correlations $\overline{n'u'}$ and $\overline{n'h'}$ as functions of the mean quantities. We use the Kesten–Papanicolaou theorem.

Theorem 5.1 (Kesten–Papanicolaou [22]). *Let u be a divergence free vector field, $f(x, t)$ a given function and $c(x, t)$ the solution of the equation*

$$\frac{\partial c}{\partial t} + u \cdot \nabla c = f. \tag{5.4}$$

Assume that u is a random variable, independent of t and mixing (see Remark 5.2) with a mean denoted by $\bar{u}$. Moreover, suppose that $|\bar{u}| \gg |u'|$. Then $\bar{c}$ satisfies the convection diffusion equation

$$\frac{\partial \bar{c}}{\partial t} + u \cdot \nabla \bar{c} - \nabla \cdot (M \nabla \bar{c}) = \bar{f}, \tag{5.5}$$

where the matrix M is the autocorrelation matrix of u':

$$M_{ij} = \frac{1}{2} \int_{-\infty}^{+\infty} \overline{u'_i(x + \bar{u}t) u'_j(x)}\, dt. \tag{5.6}$$

Remark 5.2. A random vector field u is mixing if, for any two sets of points (x^i), $i \in [1, m]$, and (y^j), $j \in [1, n]$ with $\|x^i - y^j\| > d$, for all i, j, and any functions $F(u(x^1), \ldots, u(x^m))$, $G(u(y^1), \ldots, u(y^n))$ there exists $\alpha(d)$ decaying sufficiently fast to zero with d such that

$$|\overline{FG} - \overline{F}\,\overline{G}| \leq \alpha(d) |F|_\infty |G|_\infty,$$

independent of m and n.

Assuming that all the Kesten–Papanicolaou theorem hypotheses are satisfied, we may write

$$\overline{n'u'} = -D_u \nabla \bar{n} \quad \text{and} \quad \overline{n'h'} = -D_h \nabla \bar{n}, \tag{5.7}$$

where D_u and D_h are diffusion coefficients related to the autocorrelation of u and h respectively .

To simplify the model, we assume that $D_h = 0$ and define $D = D_u$. The turbulent striation model is (we remove the bars on the mean values)

$$\frac{\partial n}{\partial t} + \nabla \cdot (nu) - \nabla \cdot (D\nabla n) = 0, \tag{5.8}$$

$$u = -\frac{\nabla\phi \times B}{|B|^2}, \tag{5.9}$$

$$\nabla \cdot (n(-\nabla\phi - u_n \times B)) = 0. \tag{5.10}$$

All the difficulty is deferred to the computation of an expression of the diffusion coefficient D. An analytical expression is hard to obtain. We proceed to its determination with a phenomenological analysis. Since the small scales contributes to increase the turbulence and the instability, we choose a characteristic length scale l such that wavelengths less or equal to l be stable. Therefore, we have to determine a minimum diffusion constant D, which guarantees the stability. In a practical point of view, l will be associated to the mesh size Δx. The stability analysis is a key point to determine empirically the diffusion constant D.

6 Stability analysis of the turbulent striation model

We again choose a steady state characterized by an exponential density profile. We assume that D is constant. Since the lack of diffusion is associated to the Pedersen mobilities, we take a diffusion coefficient equal to

$$D = \begin{pmatrix} D_1 & 0 \\ 0 & D_2 \end{pmatrix}. \tag{6.1}$$

The unperturbed state is defined by

$$n_0 = Ne^{y/\lambda}, \quad \nabla n_0 = \left(0, \frac{n_0}{\lambda}\right), \tag{6.2}$$

$$u_0 = \left(V, \frac{D_2}{\lambda}\right) = \left(-\frac{1}{|B|}\frac{\partial\phi_0}{\partial y}, \frac{1}{|B|}\frac{\partial\phi_0}{\partial x}\right), \tag{6.3}$$

$$\phi_0 = \frac{|B|D_2}{\lambda}x - V|B|y, \tag{6.4}$$

$$u_n = (V, U). \tag{6.5}$$

Let $D = |U|\lambda\bar{D}$. We proceed as in the previous sections and we get the dispersion relation

$$\omega = -\frac{V}{|U|}\xi_1 - i(\bar{D}_1\xi_1^2 + \bar{D}_2\xi_2^2) - i\frac{(\sigma - \bar{D}_2)\xi_1^2}{(\xi_1^2+\xi_2^2)^2 + \xi_2^2}((\xi_1^2+\xi_2^2) + i\xi_2). \tag{6.6}$$

We deduce that

$$\mathrm{Im}(\omega) = \frac{\mathcal{N}}{\mathcal{D}}, \tag{6.7}$$

$$\mathcal{N} = -(\bar{D}_1\xi_1^2 + \bar{D}_2\xi_2^2)((\xi_1^2 + \xi_2^2)^2 + \xi_2^2) - (\sigma - \bar{D}_2)\xi_1^2(\xi_1^2 + \xi_2^2), \tag{6.8}$$

$$\mathcal{D} = (\xi_1^2 + \xi_2^2)^2 + \xi_2^2. \tag{6.9}$$

Since $\mathcal{D} \geq 0$, we just have to discuss the sign of $\mathcal{N}$. Let $x = r\cos(\theta)$ and $y = r\sin(\theta)$. This is equivalent to the sign study of the function

$$f(r,\theta) = -r^4((\bar{D}_1\cos^2\theta + \bar{D}_2\sin^2\theta)(r^2 + \sin^2\theta) + (\sigma - \bar{D}_2)\cos^2\theta). \tag{6.10}$$

We analyse the sign of f for some values of $\bar{D}_1$ and $\bar{D}_2$ rather than the general case. We summarize the results in Table 3.

Table 3. Stability result for the turbulent striation model.

	$\sigma = +1$	$\sigma = -1$
$\bar{D} = \bar{D}_1 = \bar{D}_2$ or $\bar{D} = \bar{D}_2$ and $\bar{D}_1 = 0$	$\bar{D} \leq 1$ stable $\bar{D} > 1$ unstable if $r(\theta) < \sqrt{1 - \bar{D}^{-1}}$	$\bar{D} \leq 1$ stable $\bar{D} > 1$ unstable if $r(\theta) < \sqrt{1 + \bar{D}^{-1}}$
$\bar{D} = \bar{D}_1$ and $\bar{D}_2 = 0$	stable	unstable if $r(\theta) < 1/\sqrt{\bar{D}}$

Consequently, the turbulent striation model is more stable with an anisotropic diffusion acting only in the x direction. As the gradient drift, the diffusion in the y direction can destabilize the model in some cases. We choose therefore an anisotropic diffusion in the x-direction.

We look for a wavelength scale l such that all waves with a wavelength ζ less than l be stable. The corresponding wave numbers are respectively $1/l$ and $1/\zeta$ and the relation becomes

$$\text{if } \frac{1}{\zeta} > \frac{1}{l}, \text{ then the wave is stable.}$$

In order to achieve this property, we must have

$$\frac{1}{l} \geq \frac{1}{\sqrt{\bar{D}}}.$$

Since $\bar{D} = \frac{\lambda}{|U|}D$, this relation leads to

$$D \geq l^2\frac{|U|}{\lambda}.$$

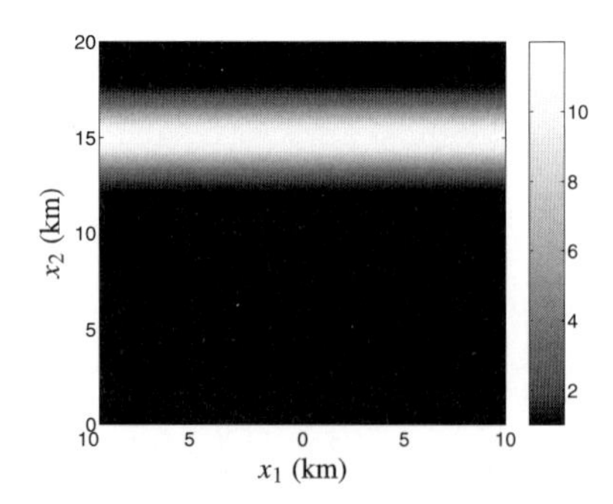

Figure 7. Initial plasma density (m^{-3}).

Table 4. Number of cells and mesh sizes.

	Nb. of cells	Δ_x, Δ_y (m)
Mesh 1	200×200	$0.1 \ 10^3$
Mesh 2	400×400	$0.05 \ 10^3$
Mesh 3	800×800	$0.025 \ 10^3$

7 Numerical experiments

In this section, we present some numerical simulations of the mono layer striation model (2.39), (2.42) and of the turbulent striation model (5.8), (5.10). We present also some simulations of the multi layers striation model in the framework of a non-uniform magnetic field.

The elliptic equation (2.42) or (5.10) is discretized by a conservative finite difference method. The plasma velocity is computed by means of finite differences applied to the second equation of (2.39) or (5.8) on staggered grids. The transport equation (first equation of (2.39) or (5.8)) is discretized thanks to a classical TVD-scheme ([15], [34], [23]). In order to deal with steep density gradients, the diffusion operator in (5.8) is implicitly discretized and we make use of a Strang splitting for the overall time discretization of this equation. A preconditioned gradient method [32] is applied to solve the linear systems resulting from the discretization of the elliptic equation (2.42) or (5.10) and from the implicit discretization of the diffusion equation (5.8).

The initial density is a random perturbation of a uniform density in the x_1-direction with a Gaussian profile in the x_2-direction (see Figure 7). The neutral wind u_n is directed along the x_2-axis and has a value of 45 ms^{-1}. Different mesh sizes listed in Table 4 are considered.

We first consider the original striation model (2.39), (2.42). In Figure 8 we represent the electronic density computed on the different meshes (see Table 4) at time $t = 804\,\text{sec}$. One can notice that the number of persisting structures grows with the number of cells while their typical size decreases with the mesh-size. This behavior can be related to the instability of the model. Indeed, numerical diffusion is the only damping mechanism and the numerical diffusivity is proportional to the mesh size [15], [34], [23]. According to the stability analysis in Section 6, the diffusive striation model becomes stable for wave-vectors of the order of $1/\sqrt{D}$, which is proportional to $1/\sqrt{\Delta x}$. Therefore, the size of the typical persisting structures must be divided by a factor $\sqrt{2}$ each time the mesh-size is divided by 2.

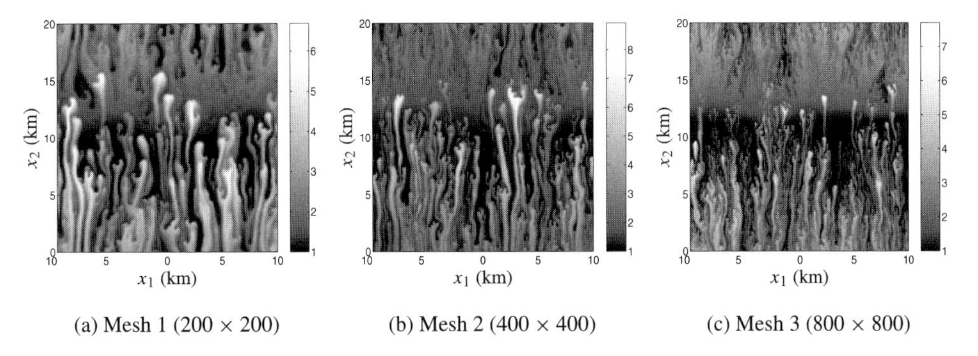

(a) Mesh 1 (200 × 200) (b) Mesh 2 (400 × 400) (c) Mesh 3 (800 × 800)

Figure 8. Electronic density after 804 sec, with the striation model.

We next consider the turbulent striation model (5.8), (5.10). Figure 9 demonstrates the stability brought by the diffusion: the number and size of the structures remain almost the same when the mesh resolution increases.

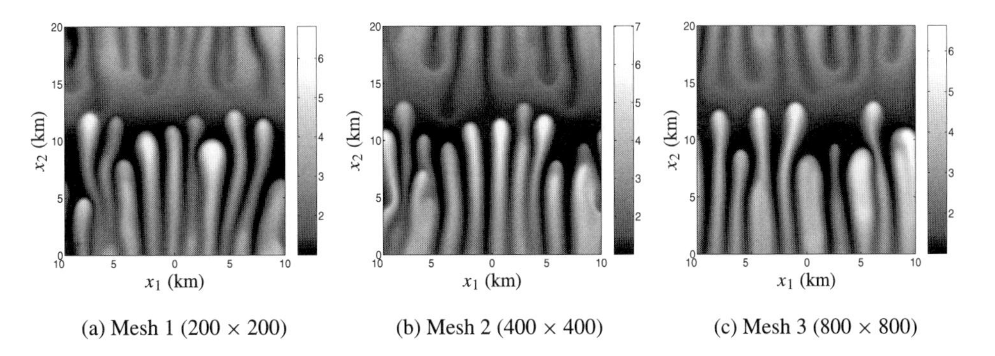

(a) Mesh 1 (200 × 200) (b) Mesh 2 (400 × 400) (c) Mesh 3 (800 × 800)

Figure 9. Electronic density after 804 sec, with the turbulent striation model.

The treatment of the multi layers striation model in the framework of a non-uniform magnetic field is performed in a similar way. The major difference is that the divergence of the velocity u is not equal to zero anymore. Therefore, a first order splitting scheme is used to discretize the transport equation (2.43). To simplify the resolution, we take a neutral wind with no component in the γ direction and the temperature is set to zero. The initial density is a random perturbation of a uniform density in the α-direction with a Gaussian profile in the β-direction. The neutral wind u_n is directed along the α-axis and has a value of 45 ms^{-1}. We consider a mesh of 300×300 cells with $\Delta_\alpha = \Delta_\beta = 4\,\text{km}$. The field tube is discretized with 60 slices. The central slice is located at an altitude of 700 km. A transverse slice is added to show the evolution of the striations along the magnetic field tube.

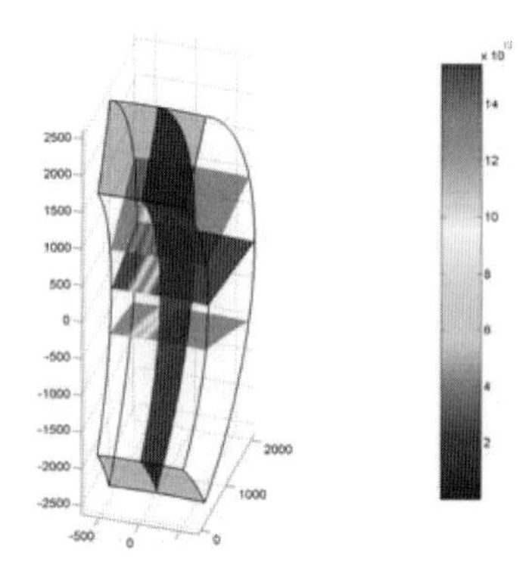

Figure 10. Initial plasma density (m^{-3}).

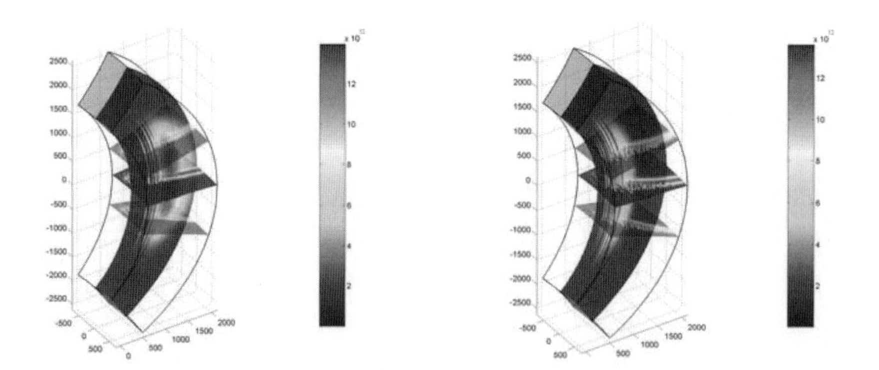

Figure 11. Plasma density at 1250 sec. Figure 12. Plasma density at 2250 sec.

The effect of the non-uniform magnetic field is important. Actually, in the classical striation model, we see that the density of the plasma is only stretched along x_2 direction (corresponding to β for the non-uniform model). In the simulations in Figures 11 and 12 we see that a new effect is added. The plasma is also bended in β direction. Additionally, the plasma is subject to the striation process along the magnetic field lines.

8 Conclusion

We have presented a hierarchy of models to describe the evolution of ionospheric plasma in a uniform and a non-uniform magnetic field. The high-level model, the so-called striation model, is proved to be linearly unstable. A statistical approach is proposed to model the instabilities. Numerical simulations are provided to confirm the validity of these modelings.

References

[1] W. G. Baker, D. F. Martyn, Electric currents in the ionosphere, I. The conductivity, *Phil. Trans. Roy. Soc. London* A246 (1953), 295–305 (see also parts II et III in the same issue).

[2] J. J. Berthelier, L'ionosphère, in *Environnement spatial: prévention des risques liés aux phénomènes de charge* (J. P. Catani et M. Romero, eds.), Cépaduès éditions, Toulouse 1996.

[3] C. Besse, J. Claudel, P. Degond, F. Deluzet, G. Gallice, C. Tessieras, A model hierarchy for ionospheric plasma modeling, *Math. Models Methods Appl. Sci.* 14 (2004), 393–415.

[4] C. Besse, J. Claudel, P. Degond, F. Deluzet, G. Gallice, C. Tessieras, Instability of the ionospheric plasma: modeling and analysis, submitted.

[5] C. Besse, J. Claudel, P. Degond, F. Deluzet, G. Gallice, C. Tessieras, Numerical simulations of the ionospheric dynamo model in a non-uniform magnetic field, in preparation.

[6] C. Besse, P. Degond, H-J. Hwang, R. Poncet, A rigorous proof of the nonlinear gradient-drift instability in the framework of the ionospheric dynamo model, to appear in *Comm. Partial Differential Equations*.

[7] E. Blanc, B. Mercandelli, E. Houngninou, Kilometric irregularities in E and F regions of the daytime equatorial ionosphere observed by a high resolution HF radar, *Geophysical Research Letters* 23 (1996), 645–648.

[8] J. W. Chamberlain, D. W. Hunter, *Theory of planetary atmospheres*, Academic Press, New York 1987.

[9] S. Chandrasekhar, *Hydrodynamic and hydromagnetic stability*, Dover Publications, 1981.

[10] B. Desjardins, E. Grenier, Linear instability implies nonlinear instability for various types of viscous boundary layers, *Ann. Inst. H. Poincaré Anal. Non Linéaire* 20 (1) (2003), 87–106.

[11] J. H. Doles III, N. J. Zabuski, F. W. Perkins, Deformation and striation of plasma clouds in the ionosphere. 3. Numerical simulations of a multilevel model with recombination chemistry, *J. Geophys. Res.* 81 (1976), 5987–6004.

[12] D. T. Farley, Theory of equatorial eletrojet plasma waves, new developments and current status, *J. Atm. Terr. Phys.* 47 (1985), 729–744.

[13] B. G. Fejer, M. C. Kelley, Ionospheric irregularities, *Reviews of Geophysics and Space Physics* 18 (1980), 401–454.

[14] V. Girault, P.-A. Raviart, *Finite Element Methods for Navier Stokes Equations, Theory and Algortihms*, Springer Ser. Comput. Math. 5, Springer-Verlag, Berlin 1986.

[15] E. Godlewski, P. A. Raviart, *Numerical Approximation of Hyperbolic Systems of Conservation Laws*, Appl. Math. Sci. 118, Springer-Verlag, New York 1996.

[16] E. Grenier , On the nonlinear instability of Euler and Prandtl equations, *Comm. Pure Appl. Math.* 53 (2000), 1067–1091.

[17] C.Grimault, Caractérisation des canaux de propagation satellite-Terre SHF et EHF en présence de plasma post-nucléaire, PhD Dissertation, University of Rennes, France, 1995.

[18] Y. Guo, W. Strauss , Nonlinear instability of double-humped equilibria, *Ann. Inst. H. Poincaré Anal. Non Linéaire* 12 (1995), 339–352.

[19] J. K. Hargreaves, *The solar-terrestrial environment*, Cambridge University Press, Cambridge 1992.

[20] H. J. Hwang, Y. Guo, On the dynamical Rayleigh-Taylor instability, *Arch. Ration. Mech. Anal.* 167 (3) (2003), 235–253.

[21] M. J. Keskinen, S. L. Ossakow, B. G. Fejer, Three-dimensional nonlinear evolution of equatorial ionospheric spread-F bubbles, *Geophys. Res. Lett.* 30 (2003), 1855.

[22] H. Kesten, G. C. Papanicolaou, A limit theorem for stochastic acceleration, *Comm. Math. Phys.* 78 (1980), 19–63.

[23] R. J. LeVeque, *Numerical methods for conservation laws*, Birkhäuser Verlag, Basel 1992.

[24] S. Matsushita On artificial geomagnetic and ionospheric storms associated with high-altitude explosions, *J. Geophys. Res.* 64 (1959), 1149–1161.

[25] B. Mohammadi, O. Pironneau, *Analysis of the K-Epsilon Turbulence Model*, Research in Applied Mathematics, Masson, Paris 1994; John Wiley & Sons, Ltd., Chichester 1994.

[26] S. L. Ossakow, P. K. Chatuverdi Morphological Studies of Rising Equatorial Spread F Bubbles *J. Geophys. Res.* 83 (1978), 2085–2090.

[27] F. Poupaud, A. Vasseur, Classical and quantum transport in random media, *J. Math. Pures Appl.* 82 (2003), 711–748.

[28] G. C. Reid, The Formation of Small-Scale Irregularities in the Ionosphere, *J. Geophys. Res.* 73 (1968), 1627–1640.

[29] H. Rishbeth, O. K. Garriott, Introduction to ionospheric physics, Academic Press, 1969.

[30] C. Ronchi, R. N. Sudan, D. T. Farley, Numerical simulations of large-scale plasma turbulence in the daytime equatorial electrojet, *J. Geophys. Res.* 96 (1991), 21263–21279.

[31] C. Ronchi, R. N. Sudan, P. L. Similon, Effect of short-scale turbulence on kilometer wavelength irregularities in the equatorial electrojet, *J. Geophys. Res.* 95 (1990), 189–200.

[32] Y. Saad, SPARSKIT: a basic tool kit for sparse matrix computations - Version 2, Tech. Rep. Computer Science Department, Univ. of Minnesota, Minneapolis, MN, 1994.

[33] R. N. Sudan, J. Akinrimisi, D. T. Farley, Generation of small scales irregularities in the equatorial electrojet, *J. Geophys. Res.* 78 (1973), 240.

[34] E. F. Toro, *Riemann Solvers and Numerical Methods for Fluid Dynamics*, Springer-Verlag, Berlin 1999.

[35] W. White, An overview of high-altitude nuclear wepaons phenomena, Heart conference short course, March 1998, preprint.

[36] S. T. Zalesak, S. L. Ossakow Nonlinear equatorial spread F: the effect of neutral winds and background Pedersen conductivity, *J. Geophys. Res.* 87 (1982), 151–166.

A case study on the reliability of multiphase WKB approximation for the one-dimensional Schrödinger equation

L. Gosse

Istituto per Ricerche di Matematica Applicata "Mauro Picone"
Sezione di Bari, Via G. Amendola 122-I, 70126 Bari, Italy
email: `l.gosse@ba.iac.cnr.it`

Abstract. We present a short overview of high-frequency asymptotics for the 1D Schrödinger equation while emphasizing the computational aspects. We show that presumably different techniques like stationary phase methods, WKB or Wigner analysis give essentially the same macroscopic behaviour. Moment systems and K-branch entropy solutions are introduced in order to derive well-suited numerical methods. Finally, we display some computations supporting these ideas on a classical example.

1 Introduction

We are interested in computing efficiently the high frequency asymptotics of the Cauchy problem for the following linear equation:

$$i\hbar\partial_t\psi + \frac{\hbar^2}{2m}\partial_{xx}\psi = 0, \quad \psi(t=0,.) = \psi_0; \quad x \in \mathbb{R}. \tag{1.1}$$

The unknown ψ is the quantum mechanical wave-function of some particle of mass $m > 0$; $\hbar$ stands of course for the Planck constant. It is customary to work with dimensionless variables and one is led to introduce a *scaled Planck constant* $\varepsilon > 0$; problem (1.1) rewrites

$$i\varepsilon\partial_t\psi + \frac{\varepsilon^2}{2}\partial_{xx}\psi = 0, \quad \psi(t=0,.) = \psi_0; \quad x \in \mathbb{R}. \tag{1.2}$$

In order to study the wave-particle transition (or *classical limit*), one is especially interested in a class of initial data of the WKB (or *monokinetic*) type:

$$\psi(t=0,.) = A_0(x)\exp(i\varphi_0(x)/\varepsilon); \quad x \in \mathbb{R}. \tag{1.3}$$

The slowly-varying quantities $A_0 \geq 0$ and φ_0 appearing in this last expression are called respectively the amplitude and the phase of the wave ψ_0. Data of the form (1.3) with A_0 compactly supported are usually called *wave packets*. We stress that the equation (1.2) may arise when considering areas of application different from quantum mechanics, see [26].

2 Asymptotics of ψ and rays

In this section, we aim at setting up techniques in order to compute the asymptotics of ψ satisfying (1.2)–(1.3) for $\varepsilon \to 0$ without solving numerically the equation in order to bypass the meshing constraints explained in [23].

2.1 The stationary phase method

This procedure is extensively exposed in [8]; we have in mind to present it briefly in our rather simple context. A first and elementary observation is that the scaled Schrödinger equation (1.2)–(1.3) admits an explicit solution in terms of a so-called *oscillatory integral*:

$$\psi(t,x) = \int_{\mathbb{R}} \int_{\mathbb{R}} A_0(y) \exp\left(\frac{i}{\varepsilon}\left(\varphi_0(y) + \xi(x-y) - t\xi^2/2\right)\right).dy.d\xi. \qquad (2.1)$$

We call "generalized phase" the function $S(\xi, y) = \varphi_0(y) + \xi(x-y) - t\xi^2/2$ in order to avoid any confusion with the aforementioned WKB phase φ. Then the stationary phase lemma ensures that provided A_0, φ_0 are C^∞ functions,

$$\psi(t,x) = \sum_j \frac{A_0(y_j)}{\sqrt{1+t\varphi_0''(y_j)}} \exp\left(\pm\frac{i\pi}{4}\right) \exp\left(\frac{i}{\varepsilon}\left(\varphi_0(y_j) + t\varphi_0(y_j)^2/2\right)\right) + O(\varepsilon). \qquad (2.2)$$

Loosely speaking, one goes from a continuous superposition of waves in (2.1) to a discrete one (2.2) discarding all the couples of points ξ, y except those around which the generalized phase S is stationary thanks to the numerous cancellations inside the integral. The selected points are therefore critical points of S hence

$$\nabla_{\xi,y} S = 0 \Leftrightarrow \varphi_0'(y) - \xi = 0, \quad x - y - t\xi = 0.$$

So, for any $x \in \mathbb{R}$ and at a given time $t > 0$, one has to find all the possible values y_j, $j = 1, 2, 3, \ldots$ satisfying the *ray equations* (also called bicharacteristics)

$$\xi = \varphi_0'(y), \quad x = y + t\varphi_0'(y). \qquad (2.3)$$

In the language of conservation laws, one is more used to speak about characteristic curves; in this case, φ_0' is related to the initial velocity, as we shall see later.

A central problem in this approach is that the approximation (2.2) seems to blow up for $1 + t\varphi_0''(y_j) \to 0$; we shall call this blow-up locus the *caustic curve*.

2.2 Wigner measure analysis

A more modern tool to investigate the classical limit of (1.2)–(1.3) is the Wigner transform, which is defined as an "angularly resolved energy density":

$$W[\psi](t,x,\xi) = \int_{\mathbb{R}} \psi(t, x + \varepsilon z/2)\bar{\psi}(t, x - \varepsilon z/2) \exp(-i\xi z).dz. \tag{2.4}$$

Let us define a sequence of solutions ψ^ε to (1.2)–(1.3) and W^ε its associated family of Wigner transforms; then a well-known compactness lemma ([21]) states that

$$\|\psi^\varepsilon\|_{L^2} \leq C \text{ uniformly } \Rightarrow W^\varepsilon \rightharpoonup W \in \mathcal{M}^+, \quad \varepsilon \to 0,$$

where $\mathcal{M}^+$ stands for the cone of nonnegative measures in $t, x, \xi \in \mathbb{R}^+ \times \mathbb{R} \times \mathbb{R}$. Moreover, the limit Wigner measure satisfies the following free transport equation:

$$\partial_t W + \xi \partial_x W = 0, \quad W(t = 0, x, \xi) = A_0(x)^2 \delta(x - \partial_x \varphi_0(x)), \tag{2.5}$$

($\delta(\cdot)$ the standard Dirac measure) which admits the explicit solution

$$W(t,x,\xi) = W(t = 0, y, \xi),$$

away from caustics, as soon as for any $x \in \mathbb{R}$ and at a given time $t > 0$, y satisfies the ray equations (2.3). Of course, one should not expect a unique solution in the general case; thus several values y_j, $j = 1, 2, 3, \ldots$ are to be found. Accordingly, the initial monokinetic density will split into a more complex object of the type (2.2) beyond a certain *breakup time*.

A major interest in this technique lies in the fact that quadratic observables can be easily obtained by taking the moments of W; for instance, the position density reads:

$$\rho(t,x) = \lim_{\varepsilon \to 0} |\psi^\varepsilon(t,x)|^2 = \int_{\mathbb{R}} W(t,x,\xi).d\xi. \tag{2.6}$$

2.3 WKB and the evolution of the position density

This technique is based on the assumption that for any value of the scaled Planck constant ε below a certain threshold, the solution of (1.2)–(1.3) is well approximated by an ansatz,

$$\psi(t,x) \simeq A(t,x) \exp(i\varphi(t,x)/\varepsilon), \quad t > 0.$$

Another convenient way to motivate this ansatz is to trace back to (2.2) while introducing a (generally multivalued) mapping $Y\colon (t,x) \mapsto y_j$, $j = 1, 2, \ldots$ such that equations (2.3) hold. Finally, one defines a phase-amplitude pair $\varphi(t,x)$, $A(t,x)$ relying on (2.2) and the quantities $y_j = Y(t,x)$. In modern language, working with an ansatz amounts to guessing the microstructure in the problem (1.2)–(1.3); it says that the fine scale structure of the initial data stands still forever. Indeed, the scaling in (2.4) is adapted to wave functions ψ endowed with oscillations of $O(\varepsilon)$ wave-length.

Thus plugging this ansatz and splitting between real and imaginary parts inside (1.2) leads to

$$\partial_t \varphi + \frac{1}{2}(\partial_x \varphi)^2 = \frac{\varepsilon^2}{2A}\partial_{xx} A, \quad \partial_t (A^2) + \partial_x (A^2 \partial_x \varphi) = 0. \tag{2.7}$$

The "act of faith" that leads to the classical WKB system is that $\frac{\varepsilon^2}{2A}\partial_{xx} A \to 0$ when $\varepsilon \to 0$; this *dispersive limit* is essentially supported beyond breakup time by the fact that the emerging eikonal equation on φ admits the same rays as (2.3). So there is a consistency with the two preceding methods. Let us also recall that for small time, this kind of asymptotics are fully justified, see [16]. A survey is also given by Keller [20]

In the limit, system (2.7) becomes weakly coupled as the eikonal equation decouples and can be solved independently; of course, in order to remain consistent with the two aforementioned procedures, one must give up the idea of solving in the context of *viscosity solutions*, [7]. See however [12] for rather complete results in this (wrong) direction. If one introduces a velocity variable $u = \partial_x \varphi$, then the first equation becomes the classical Burgers' equation

$$\partial_t u + u \partial_x u = 0, \quad u_0 = \partial_x \varphi_0,$$

for which the multivalued (or geometric) solution is to be sought through the rays (2.3), [4], [17]. If one can complete this program, then the intensity $A^2(t,x)$ can be easily recovered; indeed, at any time $t > 0$, one has

$$A^2(t,x) = A_0^2(y)\left|\frac{\partial y}{\partial x}\right|.$$

One can deduce from (2.3) that $x = y + tu_0(y) = y + tu(t,x)$, so the last expression boils down to

$$\left|\frac{\partial y}{\partial x}\right| = 1 - t\partial_x u(t,x),$$

with $u(t,x)$ supposedly known. However, in the homogeneous case, the most accurate way to derive the intensity follows from

$$\left|\frac{\partial y}{\partial x}\right| = \left|\frac{\partial x}{\partial y}\right|^{-1} = \frac{1}{|1 + tu_0'(y)|},$$

which leads to the expression

$$A^2(t,x) = \frac{A_0(y)^2}{|1 + tu_0'(y)|}, \quad y = x - tu(t,x). \tag{2.8}$$

This formula will be of constant use in the numerical computations; it is of course equivalent to the one written in (2.2) since $\varphi_0''(y) = u_0'(y)$ and $\rho(t,x) = A^2(t,x)$.

2.4 The ray-tracing method

For the sake of completeness, we shortly recall the ray-tracing method as a numerical tool. It consists in sampling a compact interval of the real axis into a collection of abscissas y_l, $l = 1, 2, \dots$ and then shoot rays forward in time using the equations (2.3). The velocity $u(t,x)$ is found through the conservation law $u(t,x) = u_0(y)$ and the intensity is deduced by means of (2.8). One of its main shortcomings is the loss of accuracy in the rarefaction fans, see Figure 4 in [11] or [2], [9], [29], which imposes repetitive regridding processes.

3 Two numerical strategies based on moments

For purely computational reasons, the simulation of a transport equation of the type (2.5) may be considered as too expensive since the velocity variable ξ would have to be sampled too. So a classical idea to move back to a less big computational space is to take moments in ξ in order to work in the physical space t, x only.

3.1 The Delta-closure (Wigner analysis)

This is the most direct way to proceed; namely, one considers (2.5) and integrate against ξ to derive an infinite hierarchy:

$$\partial_t m_i + \partial_x m_{i+1} = 0, \quad m_i(t,x) = \int_{\mathbb{R}} \xi^{i-1} W(t,x,\xi).d\xi, \quad i = 1, 2, \dots \tag{3.1}$$

However, it is not always necessary to simulate such a big system; as an example, if $u_0'(x) \geq 0$, $W(t,x,\xi)$ keeps on being monokinetic and (3.1) reduces to the pressureless gas system, [6],

$$\partial_t \rho + \partial_x q = 0, \quad \partial_t q + \partial_x (q^2/\rho) = 0, \quad q/\rho = u.$$

In this last case, ρ and u remain smooth. When breakup occurs, for some $t^* \in \mathbb{R}^+$, equations (2.3) admit several roots y_j and a more intricate kinetic density to be inserted inside (3.1). Following [28], [25], [18], [9], [6], we can assume an upper bound $j \leq K \in \mathbb{N}$ and postulate that

$$W(t,x,\xi) = \sum_{j=1}^{K} A_j(t,x)^2 \delta(\xi - u_j(t,x)), \quad j = 1, 2, \dots, K.$$

Then one has to consider $2K$ moments in order to close (3.1):

$$\partial_t m_1 + \partial_x m_2 = 0, \quad \partial_t m_2 + \partial_x m_3 = 0, \quad \dots \quad \partial_t m_{2K} + \partial_x m_{2K+1} = 0, \quad (3.2)$$

where $m_i(t,x) = \sum_{j=1}^{K} \rho_j(t,x) u_j(t,x)^{i-1}, i = 1, 2, \dots, 2K$. Closing (3.2) amounts to expressing m_{2K+1} as a function of $m_1, m_2, \dots, m_{2K}$. Despite the fact it is theoretically doable, it remains a difficult task for $K > 4$. Even worse, systems (3.2) are generally only *weakly hyperbolic* and admit measure-solutions. This makes numerical simulations delicate; results are available in [3], [9], [13], [18].

3.2 The Heaviside-closure (K-branch solutions)

It has been observed ([4]) that the geometric solutions of Burgers' equation with $u_0(x) \geq 0$ can be recovered out of a kinetic problem in the flavour of (2.5),

$$\partial_t f + \xi \partial_x f = 0, \quad f(t = 0, x, \xi) = H(u_0(x) - \xi) H(\xi),$$

with H the Heaviside function. Beyond breakup time, f ceases to be of this mono-kinetic-like form and a correct representation would be

$$f(t,x,\xi) = \sum_{j=1}^{K} (-1)^{j-1} H(u_j(t,x) - \xi), \quad (3.3)$$

as long as no more than K folds appear. A remarkable feature is that (3.3) can be obtained from an entropy minimization process; this eventually led to the definition of *K-multivalued solutions* in [5] (the classical entropy solutions [22] correspond to $K = 1$). These K-multivalued solutions admit a kinetic formulation:

Definition 3.1. We call K-multivalued solution any measurable function $f(t,x,\xi) \in \{0,1\}$ on $\mathbb{R} \times \mathbb{R}^+ \times \mathbb{R}^+$ satisfying the following equation in the sense of distributions:

$$\partial_t f + \xi \partial_x f = (-1)^{K-1} \partial_\xi^K \tilde{m}, \quad f(t,x,\xi) = \sum_{j=1}^{K} (-1)^{j-1} H(u_j(t,x) - \xi), \quad (3.4)$$

where $\tilde{m}$ is a nonnegative Radon measure on $\mathbb{R} \times \mathbb{R}^+ \times \mathbb{R}^+$.

The set of $u_j(t,x)$'s is called the *K-branch entropy solution* of Burgers' equation, [11]. The same way as in the preceding section, one considers moments $m_i(t,x) = \frac{1}{i} \sum_{j=1}^{K} (-1)^{j-1} u_j(t,x)^i, i = 1, 2, \dots, K$ and an equivalence result holds:

Theorem 3.2 (Brenier & Corrias, [5]). *A measurable function*

$$f(t,x,\xi) = \sum_{j=1}^{K} (-1)^{j-1} H(u_j(t,x) - \xi)$$

is a K-multivalued solution if and only if all the following entropy inequalities hold for any θ, $\partial_\xi^K \theta \geq 0$:

$$\partial_t \int_{\mathbb{R}^+} \theta(\xi) f(t,x,\xi).d\xi + \partial_x \int_{\mathbb{R}^+} \xi\theta(\xi) f(t,x,\xi).d\xi \leq 0. \tag{3.5}$$

Equality holds in case $\partial_\xi^K \theta \equiv 0$.

A beautiful property is that this construction is consistent with the Delta-closure coming from Wigner analysis:

Theorem 3.3 (Equivalence of moment systems, [13]). *Let $0 \leq u_0 = \partial_x \varphi_0$ be a smooth function and $\tilde{w}(t=0,x,\xi)$ be the solution of the Liouville equation* (2.5) *with initial condition, $\tilde{w}(t=0,x,\xi) = H(u_0(x) - \xi)$. Consider the set*

$$\mathfrak{C} = \{\xi \in \mathbb{R}^+ \textit{such that } \tilde{w}(t,x,\xi) = 1 \textit{ for some } (t,x) \in \mathbb{R}^+ \times \mathbb{R}\}.$$

Assume that $\mathfrak{C}$ has only M connected components. If $M = \frac{1}{2}(K+1)$ (K odd) or $M = K/2$ (K even), then the moment systems (3.2) *and* (3.5) *produce the same velocities $u_j(t,x)$, $j = 1, 2, \ldots K$.*

Of course, formula (2.8) is to be used in order to deduce the intensities in a consistent way from the K-branch entropy solution. We close this section mentioning that the choice of the upper bound $K \in \mathbb{N}$ can be done relying on the rigorous results of [30]; see [14].

3.3 About multivalued intensities

As one may be especially interested in recovering physical observables which are single-valued, we want to explain how to work them out of the $u_j(t,x)$'s and the $A_j(t,x)$'s, $j = 1, 2, \ldots, K$. For instance, the position density associated to (2.2) is an oscillatory object since cross-terms will exhibit small-scale behaviour. However, following [14], we observe that they can be expected to become negligible as $\varepsilon \to 0$ relying on a stationary phase argument. Let us pick up a smooth test function $\phi(t,x)$ and consider, for $j, j' \leq K$,

$$\int_{\mathbb{R}} A_j(t,x) A_{j'}(t,x) \exp(i(\varphi_j - \varphi_{j'})(t,x)/\varepsilon)\phi(t,x).dx \quad \text{for all } t > 0.$$

Thus the stationary phase lemma ensures that if ϕ is supported in a domain where (A_j, φ_j) and $(A_{j'}, \varphi_{j'})$ are C^∞, this integral is $O(\varepsilon)$ except on points where $\partial_x(\varphi_j - \varphi_{j'})(t,x) = (u_j - u_{j'})(t,x) = 0$, that is, on caustics.

All in all, we have obtained that a very reasonable approximation of the first quadratic observable for small $\varepsilon \geq 0$ is given by the simple expression:

$$\psi\overline{\psi}(t,x) = |\psi(t,x)|^2 \simeq \sum_{j=1}^{K} |A_j(t,x)|^2. \tag{3.6}$$

This will be checked numerically in the next section; other quadratic observables could be derived similarly.

4 Numerical results: a Gaussian pulse

We give as initial data for (1.2)–(1.3) the following WKB pulse:

$$\psi_0(x) = \exp\Big(-\frac{1}{2}(x-\pi)^2\Big).\exp(-i\cos(x)/\varepsilon), \quad x \in [0, 2\pi]. \tag{4.1}$$

4.1 K-branch solutions and the corresponding intensity

This corresponds to $\varphi_0(x) = -\cos(x)$, $u_0(x) = \sin(x)$, $A_0(x) = \exp\big(-\frac{1}{2}(x-\pi)^2\big)$ which are all C^∞. We shall not give details about the numerical schemes we used for simulating the K-branch solutions related to this problem; everything is to be found in [11], [13], [14]. The meshing was $\Delta x = 2\pi/512$, $\Delta t = \Delta x/1.05$ and the solutions are shown in $t = 3$ in Figure 1. The rays can be seen in Figure 2 in [13]; a so-called

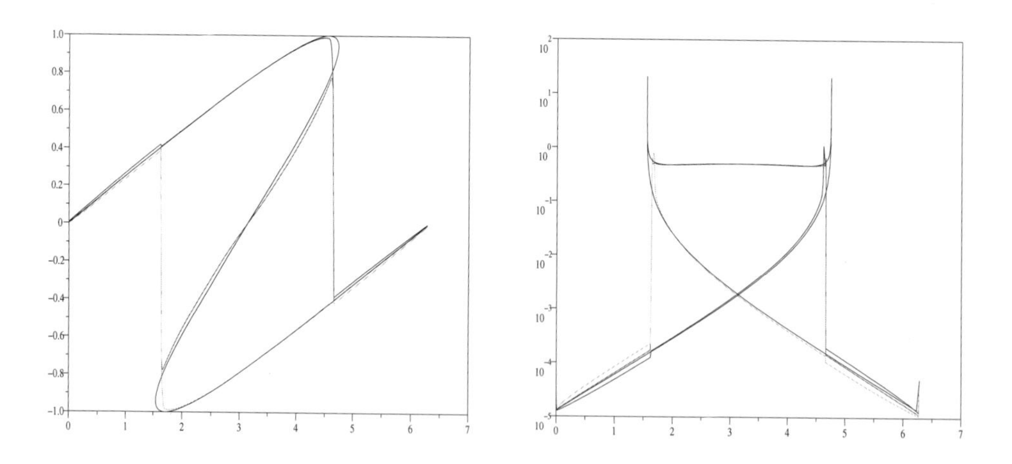

Figure 1. Comparison between the WKB solution and the ray-tracing one in $t = 3$.

cusp singularity develops and the geometric solution of Burgers' equation admits 3 values inside the central fan. The K-branch solution matches the ray-tracing profile with very good accuracy.

4.2 Comparison with the Schrödinger equation

Relying on (3.6), (2.8), we derived the approximate position density we compared with the one coming out of a standard Fourier scheme for (1.2) (see [1]) with 2^{12} modes. One can see the agreement as ε is decreased in Figure 2; the Schrödinger solution's

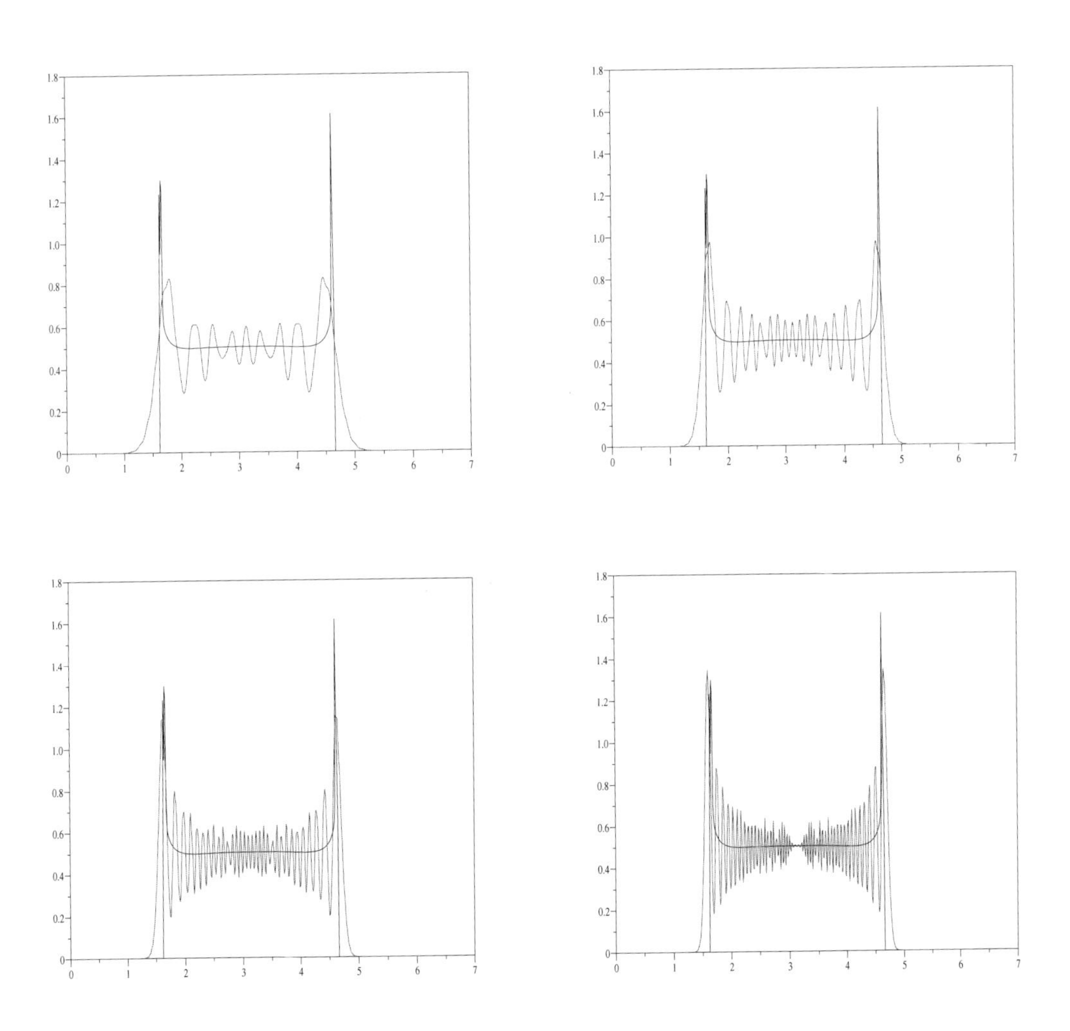

Figure 2. Comparison between the WKB position density (3.6), (2.8), and a direct Schrödinger simulation in $t = 3$.

density oscillates at higher and higher frequencies around the value which has been obtained out of the WKB computation. We took $\varepsilon = 1/50, 1/85, 1/145, 1/220$.

5 Conclusion

We tried to give an overview on some efficient computational methods for the semi-classical approximation of (1.2)–(1.3). Apart from stationary phase arguments, we emphasized most of all kinetic-based methods. Two main classes exist nowadays: the one based on Wigner equation (2.5) and the other coming from the K-multibranch solutions of [5]. When it is applicable, the second one leads usually to lighter numerical algorithms. However, open problems subsist: the hardest one being the passage from moments coordinates $m_i(t, x)$ to K-branch solutions $u_j(t, x)$. This is a well-known inverse problem called the *Markov moment problem*; progress are expected to be made in this direction, [15, 27]. Other strategies are available, see [2], [9], [19], [24, 29].

References

[1] W. Z. Bao, Shi Jin and P.A. Markowich, On time-splitting spectral approximations for the Schrödinger equation in the semiclassical regime, *J. Comput. Phys.* 175 (2002), 487–524.

[2] J. D. Benamou, Direct computation of multivalued phase space solutions for Hamilton-Jacobi equations, *Comm. Pure Appl. Math.* 52 (1999), 1443–1475.

[3] F. Bouchut, S. Jin, X. Li, Numerical Approximations of Pressureless and Isothermal Gas Dynamics, *SIAM J. Numer. Anal.* 41 (2003), 135–158.

[4] Y. Brenier, Averaged multivalued solutions for scalar conservation laws, *SIAM J. Numer. Anal.* 21 (1984), 1013–1037.

[5] Y. Brenier and L. Corrias, A kinetic formulation for multibranch entropy solutions of scalar conservation laws, *Ann. Inst. H. Poincaré Anal. Non Linéaire* 15 (1998), 169–190.

[6] Y. Brenier, E. Grenier, Sticky particles and scalar conservation laws, *SIAM J. Numer. Anal.* 35 (1998), 2317–2328.

[7] M. Crandall, P.L. Lions, Viscosity solutions of Hamilton-Jacobi equations, *Trans. Amer. Math. Soc.* 282 (1984), 487–502.

[8] J. J. Duistermaat, *Fourier Integral Operators*, Progr. Math. 130, Birkhäuser, Boston 1996.

[9] B. Engquist, O. Runborg, Computational high frequency wave propagation, *Acta Numerica* 12 (2003), 181–266.

[10] S. Filippas and G. N. Makrakis, Semiclassical Wigner function and geometrical optics, *SIAM Multiscale Model. Simul.* 1 (2003), 674–710. .

[11] L. Gosse, Using K-branch entropy solutions for multivalued geometric optics computations, *J. Comput. Phys.* 180 (2002), 155–182.

[12] L. Gosse and F. James, Convergence results for an inhomogeneous system arising in various high frequency approximations, *Numer. Math.* 90 (2002), 721–753

[13] L. Gosse, S. Jin and X. Li, Two moment systems for computing multiphase semiclassical limits of the Schrödinger Equation, *Math. Models Meth. Applied Sci.* 13 (2003), 1689– 1723.

[14] L. Gosse and P. A. Markowich, Multiphase semiclassical approximation of an electron in a one-dimensional crystalline lattice I: Homogeneous problems, *J. Comput. Phys.* 197 (2004), 387–417.

[15] L. Gosse and O. Runborg, Finite moment problems and applications to multiphase geometric optics, submitted to *J. Comp. Appl. Math.*

[16] E. Grenier, Semiclassical limit of the nonlinear Schrödinger equation in small time, *Proc. Amer. Math. Soc.* 126 (1998), 523–530.

[17] S. Izumiya and G. T. Kossioris, Geometric singularities for solutions of single conservation laws, *Arch. Rational Mech. Anal.* 139 (1997), 255–290.

[18] Shi Jin and X. Li, Multi-phase computations of the semiclassical limit of the Schrödinger equation and related problems: Whitham vs. Wigner, *Physica D* 182 (2003), 46–85.

[19] Shi Jin and S. Osher, A level set method for the computation of multivalued solutions to quasi-linear hyperbolic PDEs and Hamilton-Jacobi equations, *Commun. Math. Sci.* 1 (3) (2003), 575–591.

[20] J. B. Keller, Semiclassical mechanics, *SIAM Review* 27 (1985), 485–504.

[21] P. L. Lions and T. Paul, Sur les measures de Wigncr, *Rev. Mat. Iberoamericana* 9 (1993), 553–618.

[22] P. L. Lions, B. Perthame, E. Tadmor, A kinetic formulation of multidimensional scalar conservation laws and related equations, *J. Amer. Math. Soc.* 7 (1994), 169–191.

[23] P. A. Markowich, P. Pietra and C. Pohl, Numerical approximation of quadratic observables of Schrödinger-type equations in the semiclassical limit, *Numer. Math.* 81 (1999), 595–630.

[24] J. Qian, S. Leung, A level set based Eulerian method for paraxial multivalued traveltimes, A level set based Eulerian method for paraxial multivalued traveltimes, *J. Comput. Phys.* 197 (2004), 711–736.

[25] O. Runborg, Some new results in multiphase geometrical optics, *Math. Mod. Numer. Anal.* 34 (2000), 1203–1231.

[26] G. Papanicolaou and L. Ryzhik, Waves and transport, in *Hyperbolic equations and frequency interactions*, IAS/Park City Math. Ser. 5, Amer. Math. Soc., Providence, RI, 1999, 305–382.

[27] G. M. Sklyar, L. V. Fardigola, The Markov power moment problem in problems of controllability and frequency extinguishing for the wave equation on a half-axis, *J. Math. Anal. Appl.* 276 (2002), 109–134.

[28] C. Sparber, P. Markowich, N. Mauser, Multivalued geometrical optics: Wigner functions vs. WKB methods, *Asymptot. Anal.* 33 (2003), 153–187.

[29] W.W. Symes and J. Qian, A slowness matching Eulerian method for multivalued solutions of Eikonal equations, *J. Sci. Comput.* 19 (2003), 501–526.

[30] E. Tadmor, T. Tassa, On the piecewise regularity of entropy solutions to scalar conservation laws, *Comm. Partial Differential Equations* 18 (1993), 1631–1652.

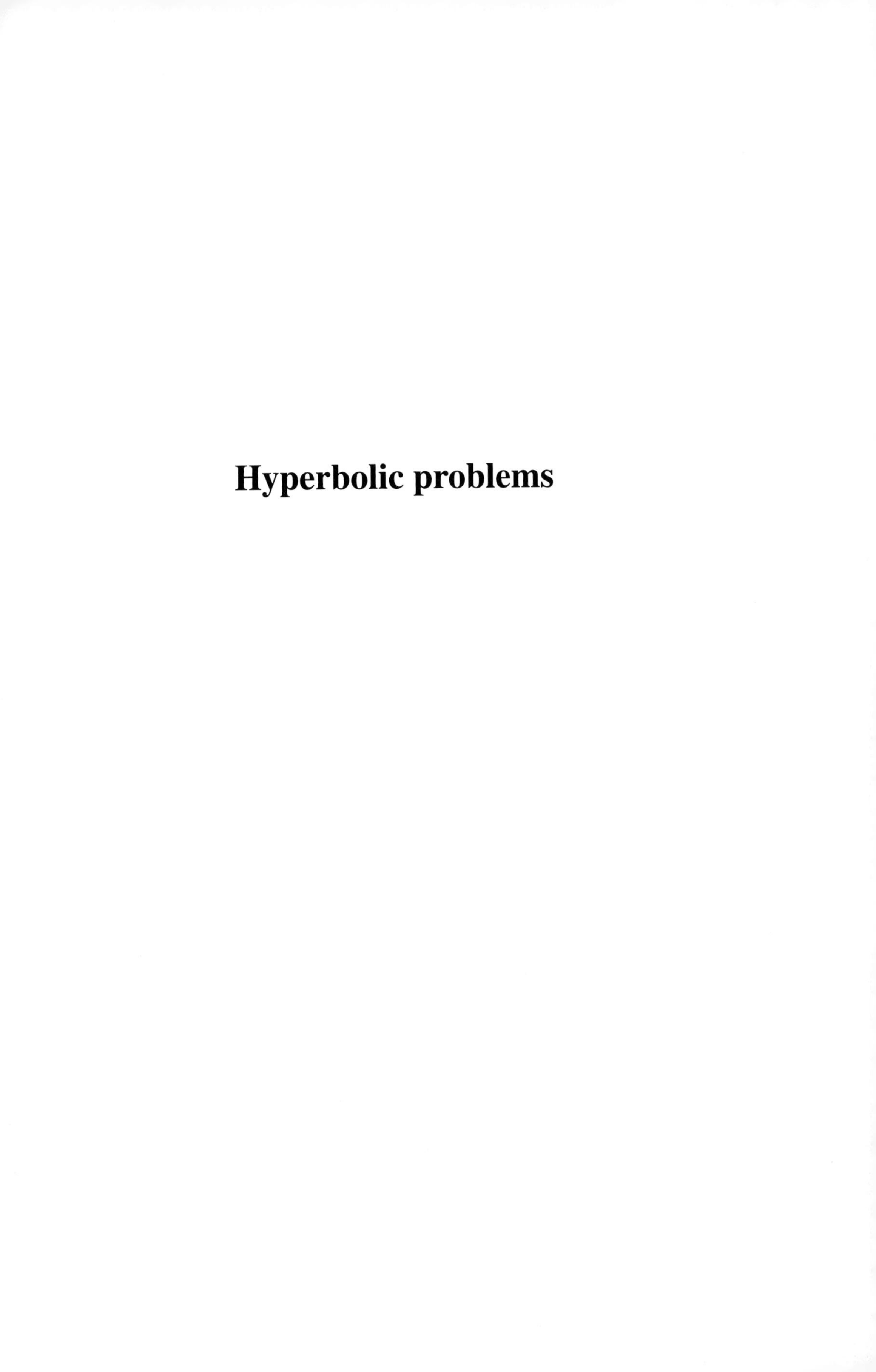

Hyperbolic problems

Introduction

by B. Després

A common feature of all the works that have been done during the CEMRACS event is that they are concerned with modelization, discussion of the mathematical correctness of the models in relation with the physical foundations, and numerical methods for the approximate solution of the models. The difference is more in the balance of the three ingredients. However the success reached at the end is in all cases due to the fact that none of these three ingredients has been neglected, and that all three of them interact with the other two.

The GENJET project was proposed by G. Baudin and E. Lapébie (both from the DGA). The group was constituted with seven members, including C. Baranger, G. Baudin, L. Boudin, B. Despres, F. Lagoutière, E. Lapébie and T. Takahashi. The large number of participants was motivated by the importance of the task: essentially this project can be subdivided into a "hyperbolic" project and a "kinetic" project. The physical motivation was the modelization and numerical solution of the generation and break-up of droplets after the impact of a rigid body on a tank filled with a compressible fluid. As a consequence the physics runs from kinetic modelization of droplets to the coupling of a compressible fluid with a non deformable solid. In order to obtain substantial results for the duration of the CEMRACS, it has been decided to simplify the study. This is why both regimes are considered separately. In the kinetic part, a model of Reitz wave was preferred to the TAB (Taylor Analogy Break-up) model because the Weber number was relatively low. The conservativity of the model was checked: it gives a indication of its correctness. The numerical tests show qualitative agreement with the physics of droplets in this regime. The hyperbolic part of the study was about the coupling of a compressible fluid with a rigid body, but limited in dimension one. The method consists in the use of a mixed cell approach (i.e. an artificial multimaterial model) with an exact calculation of the balance of forces at the boundary between the fluid and the solid body. It gives quite satisfactory numerical results. As is usual for such problems, the conservation of total energy is not exact. More research has to be done, in particular for the multidimensional extension.

The MHYMOD project has concerned B. Despres, S. Jaouen, C. Mazeran and T. Takahashi. It was about numerical methods for multivelocity compressible modeling. The originality was the interpenetration of the fluid: at $t = 0$ an interface may exist that separates the different fluids; then due to some small scale phenomena, mixing and/or interpenetration occur. This study met many of the well-known difficulties inherent in this subject: possible loss of hyperbolicity, degeneracy of the model where one of the material/phase disappears, etc. One of the goals of the study was to propose new lagrangian numerical methods for this problem. Here lagrangian means that the scheme is two steps: in the first step the system is solved in the Lagrangian reference frame that moves with the average fluid; the second step is usually called remapping or projection. This study was also the occasion of a collaboration with S. Gavrilyuk (Université Marseille). It helped to understand the similarities between the multivelocity model that was at the base of the MHYMOD project and a model that appears in two fluids modelization in pipes. At the numerical level, the famous Ransom test case was solved with the new method with convincing results.

The DINMOD project has mobilized F. Caro, F. Coquel, D. Jamet and S. Kock around diffusive interface methods for two-phase flows modeling. Other projects of this CEMRACS event are about multiphase or multivelocity flows: this is still an active field of research, where modelization is perhaps half of the hard work. Let us recall that no all purpose multiphase or multivelocity partial differential equation model exists. On the contrary one has to adapt the modelization to the physical situation encountered. One difficulty is that many of the models, if not all, display strong deviations from the standard theory of hyperbolic systems of conservation laws. This motivates multidisciplinary research where modelization (with mechanical and thermodynamical aspects), the study of the well-posedness of the PDE model and the discretization issues interlace continuously. The study done in the DINMOD project is highly representative of this tendency. It begins with the modelization, making the parallel with grad-two fluids. The Hamilton least action principle is used to derive a model. The authors stress themselves that the use of the Hamilton principle has its own part of arbitrariness, even if most of the classical physics is based on it. Once the model has been established, the authors study the Mechanical Equilibrium together with the Thermodynamical Equilibrium of the model. It consists in studying various limits of the coefficients of relaxation. An important result is that these equilibriums are well posed. Here the well-posedness is understood in a very weak sense. Nevertheless it gives a sound basis that helps to design numerical methods based on a relaxation or projection method. The scheme is made of a classical hyperbolic stage, and of a less classical relaxation to equilibrium stage. Numerical results show the efficiency of the approach.

The work done by F. Coquel, D. Diehl, C. Merkle and C. Rhode is about the modelization and numerical capture of phase transitions solutions for liquid vapor flows, with an approach referred to as the sharp/diffusive interface method. Since the pioneering work of van der Walls, many researchers have realized that the solutions of these problems need an advanced modelization for which an appropriate mathematical theory is still to be completed. For instance most of the models display an elliptic instable region and the selection of shock solutions (by a selection principle) is needed to get uniqueness of the solution of the Riemann problem. Among all the contributions of this work, let us just mention that it was proposed to use the kinetic relation as the selection principle: all this is absolutely crucial at the numerical level. Examples are given to motivate the study: for instance spurious oscillations that one may encounter in bubble-in-a-liquid computations are related to the issue of the determination of the selection principle and of its application in numerical methods. A solution with the relaxation approach is proposed, with very nice numerical results. The method is based on an analysis that insures the decrease of the energy of the system. This study definitely collects many aspects of the current research about conservation laws for complex flows.

The work of J. Cartier and A. Munnier is motivated by the numerical solution of radiative transfer problems in view of Inertial Confinement Fusion (ICF) applications. The model problem is the radiative transfer equation for $I(x, t; \vec{\Omega})$ where $\vec{\Omega}$ is the direction of photons. The coupling with the matter leads to an integrodifferential equation. An important difficulty is to get a discretization compatible with the diffusion limit of the model. This diffusion limit is of parabolic type. This is also related to the well-known Eddington factor γ approximation. The flux limiter λ is another parameter that controls the physical quality of the diffusion approximation. This is based upon a work of Levermore that gives four conditions that γ and λ should respect to be compatible. A numerical comparison of some pairs (γ, λ) is made: the numerical results show a preference for the geometric model. However the authors stress in their conclusion that even with this approach the results are not completely satisfactory. This work is representative of the theory of diffusion approximation for some hyperbolic equations.

The work presented by M. Dumbser and C.-D. Munz is about arbitrary high order Discontinuous Galerkin schemes for linear and non linear hyperbolic systems of PDE's. Compressible gas dynamics is one example treated during the CEMRACS event. Recall that Discontinuous Galerkin Methods (DGM) is a very promising approach for the discretization of PDE's. The most important feature is probably that any order of accuracy is theoretically reachable on arbitrary grids even for complicated equations, provided that these equations can be written in divergence form. The particular DGM studied in this project is based on an idea of TORO: it consists in a Lax–Wendroff time integration, but at any order. Moreover the CPU time needed for the ADER scheme is lower with respect to standard Runge–Kutta time integration. This is important because it has been stressed by many researchers that reducing the cost of DGM is an issue. Of course it depends also on the implementation. The method proposed in this work gives a positive answer to the problem of reducing costs. Many numerical experiments show that the accuracy is also at the *rendez-vous*. Another class of applications that needs very high order schemes is long time simulation. This kind of problem appears in aeroacoustics. It requires to push standard methods to their limits in order to get satisfactory results. P^1 approximations are usually not enough. P^{10} is shown to be promising. This study is highly representative of the research done in the world to promote DGM for various hyperbolic problems. Some open points remain of course. For instance it seems difficult to merge the elegant mathematical presentation of the foundations of DGM with the limiter techniques that one needs to stabilize DG schemes. Those interested in all these approaches and issues will find a very complete presentation in this paper.

The work done by C.-D. Munz, M. Dumbser and M. Zucchini is about the modeling of Low Mach number flows and their numerical approximation. Apart from the report of the specific work that has been done at the CEMRACS, the paper gives a quite extended presentation of Low Mach number approximation of compressible gas dynamics. In the case $M = \frac{u_{\text{ref}}}{c_{\text{ref}}}$ is small, acoustic waves of long range and low energy may interact with small structures containing high energy. The idea of the method is to perform an asymptotic expansion in terms of the Mach parameter M. Different Ansatz leads to different models: many of them are very demanding at the numerical level. The method used in this work has recently been proposed by Klein and Munz: it is called the multiple pressure approximation. Numerical computations have been performed with this approach. For instance the acoustic pressure of a co-rotating vortex is computed using the ADER high order method described in another report of the CEMRACS 2003. An calculation is about the noise generated by a free jet. The method consists in feeding the ADER scheme for the computation of the acoustic waves with the data generated by a commercial software used to compute the large scale averaged flow. The results may be evaluated in view of the multiple pressure modelization. Quite impressive numerical results are provided. In summary this study is representative of new tendencies about the modelization and numerical approximation of the acoustic response of a compressible flow. We are convinced that this kind of work will inspire many others in the next years.

Liquid jet generation and break-up

C. Baranger[1], *G. Baudin*[1], *L. Boudin*[2], *B. Després*[2], *F. Lagoutière*[2], *E. Lapébie*[1] *and T. Takahashi*[3]

[1]*CMLA, CNRS UMR 8536, Centre d'Études de Gramat, 46500 Gramat*
ENS Cachan and DGA, 61, av. du Président Wilson, 94235 Cachan Cedex, France
email: `baranger@cmla.ens-cachan.fr`
`gerard.baudin@dga.defense.gouv.fr`
`emmanuel.lapebie@dga.defense.gouv.fr`

[2]*Laboratoire J.-L. Lions, CNRS UMR 7598, Université Paris-VI*
Boîte courrier 187, 75252 Paris Cedex 05, France
email: `boudin@ann.jussieu.fr`, `despres@ann.jussieu.fr`
`lagoutie@ann.jussieu.fr`

[3]*Institut Élie Cartan, Université Nancy I*
B.P. 239, 54506 Vandœuvre-lès-Nancy Cedex, France
email: `takeo.takahashi@iecn.u-nancy.fr`

Abstract. This work is motivated by the numerical simulation of the generation and break-up of droplets after the impact of a rigid body on a tank filled with a compressible fluid. This paper splits into two very different parts. The first part deals with the modeling and the numerical resolution of a spray of liquid droplets in a compressible medium like air. Phenomena taken into account are the breakup effects due to the velocity and pressure waves in the compressible ambient fluid. The second part is concerned with the transport of a rigid body in a compressible liquid, involving reciprocal effects between the two components. A new one-dimensional algorithm working on a fixed Eulerian mesh is proposed.

The GENJET (GENeration and breakup of liquid JETs) project has been proposed by the Centre d'Études de Gramat (CEG) of the Délégation Générale de l'Armement (DGA). It concerns the general study of the consequences of a violent impact of a rigid body against a reservoir of fluid. Experiments show that once the solid has pierced the shell of the reservoir, it provokes a dramatic increase of the pressure inside the reservoir, whose effect is the ejection of some fluid through the pierced hole. The generated liquid jet then expands into the ambient air, where it can interact with some air pressure waves, leading to a fragmentation of the jet into small droplets. These experiments show that after having pierced the shell, the projectile behaves as a rigid body. They also show that the liquid inside the reservoir behaves as a compressible fluid (indeed, the projectile velocity, around 1000 $\mathrm{m.s^{-1}}$, is in the same order of magnitude than the sound speed in the liquid).

The modeling of such a complex flow requires to take into account very different regimes, from the pure compressible and/or incompressible flow condition to a droplet regime (such a regime sharing some similarities with kinetic modeling of particles). Moreover many scales are needed to correctly describe the complete experiments, from the large hydrodynamic scale to the small droplet scale. The study done during CEMRACS 2003 focused on the fluid regime and on the droplet regime, since some important difficulties are still there for both regimes separately.

- Concerning the breakup of droplets in the air, we have focused on physical and numerical modeling issues.
- Concerning the fluid regime, an important difficulty at the numerical level is that we want to get an accurate numerical description of the transport of a rigid body inside a compressible fluid. Even if the rigid body is of course not a fluid, the situation shares at the numerical level a lot of similarities with the coupling an incompressible fluid with a compressible one. Thus this part of the study concerns more numerical algorithms than the modeling.

The present paper follows this cutout of the study. Section 1 presents the modeling of the breakup of droplets, whereas Section 2 treats the coupling of the rigid body and the fluid. In both sections, numerical results are reported.

In view of the main goal of the GENJET project, a natural perspective of the work described below would be the coupling of the models, algorithms and numerical methods.

1 A kinetic modeling of a breaking up spray with high Weber numbers

In this section, we aim to model a spray of droplets which evolve in an ambient fluid (typically the air). That kind of problem was first studied by Williams for combustion issues [32]. The works of O'Rourke [20] helped to set the modeling of such situations and their numerical simulation through an industrial code, KIVA [1].

The main phenomenon that occurring in the spray is the breakup of the droplets. Any other phenomena, such as collisions or coalescence, will be neglected in this work, but they are reviewed in [3] for example. Instead of using the TAB model (see [2]), which is more accurate for droplets with low Weber numbers, we choose the so-called Reitz wave model [27], [21], [4]. Then this breakup model is taken into account in a kinetic model [14], [2].

The question of the spray behavior with respect to the breaking up has arisen in the context of the French military industry. One aims to model with an accurate precision the evolution of a spray of liquid droplets inside the air. In that situation, the droplets of the spray are assumed to remain incompressible (the mass density ρ_d is a constant of the problem) and spherical. We also assume that the forces on the spray are negligible with respect to the drag force, at least at the beginning of the computations. After a few seconds, the gravitation may become preponderant. Note that the aspects of energy transfer will not be tackled in this report.

In Subsection 1.1, we derive the equations of the model and check the properties of conservation of mass and momentum. Then we briefly present the scheme for the numerical resolution of our problem. Eventually, we show some numerical results.

The notations used in the first section are presented in Table 1.

Table 1. Notations.

$t \geq 0$	time
$x \in \mathbb{R}^{d_x}$	position
$v \in \mathbb{R}^{d_v}$	droplet velocity
$r \in]0, +\infty[$	droplet radius
σ	surface tension of the droplet
$f(t, x, v)$	spray probability density function (PDF)
Q_{bup}	breakup operator
ρ_d	droplet mass density (constant)
$m = \dfrac{4}{3}\pi r^3 \rho_d$	droplet mass
$\rho_g(t, x)$	fluid density
$u(t, x) \in \mathbb{R}^{d_v}$	fluid flow velocity
$\alpha(t, x)$	fluid volume fraction
C_d	drag coefficient
Re	Reynolds number associated to the fluid near a droplet
Re_d	Reynolds number of a droplet
We	Weber number associated to the gas near a droplet
We_d	Weber number of a droplet
Oh	Ohnesorge number
Ta	Taylor number

1.1 Presentation of the model

The spray is described by the probability density function (PDF) f, which depends on time t, position x, velocity v and radius r. The number of droplets located in the volume (of the phase space) $[x, x + dx] \times [v, v + dv] \times [r, r + dr]$ is the quantity $f(t, x, v, r)dxdvdr$.

The PDF f satisfies the kinetic equation

$$\partial_t f + v \cdot \nabla_x f + \nabla_v \cdot (f\gamma) + \partial_r(f\chi) = Q_{\text{bup}}(f). \quad (1.1)$$

The acceleration γ of one droplet is given by

$$\gamma(t,x,v,r) = -\frac{1}{m(r)} D(t,x,v,r)(v-u(t,x)), \tag{1.2}$$

where m is the droplet mass, u the fluid velocity and

$$D = \frac{\pi}{2} r^2 \rho_g C_d |v-u|.$$

The coefficient C_d is the drag coefficient, and its value is given in the appendix.

In order to give an explicit form of χ and Q_{bup}, we here need to explain how the breakups are handled by the Reitz wave model. We here use the classical Reynolds (Re, Re_d), Weber (We, We_d), Ohnesorge (Oh) and Taylor (Ta) numbers, whose values are described below (see [26], for instance). One can use two different Reynolds numbers: the Reynolds number associated to the fluid near a droplet, and the Reynolds number of the droplet itself. We have

$$\text{Re} = \frac{r\rho_g|v-u|}{\mu_g}, \qquad \text{Re}_d = \frac{r\rho_d|v-u|}{\mu_d},$$

where μ_g (resp. μ_d) is the fluid (resp. droplet) viscosity. In the same way, we consider two different Weber numbers

$$\text{We} = \frac{r\rho_g|v-u|^2}{\sigma_d}, \qquad \text{We}_d = \frac{r\rho_d|v-u|^2}{\sigma_d},$$

where σ_d is the surface tension of the droplet. Then it is easy to define the Ohnesorge and Taylor numbers of the droplet

$$\text{Oh} = \frac{\sqrt{\text{We}_d}}{\text{Re}_d}, \qquad \text{Ta} = \text{Oh}\sqrt{\text{We}}.$$

We refer to [5] for instance, for the definition of the drag coefficient C_d.

A breaking up droplet looses some mass by creating small children droplets, but does not disappear, as it occurs in most breakup models. Hence, there is no disappearance of droplets, only production, since only the mother droplet's radius changes: we must be able to compute the new radius of the mother droplet (thanks to χ) and to evaluate the characteristics of the children droplets (number, position, velocity, radii). The phenomenon taken into account is the main disturbance on the surface of the droplet, which is most likely to result in breakup. The wavelength of this disturbance is

$$\Lambda = \frac{9.02\left(1+0.45\sqrt{\text{Oh}}\right)\left(1+0.4\text{Ta}^{0.4}\right)}{\left(1+0.865\text{We}^{5/3}\right)^{0.6}}\, r. \tag{1.3}$$

We also need the growth rate of the wavelength

$$\Omega = \frac{0.34+0.385\text{We}^{3/2}}{(1+\text{Oh})\left(1+1.4\text{Ta}^{0.6}\right)}\sqrt{\frac{\sigma}{\rho_d r}}, \tag{1.4}$$

where σ is the surface tension of the droplet.

We can then provide a characteristic breakup time

$$\tau = \frac{C_\tau r}{\Omega \Lambda}, \tag{1.5}$$

where C_τ mainly depends on the characteristic lengths of the spray. We here propose to choose $C_\tau = 37.88$. Note that (1.3)–(1.5) come from [4], as well as this value of C_τ. This choice seems to be significant, since the physical setting of [4] is close to ours.

Eventually, the variation χ of the droplet radius satisfies

$$\chi(t, x, v, r) = -\nu(t, x, v, r)(r - R(t, x, v, r)), \tag{1.6}$$

where $R = C_R \Lambda$ (we here choose $C_R = 0.61$). The breakup frequency ν is given by

$$\begin{aligned} \nu &= 0 \quad \text{if } r \le \Lambda \text{ or We} \le \frac{1}{2}\sqrt{\text{Re}}, \\ \nu &= \frac{1}{\tau} \quad \text{in any other cases.} \end{aligned} \tag{1.7}$$

Note that (1.6) cannot be seen as a linear ODE on r: even if R linearly depends on Λ, the wavelength itself does not linearly depend on r, since in (1.3) the Ohnesorge, Taylor and Weber numbers also depend on r.

The breakup operator only deals with the small children droplets. We then propose the following definition of Q_{bup}, where we only keep the dependencies on the velocity and the radius, for the sake of simplicity,

$$\begin{aligned} Q_{\text{bup}}(f)(v, r) = \int_{v_*, r_*} \int_{\mathcal{S}_1(v_*)} & \nu(v_*, r_*) \frac{3{r_*}^2 (r_* - R_*)}{{R_*}^3} \\ & \delta_{V_*}(v)\, \delta_{R_*}(r)\, f(v_*, r_*)\, \frac{ds}{2\pi}\, dr_* dv_*, \end{aligned} \tag{1.8}$$

where $V_* = v_* + v_\perp n$, with $v_\perp = C_V \Lambda \Omega$ ($C_v = 0.5$ in our problem) and n is a random vector of the unit disc $S_1(v_*)$ of the linear plane normal to the vector v_*.

1.2 Conservations

In this subsection, we check that the classical properties of conservations of mass and momentum are satisfied.

Proposition 1.1. *The total mass of the spray is constant (in time).*

Proof. Let us denote by $M(t)$ the total mass of the spray at time t. We immediately have

$$\begin{aligned} \frac{dM}{dt} &= \frac{4\pi\rho_d}{3} \int_{x,v,r} \partial_t f \cdot r^3 \, dr dv dx \\ &= \frac{4\pi\rho_d}{3} \left[-\int_{x,v,r} \partial_r (f\chi) r^3 \, dr dv dx + \int_{x,v,r} Q_{\text{bup}}(f) r^3 \, dr dv dx \right], \end{aligned}$$

with the disappearance of the conservative terms on x and v. Then, by a standard computation and using (1.6) and (1.8), we get

$$\frac{dM}{dt} = 4\pi\rho_d \Bigg[-\int_{x,v,r} f(v,r)\nu(v,r)(r-R)\, r^2\, dr dv dx \\ + \int_{x,v,r} \left(\int_{v_*,r_*} \int_{\mathcal{S}_1(v_*)} \nu(v_*,r_*) \frac{{r_*}^2(r_*-R_*)}{{R_*}^3}\, r^3 \delta_{V_*}(v)\, \delta_{R_*}(r) \\ f(v_*,r_*)\, \frac{ds}{2\pi}\, dr_* dv_* \right) dr dv dx \Bigg].$$

The integration along $S_1(v_*)$ gives 1 and the one on v against the Dirac mass is 1, too. It is then clear that

$$\frac{dM}{dt} = 4\pi\rho_d \Bigg[-\int_{x,v,r} f(v,r)\nu(v,r)(r-R)\, r^2\, dr dv dx \\ + \int_{x,v_*,r_*} \nu(v_*,r_*) \frac{{r_*}^2(r_*-R_*)}{{R_*}^3}\, {R_*}^3 f(v_*,r_*)\, dr_* dv_* dx \Bigg],$$

which ends the proof. □

Since the fluid acts on the spray, the total momentum of the spray is not constant, but we have the following

Proposition 1.2. *The time derivative of the total momentum of the spray equals the mean force of the fluid on the spray.*

Proof. Let us denote by $I(t)$ the total momentum of the spray at time t. We immediately have

$$\frac{dI}{dt} = \frac{4}{3}\pi\rho_d \int_{x,v,r} f\gamma r^3 dr dv dx \\ + 4\pi\rho_d \Bigg[-\int_{x,v,r} f(v,r)\nu(v,r)(r-R)\, r^2 v\, dr dv dx \\ + \int_{x,v,r} \left(\int_{v_*,r_*} \int_{\mathcal{S}_1(v_*)} \nu(v_*,r_*) \frac{{r_*}^2(r_*-R_*)}{{R_*}^3}\, r^3 v\, \delta_{V_*}(v)\, \delta_{R_*}(r) \\ f(v_*,r_*)\, \frac{ds}{2\pi}\, dr_* dv_* \right) dr dv dx \Bigg].$$

The first integral term is exactly the mean force of the fluid on the spray. Let us denote by J the term inside the brackets and show that it equals 0. By similar computations we obtain that

$$J = -4\pi\rho_d \int_{x,v,r} f(v,r)\nu(v,r)(r-R)\, r^2 v\, dr dv dx \\ + 4\pi\rho_d \int_{x,v_*,r_*} \int_{\mathcal{S}_1(v_*)} \nu(v_*,r_*) f(v_*,r_*) {r_*}^2 (v_* + v_\perp n)\, \frac{ds}{2\pi}\, dr_* dv_* dx.$$

Since

$$\int_{\mathcal{S}_1(v_*)} n \frac{ds}{2\pi} = 0,$$

it becomes clear that $J = 0$ and the proof is ended.

The well-posedness of (1.1) is not investigated in this paper. However, such a question is studied for other breakup operators, as in [14].

1.3 Numerical tests

We do not here take into account the motion of the droplets. Nevertheless, we aim to ensure the compatibility of the particle method used here, with any finite volume method fitted to the surrounding fluid.

1.3.1 Numerical scheme. We first start by a description of the numerical scheme used for solving (1.1). We assume that the particles have three-dimensional velocities.

The numerical scheme is time-split into two steps: the transport (in velocity) and the breakup. That means that we successively solve

$$\partial_t f + \partial_r (f\chi) = Q_{\text{bup}}(f) \tag{1.9}$$

and

$$\partial_t f + \nabla_v \cdot (f\gamma) = 0 \tag{1.10}$$

during a whole time step. Since we use a particle method, the PDF f is sought to be under the form

$$f(t, v, r) = \omega \sum_{p=1}^{P(t)} \delta_{v_p(t)}(v)\, \delta_{r_p(t)}(r),$$

where $(v_p(t), r_p(t))$ denote the velocity and the radius of a numerical particle p at a time t, $P(t)$ is the total number of numerical particles and ω is the representativity of the particles (constant in this work).

Transport. During the first part of a time step, we follow the movement of the particles along their characteristics (in terms of velocity, not radius), i.e. we solve

$$\frac{dv_p}{dt} = -\frac{1}{m_p} D_p(v_p - u).$$

Then the time discretization gives

$$v_p^{n+1} = v_p^n - \frac{\Delta t}{m_p} D_p^n (v_p^{n+1} - u^n),$$

where u^n is the velocity of the fluid at time $n\Delta t$. Here we use an implicit scheme for the discretization of $\frac{dv_p}{dt}$ in order to ensure the numerical stability.

Breakup. The second part of the splitting is the breakup step. This is a linear process, that is, for each particle p, it happens as follows.

1. If $r_p^n \leq C_R \Lambda_p^n$ then the particle p remains as it is.
2. If $r_p^n > C_R \Lambda_p^n$ and $\mathrm{We}_p^n \leq \frac{\sqrt{\mathrm{Re}_i^n}}{2}$ then the particle p remains as it is.
3. In the other cases a breakup occurs with the probability $\Delta t / \nu_p^n$. If a breakup occurs,
   (a) the radius of the particle evolves: $r_p^{n+1} = \frac{\tau_p^n r_p^n + \Delta t C_R \Lambda_p^n}{\tau_p^n}$,
   (b) the total number of the new particles created by the breakup of a particle p is computed in order to ensure the mass conservation $N_p = \frac{3 r_p^{n2}(r_p^n - R_p^n)}{R_p^{n3}}$, and with the radius $R_p^n = C_R \Lambda_p^n$,
   (c) N_p particles $\tilde{p}$ are created with position $x_{\tilde{p}}^{n+1} = x_p^n$, radius $r_{\tilde{p}}^{n+1} = R_p^n$ and velocity $v_{\tilde{p}}^{n+1} = v_p^n + C_V \Lambda_p^n \Omega_p^n \vec{n}$ ($\vec{n}$ unit vector normal to v_p^n).

1.3.2 Numerical simulations. The simulations that we made do not take into account the simulation of the surrounding fluid. First, we only compute the breakup of some droplets which remain motionless. Second, we compute the breakup and the transport of the droplets.

Breakup. We here aim to study the influence of the velocity of the surrounding fluid on the breakup phenomenon (number of breakups, number and size of the created droplets). We have several possibilities for the value of the velocity of the fluid: either it is a constant (equal to 0) during the whole computation, or there could be a shock (in our example, $|u|$ increases from 0 to 400 m.s^{-1} at a given time), or it could follow a linear profile in time. These three cases are described below.

1. In the first simulation the fluid is motionless ($u \equiv 0$). Initially the radii of the 100 droplets considered are uniformly distributed between 0.5 and 1 mm and the velocities are equal to 50 m.s^{-1} in one direction. The computation lasts for 0.1 s, but actually there is no fragmentation after 0.06 s. The 100 particles become 8000 with a radius around 0.17 mm (between 0.1725 and 0.1775 mm, the higher radii correspond to the "mother" droplets). We can see in Figure 1 the initial and final distributions of the radii.
2. The second simulation consists in including a shock in the fluid: the velocity of the fluid decreased from 0 to -400 m.s^{-1} at time 0.04 s. For the sake of the computation, the initial distribution of the radii of the 100 droplets is chosen to be smaller because of the high number of created droplets at this relative

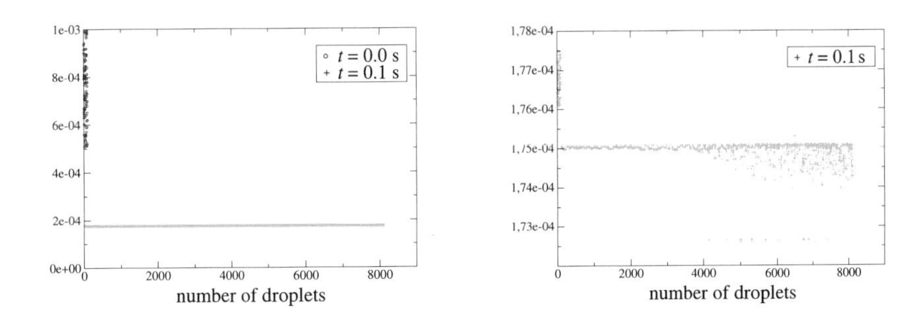

Figure 1. Radii (m) of the particles with a fixed surrounding fluid.

velocity (around 350 m.s^{-1}). So we start with radii around 5 and 10 μm. At the end, the radii of the droplets is between 2.54 and 2.62 μm.

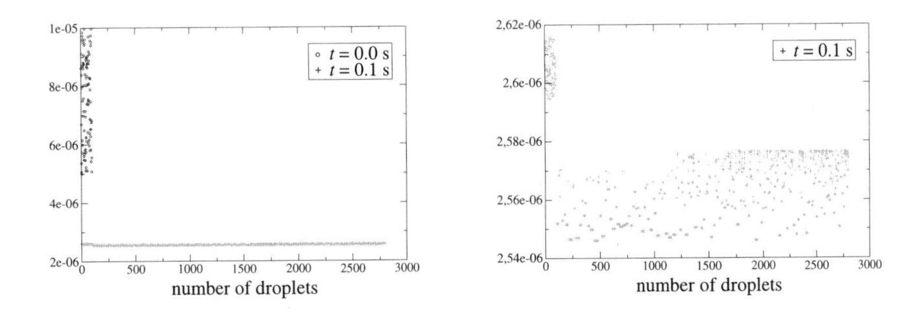

Figure 2. Radii (m) of the particles with a shock in the fluid.

3. In the last simulation the velocity of the surrounding fluid linearly decreases from 0 to -400 m.s^{-1} and from $t = 0.04$ to 0.05 s. We here keep the same initial radius distribution than in the previous simulation. There are less particles created. In Figure 3 we note that several values of radii are pointed out (2.4 μm, 2.6 μm and 2.8 μm approximately). As a matter of fact the continuous linear profile of the fluid velocity is discretized into piecewise constant values with respect to the time steps. To avoid this numerical artefact, one must refine the time steps.

To conclude this part, we could say that there is a mean radius for the created droplets which depends on the relative velocity between the droplets and the fluid (for

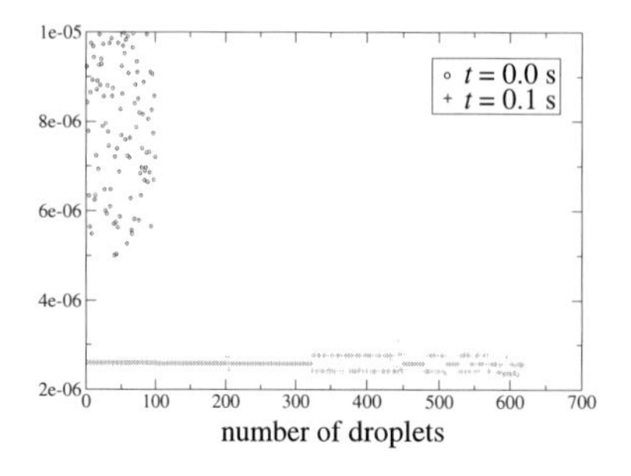

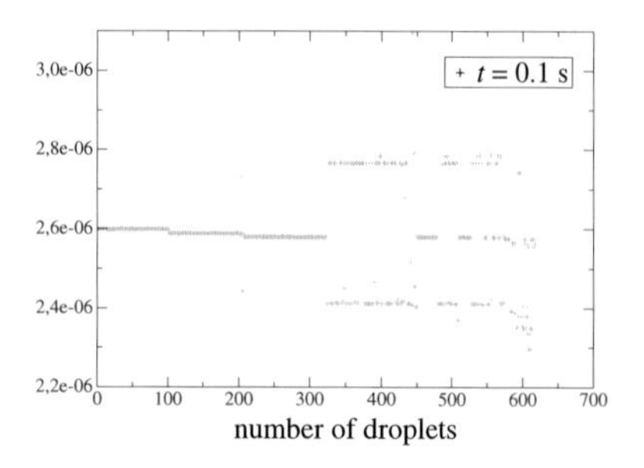

Figure 3. Radii (m) of the particles with a linearly profiled velocity for the fluid.

example, around 10^{-4} m without shock, and 10^{-6} m when a shock occurs in the fluid). The radii seem to correspond to the physical behavior of a spray when a shock occurs in the air, i.e. the 'atomization' of the spray detected during some experimentations made at the CEG.

Breakup and transport. We made some computations of the phenomena: breakup and transport of the droplets. For the time step (verifying a CFL condition) in the transport step we assume that the size of a cell is around 0.005 mm (200 cells for 1 meter). In order to take the interaction between these phenomena into account (because the breakup time is larger than the transport time), the initial radii are randomly picked between 10 and 20 mm. Of course there is some breakup in the beginning, but as the velocity of the droplets goes to 0 (the velocity of the surrounding fluid), there are less and less breakups. In Figure 4 we could see the final distribution of the droplets and in Figure 5 the successive positions of the droplets at several times.

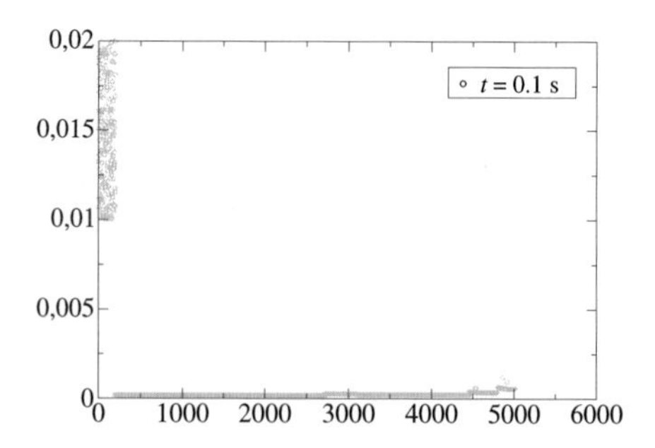

Figure 4. Final radii of the droplets during the transport-breakup computation.

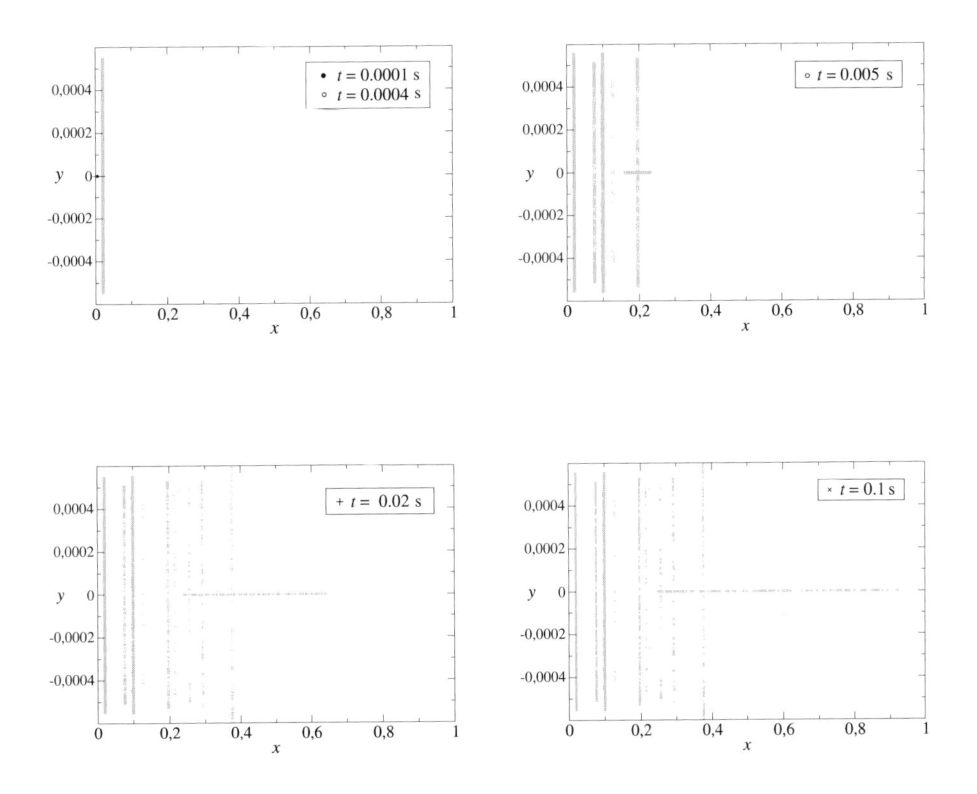

Figure 5. Positions of the droplets during the transport-breakup computation.

2 Fluid-structure coupling

The general context of this section is the transport of a rigid body in a compressible fluid, including the coupling effects.

The compressible fluid under consideration is governed by the Euler system of partial differential equations, with ideal or stiffened gas pressure law. The goal is to take into account the coupling effects between the solid and the fluid.

Let us briefly present some of the most used methods for the numerical resolution of fluid-structure coupling.

- One of the most known methods, the Immersed Boundary Method, was initiated by Peskin in [22] for blood flow in heart computations and then had a lot of improvements for other applications: see [23] for a survey, [19] for another recent development. These applications concern the coupling between an elastic or rigid solid and an *incompressible* fluid where, for the numerical stability, the velocity is assumed to be regular enough (otherwise a CFL-like condition would be very restrictive: see the discussion on the hyperbolicity and on the numerical stability in the following). This method thus cannot be envisaged for the present problem.

- Among other existing methods let us mention the Fat Boundary Method of Maury, Maury and Ismail: [18] (for the Poisson problem) and [15] (for incompressible fluid problems), also dedicated to incompressible fluids. A fictitious domain method is used in [11]. Another method, proposed in [28], uses a fixed mesh like the fictitious domain and considers a global mixed formulation (for the fluid and the solid) where the condition of rigidity of the solid is imposed on the finite element spaces. The major drawback of these last three methods for handling *compressible* fluids is their lack of conservativity.
- The method of Piperno, Farhat, etc. (see for example [9], [24], [25]) is conservative in mass and momentum. The coupling algorithm described there uses an Arbitrary Lagrangian Eulerian (ALE) formalism with a moving mesh (typically a mesh following the solid). A precise analysis of this type of methods has been done by Piperno. Nevertheless, the aim of the present CEMRACS project is to do computations on a motionless Cartesian mesh which is the same for the solid and the fluid. We thus abandon these methods (note that another ALE method is used in [17] for particle transport in an incompressible medium).
- A penalisation procedure coupled with a Volume Of Fluid (VOF) algorithm for the transport of the volume fraction has been developed in [31]. It has the little drawback of allowing artificial deformation of the solid in spite of the penalisation and of the non dissipative VOF method. Furthermore, it seems that the method has not been used for compressible fluids yet. From another point of view it has been shown in previous numerical experiments that replacing the undeformable solid with a fluid or a *very* rigid body and using a bi-fluid or a fluid-elastodynamic code (which is similar to a penalisation procedure) was not a good solution when computing strong shocks (this leads to an over-evaluation of the deceleration of the body in the fluid). Thus we decide to really consider the solid as undeformable, both in the mathematical equations and in the final code.

The method we developed is based on an explicit finite volume scheme on a Eulerian unique mesh. The presence of mixed cells (containing solid *and* fluid: this is due to the projection of the moving solid on the fixed mesh) leads to use a mixture model that takes into account the undeformable behavior of the solid.

This section is organized as follows. Subsection 2.1 gives the mathematical model governing the fluid-solid system, taking into account the compressible behavior of the fluid (ideal or stiffened gas), the undeformability of the solid, and the coupling effects. Subsection 2.2 presents a mathematical way to take into account the presence of mixed cells that necessarily appear in the numerical stage (on a fixed grid). We then propose a rapid mathematical analysis of this model in Subsection 2.3. The numerical scheme is developed in Subsection 2.4. Finally, Subsection 2.5 shows some numerical results in dimension one.

2.1 The system fluid-rigid body

Let us denote by $\Omega \subset \mathbb{R}^3$ the domain of interest. The solid body is assumed to be a connex region $S(t) \subset \mathbb{R}^3$ (for any time $t \in [0, T]$) such that $S(t) \cap \partial\Omega = \emptyset$. Thus the fluid occupies the region $\Omega \setminus S(t)$. The medium in $\Omega \setminus S(t)$ is supposed to be governed by the compressible Euler equations (conservations of mass, momentum and total energy), the region $S(t)$ evolves following a rigid solid displacement law: it can be decomposed into a translation and a rotation. These two parameters are computed with the help of the action of the fluid on the solid. The action of the solid onto the fluid is taken into account by considering some "wall-condition" on $\partial\Omega$: friction between the fluid and the solid is neglected (so as the friction inside the fluid). For the Eulerian fluid we use the variables ρ_1, u_1 and e_1 for the density, the velocity and the total energy. The density of the solid is constant in time and space. For the solid we use the variables h, ω for the center of mass and the angular velocity, respectively. If we denote by $Q(t)$ the matrix of rotation of the rigid body, i.e.

$$\dot{Q}(t)x = \omega(t) \wedge (Q(t)x) \quad \text{for all } x \in \mathbb{R}^3,$$

then we can describe the domain $S(t)$ of the solid at time t by using the domain S occupied by the solid at the initial time:

$$S(t) = \{Q(t)y + h(t),\ \ y \in S\}.$$

In the sequel we denote by $n(x,t)$ the unit normal vector to $\partial S(t)$ at the point x directed to the interior of the rigid body. We also introduce the mass M of the solid and J its inertial matrix. The equations of the solid are obtained by applying the conservation of linear and angular momentum (Newton's laws).

With the above notation the equations modeling the movement of the fluid and of the rigid body can be written in the following form:

$$\left.\begin{aligned}
&\partial_t \rho_1 + \operatorname{div}(\rho_1 u_1) = 0, && \text{in } \Omega \setminus S(t),\\
&\partial_t(\rho_1 u_1) + \operatorname{div}(\rho_1 u_1^2 + p_1) = 0, && \text{in } \Omega \setminus S(t),\\
&\partial_t(\rho_1 e_1) + \operatorname{div}(\rho_1 u_1 e_1 + p_1 u_1) = 0, && \text{in } \Omega \setminus S(t),\\
&u_1 \cdot n = h' \cdot n, && \text{on } \partial S(t),\\
&M h''(t) = \int_{\partial S(t)} p_1 n \,\mathrm{d}\sigma_x, &&\\
&(J\omega)'(t) = \int_{\partial S(t)} (x - h(t)) \wedge p_1 n \,\mathrm{d}\sigma_x, &&
\end{aligned}\right\} \tag{2.1}$$

where $d\sigma_x$ is the induced measure on $\partial S(t)$. The first 3 equations are the classical Euler system for a compressible fluid. The fourth one ensures the continuity of the normal velocity on the solid boundary, and the last two ones describe the Newton laws for the solid. In this system p_1 is the pressure inside the fluid and is given by the model of the fluid. Here we consider the case of a stiffened gas, whose law is given by $p_1 = (\gamma_1 - 1)\rho_1\left(e_1 - \frac{u_1^2}{2}\right) - \gamma_1\pi_1$ with $\gamma_1 > 1, \pi_1 \geq 0$ (the case $\pi_1 = 0$ corresponding to an ideal gas). Let us now introduce classical notation: $\varepsilon_1 = e_1 - u_1^2/2$, the internal energy, and $\tau_1 = 1/\rho_1$, the specific volume.

We now simplify this system, assuming slab symmetry (as will be done in the numerical subsections). We suppose that the domain of the whole system is $[0, 1]$ and we denote by $B(t) = [a(t), b(t)]$ the domain of the rigid body. We always assume that $0 < a(t) < b(t) < 1$ (with $b(t) - a(t) = b(0) - a(0)$). Therefore, the domain of the fluid is

$$\Omega(t) = [0, 1] \setminus [a(t), b(t)].$$

The density of the solid is constant and thus the position of the center of mass of the rigid body is $h(t) = \dfrac{a(t) + b(t)}{2}$. We also denote by M the mass of the solid.

In Eulerian coordinates the equations modeling the movement of the system can be written as follows:

$$\partial_t(\rho_1) + \partial_x(\rho_1 u_1) = 0 \qquad x \in \Omega(t),\ t \in [0, T], \tag{2.2}$$

$$\partial_t(\rho_1 u_1) + \partial_x(\rho_1 u_1^2 + p_1) = 0 \qquad x \in \Omega(t),\ t \in [0, T], \tag{2.3}$$

$$\partial_t(\rho_1 e_1) + \partial_x(\rho_1 u_1 e_1 + p_1 u_1) = 0 \qquad x \in \Omega(t),\ t \in [0, T], \tag{2.4}$$

$$u_1(t) = h'(t) \qquad x \in \{a(t), b(t)\},\ t \in [0, T], \tag{2.5}$$

$$Mh''(t) = p_1(b(t)) - p_1(a(t)) \qquad t \in [0, T], \tag{2.6}$$

$$\rho_1(x, 0) = \rho_1^0(x), \qquad x \in \Omega(0), \tag{2.7}$$

$$u_1(x, 0) = u_1^0(x), \qquad x \in \Omega(0), \tag{2.8}$$

$$e_1(x, 0) = e_1^0(x), \qquad x \in \Omega(0), \tag{2.9}$$

$$h(0) = h^0 \in \mathbb{R},\ h'(0) = h^1 \in \mathbb{R}. \tag{2.10}$$

2.2 An approximate system involving mixing

The numerical strategy that will be chosen in Subsection 2.4 involves a fixed grid. For this reason the solid and the fluid may occupy only parts of cells: the mass fractions of solid and fluid become real numbers in $[0, 1]$ once projected on the grid. This subsection is devoted to the writing of a mathematical model allowing this artificial mixing. Here we introduce the mass fractions c_1 of the fluid and c_2 of the solid. These quantities are such that $c_1 + c_2 = 1$.

For the sake of simplicity we limit the presentation to dimension 1 and do not insist on the treatment of the solid region (where $c_1 = 1, c_2 = 0$).

The conservation equations are those of each component mass, global momentum and global energy. Let us denote by ρ the global density. The partial densities of each component are then $c_1\rho$ and $c_2\rho$. The mixing between the two components being only a numerical artefact, it is possible to determine the specific volume of the components, $\tau_1 = 1/\rho_1$ and $\tau_2 = 1/\rho_2$, and the additivity of volume leads to the equation $c_1\tau_1 + c_2\tau_2 = \tau = 1/\rho$ (cf. [16]). The system then reads

$$\left.\begin{aligned} &\partial_t(c_1\rho) + \partial_x(c_1\rho u) = 0,\\ &\partial_t\rho + \partial_x\rho u = 0,\\ &\partial_t(\rho u) + \partial_x(\rho u^2 + p) = 0,\\ &\partial_t(\rho e) + \partial_x(\rho u e + p u) = 0,\\ &c_1 + c_2 = 1,\\ &c_1\tau_1 + c_2\tau_2 = \tau, \end{aligned}\right\} \tag{2.11}$$

where the unknowns are c_1, c_2, ρ, u and e. We propose to compute the pressure p with $p = p_1(\tau_1, \varepsilon_1)$, which means that the solid has no influence on the fluid pressure. Recall that τ_2 is a constant and that ε_1 is computed with the help of $e = u^2/2 + \varepsilon_1$. System (2.11) is thus formally closed.

2.3 Mathematical study

The mathematical study of this system is very difficult, and as far as we know there is no result available in the literature. For the system fluid-rigid body with a viscous incompressible fluid modelled by the Navier–Stokes equations, several papers concerning the existence have been published in the last years (see, for example, [12], [13], [6], [29] and the references in [30]). In the case of a viscous compressible fluid some results of existence can be found in [6] and [10].

This subsection is devoted to the computation of eigenvalues and eigenvectors of system (2.11) and to the analysis of its hyperbolicity. In order to simplify the analysis, we propose to use the Lagrangian formulation of (2.11). This formulation will be used for the derivation of the numerical scheme in Subsection 2.4. It relies on the definition of the convective derivative $\mathrm{D}_t = \partial_t + u\partial_x$. In the regular case standard computations lead to the equivalent system

$$\left.\begin{aligned} &\rho\mathrm{D}_t c_1 = 0,\\ &\rho\mathrm{D}_t\tau = \partial_x u,\\ &\rho\mathrm{D}_t u = -\partial_x p,\\ &\rho\mathrm{D}_t e = -\partial_x(pu), \end{aligned}\right\} \tag{2.12}$$

where τ_2 is a positive constant. This PDE system is closed via the algebraic relations $c_1+c_2 = 1, c_1\tau_1+c_2\tau_2 = \tau = 1/\rho$. Assuming the fluid has a strictly concave entropy

$S_1(\tau_1, \varepsilon_1)$ verifying $\rho \mathrm{D}_t S_1 = 1/T \mathrm{D}_t \varepsilon_1 + p/T \mathrm{D}_t \tau_1 = 0$ where T is the temperature (which is the case for a stiffened gas, see below), we can symmetrize (2.12) with respect to pressure, writing

$$\mathrm{D}_t p_1 = \frac{\partial p_1}{\partial \tau_1} \mathrm{D}_t \tau_1 + \frac{\partial p_1}{\partial S_1} \mathrm{D}_t S_1 = \frac{\partial p_1}{\partial \tau_1} \mathrm{D}_t \tau_1$$

where p_1 is viewed as a function of τ_1 and S_1. Now recalling that $c_1\tau_1 + c_2\tau_2 = \tau$ and that τ_2 is constant, we get $\mathrm{D}_t \tau = c_1 \mathrm{D}_t \tau_1$, and, finally, $\rho \mathrm{D}_t p_1 = \omega \partial_x u$, where $\omega = 1/c_1 \partial p_1/\partial \tau_1 < 0$ is defined if $c_1 > 0$. In Lagrangian coordinates, putting $\partial_m = \tau \partial_x$, the symmetrized system reads

$$\begin{aligned} &\mathrm{D}_t c_1 = 0, \\ &\mathrm{D}_t p = \omega \partial_m u, \\ &\mathrm{D}_t u = -\partial_m p, \\ &\mathrm{D}_t S_1 = 0, \end{aligned}$$

where the equation on the total energy has been replaced by the one on the entropy, and is completed with the algebraic equations $c_1 + c_2 = 1$, $c_1\tau_1 + c_2\tau_2 = 1/\rho$, $p = p_1 = (\gamma_1 - 1)\rho_1\varepsilon_1 - \gamma_1\pi_1$, $\varepsilon_1 = e - \frac{u_1^2}{2}$. This quasi-linear system is $\mathrm{D}_t U = A \partial_m U$ with

$$U = \begin{pmatrix} c_1 \\ p \\ u \\ S_1 \end{pmatrix}, \quad A = \begin{pmatrix} 0 & 0 & 0 & 0 \\ 0 & 0 & \omega & 0 \\ 0 & -1 & 0 & 0 \\ 0 & 0 & 0 & 0 \end{pmatrix}.$$

The eigenvalues of A are 0, $\sqrt{-\omega}$ and $-\sqrt{-\omega}$ with corresponding eigenvectors $(1, 0, 0, 0)^T$, $(0, 0, 0, 1)^T$, $(0, -\sqrt{-\omega}, 1, 0)^T$ and $(0, \sqrt{-\omega}, 1, 0)^T$, which form a basis of $\mathbb{R}^4$, assuming $\omega \neq 0$, which is the case under the assumption $c_1 \in]0, 1]$ and S_1 strictly concave. Hence we have the following

Proposition 2.1. *Assuming the existence of a strictly concave entropy, system* (2.12) *is hyperbolic if* $c_1 > 0$.

Usually the strict concavity of the entropy holds for $\tau_1 > 0$, $\varepsilon_1 > 0$, so that the domain of hyperbolicity of (2.12) is $\{c_1, \tau, u, e \in \mathbb{R}^4, c_1 > 0, \tau > 0, e - \frac{u^2}{2} > 0\}$.

Note that $s = \sqrt{-\omega}$ appears as the sound speed (in Lagrangian coordinates) in the medium consisting in the mixing of the compressible fluid and the undeformable solid,

$$s = \sqrt{-\omega} = \sqrt{-\frac{1}{c_1}\frac{\partial p_1}{\partial \tau_1}} = \sqrt{\frac{1}{c_1}\frac{s_1^2}{\tau_1^2}} = \frac{s_1}{\sqrt{c_1}\frac{\tau_1}{\tau}},$$

where s_1 is the sound speed (in Lagrangian coordinates) in the fluid given by $s_1 = \sqrt{\gamma_1(p_1(\tau_1, \varepsilon_1) + \pi_1)\tau_1}$. Thus the sound speed of the mixing has a singularity when

c_1 tends to 0. Let us end this subsection by giving the expression of the strictly concave classical entropy for a stiffened gas:

$$S_1(\tau_1, \varepsilon_1) = (p_1(\tau_1, \varepsilon_1) + \pi_1)\tau_1^{\gamma_1}.$$

2.4 Numerical algorithm

Here we present the numerical scheme we developed to solve the one dimensional problem. It is based on the "Lagrange-projection" algorithm analyzed by Després ([7], [8]). This is an explicit finite volume method. The general procedure consists in an operator splitting whose first step solves (2.12) (Lagrange stage) and whose second step solves $D_t V = 0$ (projection, or convection step) where V is the vector of all the transported unknowns.

In this subsection we assume a mesh with cells of constant size Δx is given and, for every $j \in \mathbb{Z}, n \in \mathbb{N}$, denote by z_j^n the approximate value of the quantity z at $j\Delta x$ and at time $n\Delta t$ (in the naive assumption that Δt remains constant). Without recalling basic definitions of finite volume schemcs, we briefly discuss how to compute $(z_j^{n+1})_{j\in\mathbb{Z}}$ from $(z_j^n)_{j\in\mathbb{Z}}$, z representing any of the variables in (2.11). In the following we skip the time index n (but not $n+1$).

The Lagrange step's role is to solve (2.12). The use of an explicit method and the equation $\rho D_t p_1 = \omega\partial_x u$ of the symmetrized system suggest that a CFL condition of order $\sqrt{-\omega}\Delta t/\Delta x \leq 1$ will have to be imposed in order that the scheme is stable. This is a constraint, because nothing plays against the rarefaction of the fluid in any cell: a dramatic decrease of the time step is predictable when the solid occupies almost a whole cell (ω is of order $1/c_1$). In order to avoid this problem, we propose a derefinement procedure (that will have to be followed by a refinement procedure). The general design of the algorithm is the following:

- derefinement (z_j becomes $\overline{z_j}$),
- Lagrange step ($\overline{z_j}$ becomes $\widetilde{z_j}$),
- projection step ($\widetilde{z_j}$ becomes $\widehat{z_j}$),
- refinement ($\widehat{z_j}$ becomes z_j^{n+1}),
- computation of the new velocity of the solid (with Newton's laws).

2.4.1 Derefinement. This part of the algorithm takes advantage of the fact that mixed cells are very localized: in dimension 1, each boundary of the solid intersects at most one cell. We thus know that any mixed cell is located near a pure fluid cell. The principle here is to consider the mixed cell with its pure fluid neighbor as a unique big cell. This allows to recover a volume fraction of fluid $c_1\tau_1/\tau > 1/2$ in the whole computational domain.

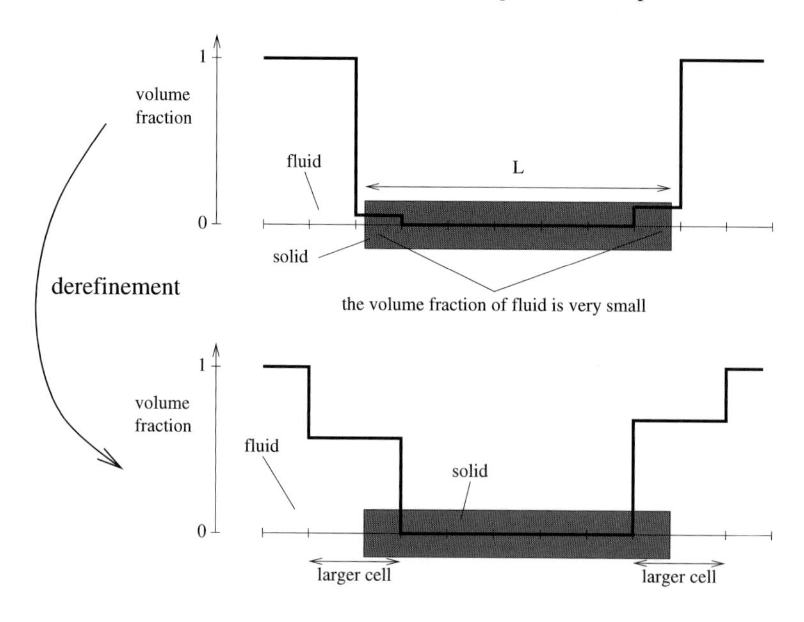

Figure 6. Derefinement.

The derefinement algorithm is the following.

- Detect cells where $c_1\tau_1/\tau < 1/2$, let us say cell j.
- For each of these cells, detect the neighbor cell where $c_1 = 1$ (in the other neighbor, $c_1 = 0$): assume for example it is cell $j-1$.
- In the mixed cell j and its neighbor $j-1$, replace all conservative quantities ρ, ρc_1, ρu, ρe by their mean values on j and $j-1$:

$$
\begin{aligned}
&\overline{\rho_{j-1}} = \overline{\rho_j} = (\rho_{j-1} + \rho_j)/2,\\
&\overline{\rho_{j-1}c_{1,j-1}} = \overline{\rho_j c_{1,j}} = (\rho_{j-1}c_{1,j-1} + \rho_j c_{1,j})/2,\\
&\overline{\rho_{j-1}u_{j-1}} = \overline{\rho_j u_j} = (\rho_{j-1}u_{j-1} + \rho_j u_j)/2,\\
&\overline{\rho_{j-1}e_{j-1}} = \overline{\rho_j e_j} = (\rho_{j-1}e_{j-1} + \rho_j e_j)/2,
\end{aligned}
$$

  (and new values of $\tau_1, \varepsilon_1, p_1, \ldots$ follow).
- In the other cells the values are unchanged: $\overline{z_k} = z_k$;

The consequence of this trick is that $c_1\tau_1/\tau > 1/2$ in each cell, so that

$$
s = \frac{s_1}{\sqrt{c_1\frac{\tau_1}{\tau}}} = \frac{\sqrt{c_1}s_1}{c_1\frac{\tau_1}{\tau}} \le 2\sqrt{c_1}s_1 \le 2s_1
$$

if $c_1 \in]0,1]$. Thus the sound speed of the mixing (where $c_1 > 0$) is bounded if this holds for the fluid.

Remark 2.2. This derefinement produces an artificial spreading (diffusion) of the solid.

2.4.2 Lagrange step. This part of the algorithm, so as the following, is directly taken from the Lagrange-projection schemes of [7] and [8] (and here we do not go further in explanations). The Lagrange stage solves (2.11). It relies on the definition of the cell edge quantities

$$\rho s^*_{j+1/2} = \sqrt{\max(\rho_j {s_j}^2, \rho_{j+1} {s_{j+1}}^2) \min(\rho_j, \rho_{j+1})},$$

$$p_{j+1/2} = \frac{p_j + p_{j+1}}{2} + \frac{\rho s^*_{j+1/2}}{2}(u_j - u_{j+1}),$$

$$u_{j+1/2} = \frac{p_j - p_{j+1}}{2\rho s^*_{j+1/2}} + \frac{1}{2}(u_j + u_{j+1}),$$

with $p_j = p_1(\tau_{1,j}, \varepsilon_{1,j})$ and $s_j = s_1(\tau_{1,j}, \varepsilon_{1,j})\tau_j / (\sqrt{c_{1,j}}\tau_{1,j})$ (s is the sound speed in the global fluid in Lagrangian coordinates).

Remark 2.3. The Lagrangian edge quantities are evaluated on the discrete solution *before* the derefinement.

These edge quantities allow to compute all the variables after the Lagrange step with formulae

$$\left.\begin{aligned} &\widetilde{c_{1,j}} - \overline{c_{1,j}} = 0, \\ &\overline{\rho_j}\frac{\widetilde{\tau_j} - \overline{\tau_j}}{\Delta t} - \frac{u_{j+1/2} - u_{j-1/2}}{\Delta x} = 0, \\ &\overline{\rho_j}\frac{\widetilde{u_j} - \overline{u_j}}{\Delta t} + \frac{p_{j+1/2} - p_{j-1/2}}{\Delta x} = 0, \\ &\overline{\rho_j}\frac{\widetilde{e_j} - \overline{e_j}}{\Delta t} + \frac{p_{j+1/2}u_{j+1/2} - p_{j-1/2}u_{j-1/2}}{\Delta x} = 0, \end{aligned}\right\} \tag{2.13}$$

with $\overline{\tau_j} = 1/\overline{\rho_j}$ for every cell (for all j).

The contact between fluid and solid are treated as wall conditions with imposed velocity of the wall (pragmatically, this is a way to obtain edge values $p_{j+1/2}, u_{j+1/2}$ at the contact with pure solid that lead to a stable scheme).

2.4.3 Projection step. This step of the algorithm can be viewed as a remapping part. It solves the advection system

$$\left.\begin{aligned} &\partial_t c_1 + u\partial_x c_1 = 0, \\ &\partial_t \rho + u\partial_x \rho = 0, \\ &\partial_t u + u\partial_x u = 0, \\ &\partial_t e + u\partial_x e = 0. \end{aligned}\right\} \tag{2.14}$$

In the chosen discrete version the fluxes $\widetilde{z_{j+1/2}}$ are computed according to the following procedure (as in [16]):

- First compute the mass fraction fluxes *exactly*, which is easy because the velocity inside the solid is constant in space and in time during the whole time step: denote it $\widetilde{c_{1,j+1/2}}$ and $\widetilde{c_{2,j+1/2}}$, of course verifying $\widetilde{c_{1,j+1/2}} + \widetilde{c_{2,j+1/2}} = 1$.

- Secondly, compute the upwinded fluxes for the fluid specific volume, the fluid internal energy, the solid specific volume and the velocity:

$$\begin{aligned}
&\widetilde{\tau_{1,j+1/2}} = \widetilde{\tau_{1,j}} && \text{if } u_{j+1/2} \geq 0,\\
&\widetilde{\tau_{1,j+1/2}} = \widetilde{\tau_{1,j+1}} && \text{if } u_{j+1/2} < 0,\\
&\widetilde{\varepsilon_{1,j+1/2}} = \widetilde{\varepsilon_{1,j}} && \text{if } u_{j+1/2} \geq 0,\\
&\widetilde{\varepsilon_{1,j+1/2}} = \widetilde{\varepsilon_{1,j+1}} && \text{if } u_{j+1/2} < 0,\\
&\widetilde{\tau_{2,j+1/2}} = \tau_2, &&\\
&\widetilde{u_{j+1/2}} = \widetilde{u_j} && \text{if } u_{j+1/2} \geq 0,\\
&\widetilde{u_{j+1/2}} = \widetilde{u_{j+1}} && \text{if } u_{j+1/2} < 0.
\end{aligned}$$

- Thirdly, compute the conservative fluxes with

$$\widetilde{\rho_{j+1/2}} = \frac{1}{\widetilde{c_{1,j}}\widetilde{\tau_{1,j+1/2}} + \widetilde{c_{2,j}}\widetilde{\tau_{2,j+1/2}}},$$

$$\widetilde{e_{j+1/2}} = \widetilde{c_{1,j}}\widetilde{\varepsilon_{1,j+1/2}} + \frac{\widetilde{u_{j+1/2}}^2}{2};$$

$$\begin{aligned}
&\frac{\widehat{\rho_j}\widehat{c_{1,j}} - \widetilde{\rho_j}\widetilde{c_{1,j}}}{\Delta t} + u_{j-1/2}\frac{\widetilde{\rho_{j+1/2}}\widetilde{c_{1,j+1/2}} - \widetilde{\rho_{j-1/2}}\widetilde{c_{1,j-1/2}}}{\Delta x} = 0,\\
&\frac{\widehat{\rho_j} - \widetilde{\rho_j}}{\Delta t} + u_{j-1/2}\frac{\widetilde{\rho_{j+1/2}} - \widetilde{\rho_{j-1/2}}}{\Delta x} = 0,\\
&\frac{\widehat{\rho_j}\widehat{u_j} - \widetilde{\rho_j}\widetilde{u_j}}{\Delta t} + u_{j-1/2}\frac{\widetilde{\rho_{j+1/2}}\widetilde{u_{j+1/2}} - \widetilde{\rho_{j-1/2}}\widetilde{u_{j-1/2}}}{\Delta x} = 0,\\
&\frac{\widehat{\rho_j}\widehat{e_j} - \widetilde{\rho_j}\widetilde{e_j}}{\Delta t} + u_{j-1/2}\frac{\widetilde{\rho_{j+1/2}}\widetilde{e_{j+1/2}} - \widetilde{\rho_{j-1/2}}\widetilde{e_{j-1/2}}}{\Delta x} = 0.
\end{aligned}$$

2.4.4 Refinement. In this part we recover a constant-by-cell approximation (on the original little cells of the mesh) and to balance the artificial spreading of the derefinement phase (cf. Remark 2.2). It only concerns the cells where $\widehat{c_{1,j}}(1 - \widehat{c_{1,j}}) > 0$. Without going into the details (that are tedious), let us just mention that this last step of the time iterate globally conserves

- volume fraction of fluid and solid,

- mass fraction of fluid and solid,

- total momentum,
- total energy

on the two neighbor cells that are concerned.

2.4.5 New evaluation of the velocity of the solid. The new velocity v^{n+1} of the solid is simply computed using Newton's laws. Let us denote by L the length of the solid and by j_l, j_r the indices of the mixed cells respectively on the left and on the right of the solid. The discrete velocity evolution is

$$v^{n+1} = v - \frac{\Delta t}{L} \tau_2 (p_{j_r - 1/2} - p_{j_l + 1/2}).$$

Remark 2.4. The present scheme is by construction conservative in mass and momentum. Nevertheless, it is not conservative in total energy: this is due to the fact that the new evaluation of the solid velocity is done via the momentum fluxes on the mixed cell edges, and the new solid energy, $1/2v^{n+1^2}$, does not coincide with the ancient one updated via the energy fluxes on the mixed cell edges. This phenomenon can be compared with the so-called "wall heating" that occurs in pure fluid cells, but there the difference is reported in the internal energy (which is not taken into account in the solid). The lack of energy conservation is a general drawback of coupled fluid-solid numerical systems, even in moving meshes formulation, as reported and studied in [25].

The evolution of total energy in time is reported in Figure 13 in the next subsection.

2.5 Some numerical results

Let us now present some numerical results for the following simulation. In the interval [0, 1] limited with wall condition on both sides, a solid of density 20 and length 0.1 (thus with mass 2) is initially centered on 0.5 and has velocity 100. The fluid on both sides is an ideal gas with $\gamma_1 = 2$, $\pi_1 = 0$ and is initially at rest with pressure and density equal to 1.

This model problem is very similar to the one studied in [24].

We present three numerical results obtained for $t = 0.001$, $t = 0.0051$ (the solid being at its extremal position on the right) and $t = 0.016$ (the solid being at its extremal position on the left), with three different meshes: 100, 300 and 1000 cells. We note a relatively slow convergence for the velocity, particularly at time $t = 0.016$. What is remarkable is that although the velocity of the solid seems far to be converged, the position of the solid is good. The reason is that the acceleration of the solid at its extremal position is important, so that a small error in time gives a huge error for the velocity.

Figure 13 shows the evolution of total energy in the whole domain. It seems to converge to a conserved quantity when refining the mesh, as expected. The variation of the total energy seems to be an increasing function of the acceleration of the solid, but a more precise study should be done to allow a precise analysis of the phenomenon.

At last we present a result obtained under the same general conditions but with $\pi_0 = 100\,000$. Figure 14 (with 200 cells) shows the pressure at time $t = 0.0024$: the solid is at its extremal position on the left. Of course, this position is more on the left than on the preceding figures for time $t = 0.0051$. This is due to the fact that the fluid is now stiffer. The pressure inside the solid is by convention fixed to 1, but this does not affect the computation.

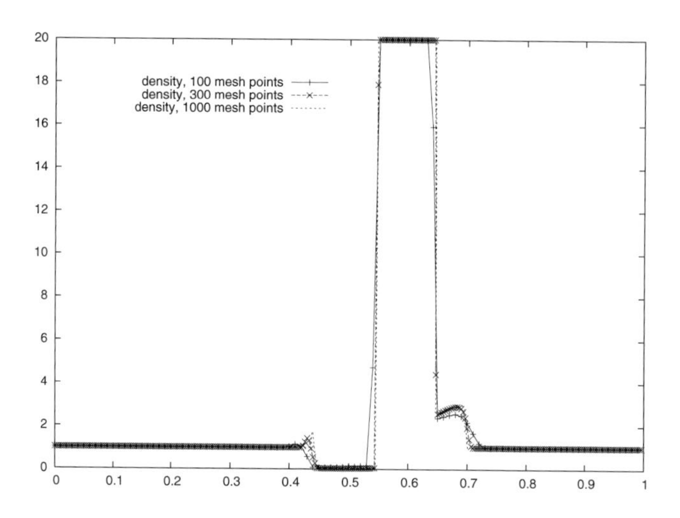

Figure 7. Density, time $t = 0.001$.

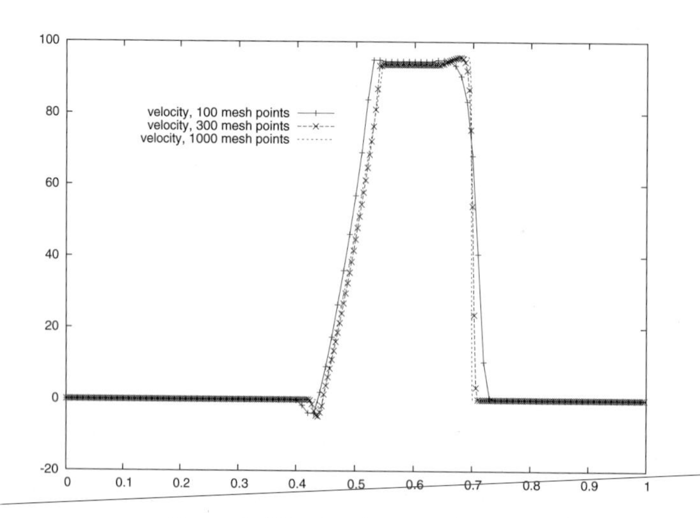

Figure 8. Velocity, time $t = 0.001$.

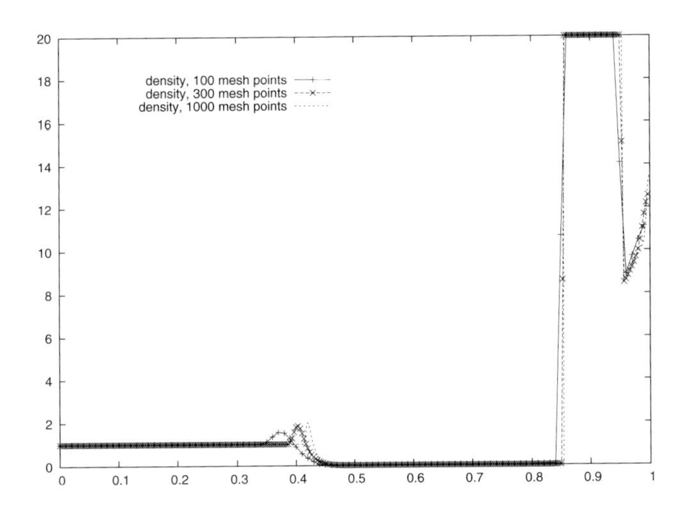

Figure 9. Density, time $t = 0.0051$.

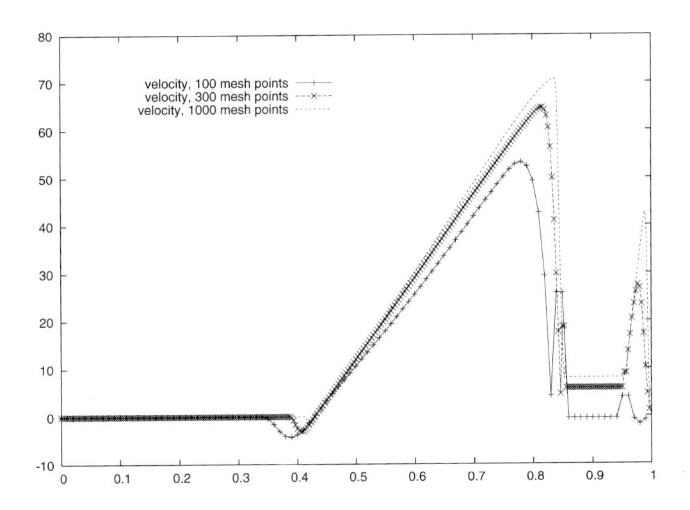

Figure 10. Velocity, time $t = 0.0051$.

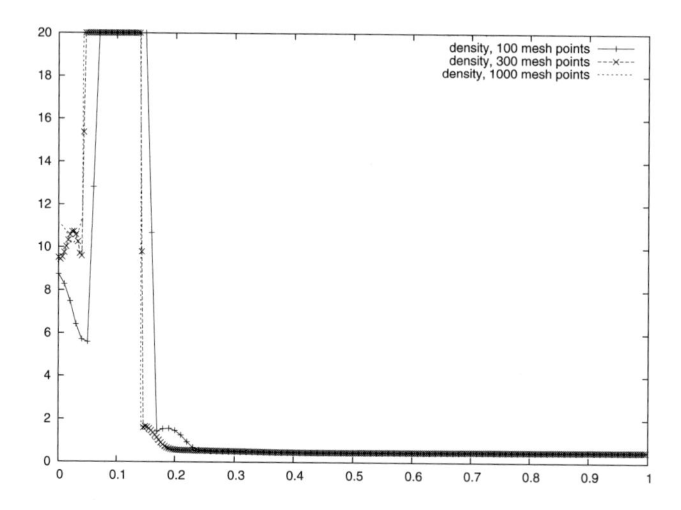

Figure 11. Density, time $t = 0.016$.

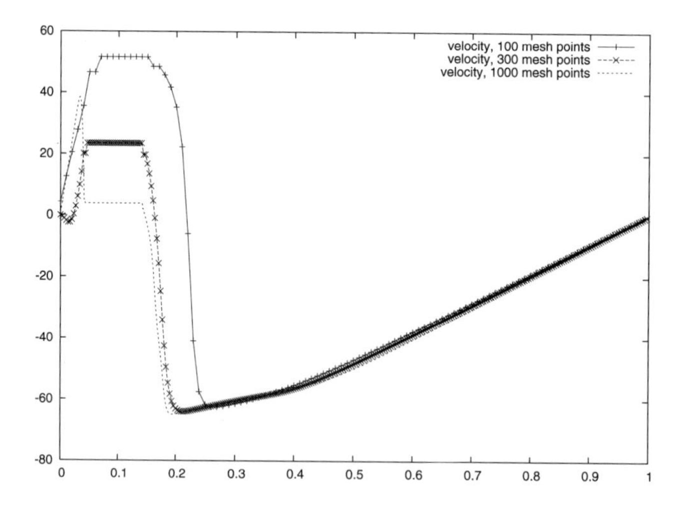

Figure 12. Velocity, time $t = 0.016$.

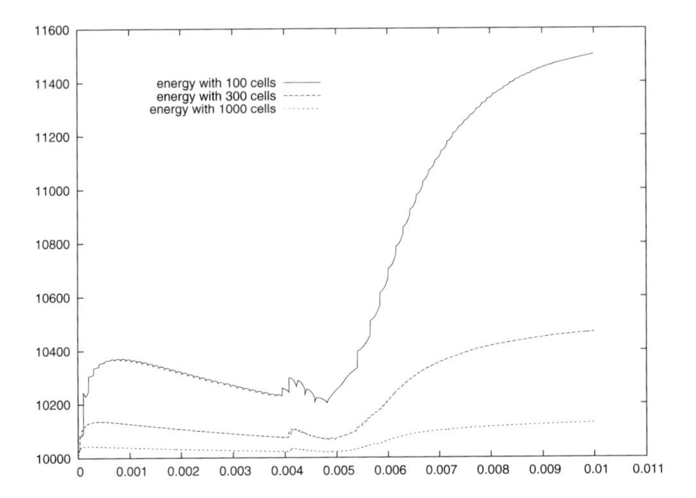

Figure 13. Total energy in [0, 1] as a function of time.

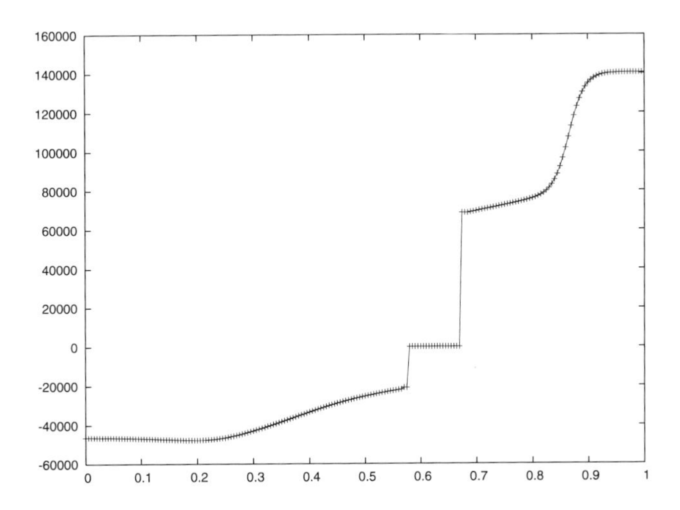

Figure 14. Pressure with $\pi_1 = 100\,000$, 200 cells, time $t = 0.0024$.

A conclusion of this part could be as follows. We developed an algorithm to solve the coupled compressible Euler system with the motion of a rigid body. One important feature of the algorithm is that it only needs a unique Cartesian mesh. For the sake of simplicity we chose an explicit solving of the problem, which is known to require a CFL-like condition that prescribes an upper bound for the time steps Δt. This upper bound is a function of the length of the fluid cell. In order not to dramatically reduce the time step due to the existence of cells that are almost empty of fluid, we developed a derefinement (or projection) procedure. The rest of the algorithm uses a classical Lagrange-Projection solver (for which some stability results, via entropy inequalities, are available). Some comparison with results obtained by ALE methods on moving meshes will be done in the immediate future to evaluate the error due to this projection on larger cells in the area of the interface between the fluid and the solid. Finally, dimensional splitting to two and three dimensional computations has to be done to demonstrate the interest of this method.

References

[1] A. A. Amsden, Kiva-3V, release 2, improvements to Kiva-3V. Report #LA-UR-99-915, Los Alamos National Laboratory, 1999.

[2] C. Baranger, Modeling of oscillations, breakup and collisions for droplets: the establishment of kernels for the T.A.B. model, to appear in *Math. Models Methods Appl. Sci.*

[3] C. Baranger and L. Desvilettes, Study at the numerical level of the kernels of collision, coalescence and fragmentation for sprays, to appear in *Proceedings of the International Workshop on Multiphase and Complex Flow Simulation for Industry, Cargèse, France, 2003*

[4] S. D. Bauman, A spray model for an adaptive mesh refinement code, PhD Thesis, University of Wisconsin-Madison, 2001.

[5] L. Boudin, L. Desvillettes and R. Motte, A modeling of compressible droplets in a fluid, *Commun. Math. Sci.* 1 (2003), 657–669.

[6] B. Desjardins and M. Esteban, On weak solutions for fluid-rigid structure interaction: compressible and incompressible models, *Comm. Partial Differential Equations* 25 (2000), 1399–1413.

[7] B. Després, Lagrangian systems of conservation laws. Invariance properties of Lagrangian systems of conservation laws, approximate Riemann solvers and the entropy condition, *Numer. Math.* 89 (2001), no. 1, 99–134.

[8] B. Després, Symétrisation en variable de Lagrange pour la mécanique des milieux continus et schémas numériques, *Matapli* 72 (2003), 45–61.

[9] C. Farhat, B. Larrouturou and S. Piperno, Partitioned procedures for the transient solution of coupled aeroelastic problems I: model problem, theory and two-dimensional application, *Comput. Methods Appl. Mech. Engrg.* 124 (1995), no. 1–2, 79–112.

[10] E. Feireisl, On the motion of rigid bodies in a viscous compressible fluid, *Arch. Ration. Mech. Anal.* 167 (2003), no. 4, 281–308.

[11] R. Glowinski, T.-W. Pan, T. I. Hesla, D. D. Joseph, and J. Périaux, A fictitious domain approach to the direct numerical simulation of incompressible viscous flow past moving rigid bodies: application to particulate flow, *J. Comput. Phys.* 169 (2001), no. 2, 363–426.

[12] C. Grandmont and Y. Maday, Existence for an unsteady fluid-structure interaction problem, *M2AN Math. Model. Numer. Anal.* 34 (2000), 609–636.

[13] M. Gunzburger, H.-C. Lee and G. Seregin, Global existence of weak solutions for viscous incompressible flows around a moving rigid body in three dimensions, *J. Math. Fluid Mech.* 2 (2000), 219–266.

[14] J.J. Hylkema, Modélisation cinétique et simulation numérique d'un brouillard dense de gouttelettes. Application aux propulseurs à poudre, PhD Thesis, Université Toulouse-III, 1999.

[15] M. Ismail, Simulation d'Écoulements fluide-particules par la FBM, PhD Thesis, Université Paris-VI, 2004.

[16] F. Lagoutière, Modélisation mathématique et résolution numérique de problèmes de fluides à plusieurs constituants, PhD thesis, Université Paris-VI, 2000.

[17] B. Maury, Direct Simulations of 2D Fluid-Particle Flows in Biperiodic Domains, *J. Comput. Phys.* 156 (1999), 325–351.

[18] B. Maury, A Fat Boundary Method for the Poisson problem in a domain with holes, *J. Sci. Comput.* 16 (3) (2001), 319–339.

[19] P. Métier, Modélisation, analyse mathématique et applications numériques de problèmes d'interaction fluide-structure instationnaires, PhD Thesis, Université Paris-VI, 2003.

[20] P. J. O'Rourke, Collective drop effects on vaporizing liquid sprays, PhD Thesis, Los Alamos National Laboratory, 1981.

[21] M. A. Patterson and R. D. Reitz, Modeling the effects of fuel spray characteristics on Diesel engine combustion and emissions, *SAE paper* #980131, 1998.

[22] C. S. Peskin, Flow patterns around heart valves: a digital computer method for solving the equations of motion, PhD Thesis, Albert Einstein College of Medicine, 1972.

[23] C. S. Peskin, The immersed boundary method, *Acta Numer.* 11 (2002), 479–517.

[24] S. Piperno, Explicit/implicit fluid/structure staggered procedures with a structural predictor and fluid sub-cycling for 2D inviscid aeroelastic simulations, *Internat. J. Numer. Methods Fluids* 25 (1997), no. 10, 1207–1226.

[25] S. Piperno, Contribution à l'étude mathématique et à la simulation numérique de phénomènes d'interaction fluide-structure, habilitation à diriger des recherches, Université Paris-VI, 2000.

[26] W. E. Ranz and W. R. Marshall, Vaporization from drops, part I–II, *Chem. Eng. Prog.* 48 (3) (1952), 141–180.

[27] R. D. Reitz, Modeling atomization processes in high-pressure vaporizing sprays, *Atom. Spray Tech.* 3 (1987), 309–337.

[28] J. San Martín, J. F. Scheid, T. Takahashi and M. Tucsnak, Convergence of the Lagrange-Galerkin method for the Equations Modelling the Motion of a Fluid-Rigid System, preprint.

[29] J. San Martín, V. Starovoitov, and M. Tucsnak, Global weak solutions for the two-dimensional motion of several rigid bodies in an incompressible viscous fluid, *Arch. Ration. Mech. Anal.* 161 (2002), 113–147.

[30] T. Takahashi, Analyse des équations modélisant le mouvement des systèmes couplant des solides rigides et des fluides visqueux, PhD Thesis, Université Nancy-I, 2002.

[31] S. Vincent, J.-P. Caltagirone, M. Azaiez and N. Randrianarivelo, Méthodes de pénalisation pour les équations de Navier-Stokes. Application à l'interaction fluide/structure, oral communication at J.-L. Lions laboratory, January 20th, 2003.

[32] F. A. Williams, *Combustion theory*, second edition, Benjamin Cummings, 1985.

Numerical study of a conservative bifluid model with interpenetration*

B. Després[1], S. Jaouen[1], C. Mazeran[1] and T. Takahashi[2]

[1] *CEA/DIF, DSSI/SNEC/LPNP, BP 12, 91680 Bruyères le Châtel, France*
email: `bruno.despres@cea.fr`, `stephane.jaouen@cea.fr`
`constant.mazeran@cea.fr`

[2] *Institut Élie Cartan, Université Nancy, BP 239, 54506 Vandoeuvre-lès-Nancy, France*
email: `takahash@iecn.u-nancy.fr`

Abstract. The present work is devoted to the study of a system of conservation laws which arises when modelling compressible bifluid flows, in view of preliminary inertial confinement fusion simulations. By opposition to usual multiphase models, the present system of PDE's turns out to be in fully conservative form, what is of great practical interest in the development of usual numerical approaches and for the capture of shocks. The study is strongly conditioned by the so-called isobar-isothermal closure. We propose a numerical scheme for the approximate resolution of the model which respects positivity of mass fractions. Numerical results illustrate the interest of the model and the method on classical bifluid test-cases.

1 Introduction

Modelling and numerical simulation of multiphase flows is a great challenge for many industrial applications. The main motivation of the present study is related to Inertial Confinement Fusion simulations, which involve hot and dense multiphase flows with shocks. Difficulty of such models and major trends about modelisation and numerics can be found in [11], in the context of nuclear reactors.

Here we are interested in the numerical resolution of a compressible bifluid model initially proposed by Després in [2]. By bifluid we mean that each fluid has its own velocity, which allows to consider complex flows, as for example interpenetration or sedimentation. In this work we restrict ourselves to only two fluids mixture in one-dimension of space. Thus the mixture will notably be characterized with two velocities $u_1(x,t)$ and $u_2(x,t)$ and with the mass fractions $\alpha_1(x,t)$, $\alpha_2(x,t)$ (or possibly the

*This work was supported by the Département Science de la Simulation et de l'Information (DSSI), Commissariat à l'Energie Atomique (CEA/DIF), BP 12, 91680 Bruyères le Châtel, France.

volume fractions $\beta_1(x,t)$ and $\beta_2(x,t)$). The interest of the considered model is that it is written in completely conservative form, by opposition of usual ones (see [1], [8], [9] for instance); analysis of shock solutions and development of numerical methods are thus *a priori* more natural.

The model has been derived in Lagrangian coordinates in accordance with general canonical structure of conservation laws coming from continuum mechanics. We refer the reader to [3] for an introduction to this general structure in the 1D case. The abstract Lagrangian form of the system is

$$D_t \boldsymbol{u} + D_m \boldsymbol{f}(\boldsymbol{u}) = 0, \tag{1.1}$$

where D_t represents the material derivative related to the global mean velocity u of the fluid: $D_t = \partial_t + u\partial_x$, and D_m stands for the mass derivative related to the global density ρ: $D_m = \frac{1}{\rho}\partial_x$. This derivation assumes the system to be endowed with a global entropy S with zero flux, *i.e.* smooth solutions satisfy

$$D_t S = 0, \tag{1.2}$$

whereas discontinuous ones satisfy

$$D_t S \geq 0 \tag{1.3}$$

in the sense of distributions. This model thus attempts to be an extension of the classical Euler system for compressible gas dynamics.

The canonical formalism [3] states that the lagrangian flux may be rewritten as

$$\boldsymbol{f}(\boldsymbol{u}) = \begin{pmatrix} \mathbb{B}\Psi \\ -\frac{1}{2}\Psi^t\mathbb{B}\Psi \end{pmatrix}, \tag{1.4}$$

where $\mathbb{B}$ is a constant symmetric matrix and Ψ a primitive variable related to the entropy

$$\Psi = \frac{1}{\left(\frac{\partial S}{\partial \boldsymbol{u}}\right)_n} \begin{pmatrix} \left(\frac{\partial S}{\partial \boldsymbol{u}}\right)_1 \\ \left(\frac{\partial S}{\partial \boldsymbol{u}}\right)_2 \\ \vdots \\ \left(\frac{\partial S}{\partial \boldsymbol{u}}\right)_{n-1} \end{pmatrix}. \tag{1.5}$$

We remind that in the case of the Euler equations, theses variables are set to

$$\boldsymbol{u} = \begin{pmatrix} \tau \\ u \\ e \end{pmatrix}, \quad \mathbb{B} = \begin{pmatrix} 0 & 1 \\ 1 & 0 \end{pmatrix} \quad \text{and} \quad \Psi = \begin{pmatrix} p \\ -u \end{pmatrix}.$$

Of course the physical justification of such a model, like others, has still to be proved. However, in our mind the conservative form of the system appears in itself as a sufficient motivation for the present study.

A first difficulty – also generally appearing in non-conservative models, see [11] – is to know whether the system is well posed or not, *i.e.* if the solutions depend continuously on the initial data. In the case of conservation laws this property is related to the hyperbolicity of the system. Unfortunately the present system with the chosen closure leads to computational difficulties which do not allow to conclude. At the numerical level an important issue is to design a positive scheme, that is a scheme which leads to positivity for the mass fractions.

The paper is organised as follows. We first recall in Section 2 the conservative model of [2] and present some links between specific and global physical quantities, necessary for further study. This strongly depends on the chosen closure. In Section 3 we show that the *non classical* part of this model is close to the shallow water system for two fluids. In Section 4 we present some algebraic results about the hyperbolicity of the system. Section 5 is devoted to the numerical resolution of the system, with a special emphasis on the positivity of the mass fractions. Then we turn to numerical experiments in Section 6. Finally in Section 7, we point out some perspectives for future developments on the same class of models but with other closure laws.

2 The isobar-isothermal conservative multiphase model

We present in this section the model proposed in [2], first introducing some notations. Each fluid indexed by the letter $i = 1, 2$ is supposed to be inviscid and non miscible. It is characterised by four physical quantities, as for example its density ρ_i, its mass fraction α_i, its velocity u_i and its total energy e_i. Of course the mass fractions are valued in $[0, 1]$ and satisfy

$$\alpha_1 + \alpha_2 = 1. \tag{2.1}$$

Then its internal energy ε_i is defined by $\varepsilon_i = e_i - \frac{1}{2}u_i^2$, and its pressure p_i as to be precised as a function of ρ_i and ε_i: $p_i = p_i(\rho_i, \varepsilon_i)$. In the following we shall also often use the specific volume $\tau_i = \rho_i^{-1}$. The entropy S_i is defined by the fundamental law of thermodynamics

$$T_i dS_i = d\varepsilon_i + p_i d\tau_i, \tag{2.2}$$

where $T_i > 0$ is the temperature, to be specified with respect to τ_i and ε_i.

The present model describes the evolution of the mixture and thus brings into play averaged quantities (without index). Extensive quantities are naturally defined by

$$\begin{aligned} \tau &= \alpha_1\tau_1 + \alpha_2\tau_2 = 1/\rho, \\ u &= \alpha_1 u_1 + \alpha_2 u_2, \\ \varepsilon &= \alpha_1\varepsilon_1 + \alpha_2\varepsilon_2, \\ e &= \alpha_1 e_1 + \alpha_2 e_2. \end{aligned}$$

The first two relations express respectively the additivity of volumes for a two fluids non miscible mixture (see [6] for more details) and the additivity of impulses. Introducing the differential velocity $w = u_1 - u_2$ allows us to rewrite the global total energy in the form

$$e = \varepsilon + \frac{1}{2}u^2 + \frac{\alpha_1\alpha_2}{2}w^2.$$

It remains to define the global entropy S. In [2] the choice is to consider that the entropy of the mixture should be greater than the sum of the partial entropies. This is modelized by

$$S = \alpha_1 S_1 + \alpha_2 S_2 + S_{\text{mix}}, \tag{2.3}$$

where S_{mix} is a concave positive function. In the following we shall mainly consider two functions

$$S_{\text{mix}}(\alpha_1, \alpha_2) = \begin{cases} -k(\alpha_1 \log \alpha_1 + \alpha_2 \log \alpha_2), \\ k\alpha_1\alpha_2, \end{cases} \quad \text{with } k \geq 0.$$

Note that this choice is of greatest importance in the present study, since the model is derived through a canonical formulation based on the global entropy (see [3] and [2]). In other words, another choice of this entropy would change the Lagrangian flux.

Finally, the model has been designed upon the so-called *isobar-isothermal* hypothesis, which also strongly conditions the study. This means that the mixture is in pressure and temperature equilibrium:

$$\left.\begin{aligned} p_1(\tau_1, \varepsilon_1) &= p_2(\tau_2, \varepsilon_2) \equiv p, \\ T_1(\tau_1, \varepsilon_1) &= T_2(\tau_2, \varepsilon_2) \equiv T. \end{aligned}\right\} \tag{2.4}$$

We can now finish to present the conservative model first proposed in [2]. In Lagrangian coordinates the system of conservation laws reads:

$$\left.\begin{aligned} & D_t(\tau) - D_m(u) = 0, \\ & D_t(\alpha_2) - D_m\left(\frac{w(1-\alpha_2)\alpha_2}{\tau}\right) = 0, \\ & D_t(u) + D_m(P) = 0, \\ & D_t(\tau w) + D_m\left(\mu_1 - \mu_2 - \frac{1-2\alpha_2}{2}w^2\right) = 0, \\ & D_t(e) + D_m\left(Pu + \frac{1}{\tau}w(1-\alpha_2)\alpha_2\left(\mu_1 - \mu_2 - \frac{1-2\alpha_2}{2}w^2\right)\right) = 0. \end{aligned}\right\} \tag{2.5}$$

In (2.5) the pressure tensor P stands for

$$P = p + \rho\alpha_1\alpha_2 w^2, \tag{2.6}$$

whereas μ_i is the generalized chemical potential

$$\mu_i = -TS_i + \varepsilon_i + p\tau_i - T\frac{\partial S_{\text{mix}}}{\partial \alpha_i}. \tag{2.7}$$

The three first equations, respectively describing the conservation of total mass, partial mass and total impulse, are common to classical models derived *via* averaging the standard Euler equations, or could at least be deduced from them. Hence they are acceptable from the modeling point of view. The originality of the present model comes from the equation on the differential velocity, which is unusual. This equation has been deduced in [2] assuming that a 1D Lagrangian multiphase model with vanishing entropy flux should share the canonical structure of general fluid model [3]. By construction, system (2.5) indeed recast on the form (1.4)–(1.5) with

$$\boldsymbol{u} = \begin{pmatrix} \tau \\ \alpha_2 \\ u \\ \tau w \\ e \end{pmatrix}, \quad \mathbb{B} = \begin{pmatrix} 0 & 0 & 1 & 0 \\ 0 & 0 & 0 & 1 \\ 1 & 0 & 0 & 0 \\ 0 & 1 & 0 & 0 \end{pmatrix} \quad \text{and} \quad \Psi = \begin{pmatrix} P \\ \mu_1 - \mu_2 - \frac{1-2\alpha_2}{2} w^2 \\ -u \\ -\frac{w(1-\alpha_2)\alpha_2}{\tau} \end{pmatrix}. \tag{2.8}$$

Notice that of course we could turn out to the Eulerian description. Motivation for writing down the system in Lagrangian coordinates is essentially related to the numerical strategy for approximation, see Section 5. For the reader's curiosity, we write the model in Eulerian coordinates:

$$\begin{aligned}
&\partial_t(\rho) + \partial_x(\rho u) = 0, \\
&\partial_t(\rho\alpha_2) + \partial_x(\rho\alpha_2 u_2) = 0, \\
&\partial_t(\rho u) + \partial_x(\rho u^2 + P) = 0, \\
&\partial_t(w) + \partial_x\big(uw + \mu_1 - \mu_2 - \tfrac{1-2\alpha_2}{2} w^2\big) = 0, \\
&\partial_t(\rho e) + \partial_x\big(\rho u e + P u + \rho w(1-\alpha_2)\alpha_2\big(\mu_1 - \mu_2 - \tfrac{1-2\alpha_2}{2} w^2\big)\big) = 0.
\end{aligned}$$

This model is very close to the one derived by Romensky in [12], [13].

To conclude the presentation of the model and enlighten on its domain of applications, let us precise the equation of states. In this study we shall restrict test-cases to stiffened gases, whose equation is

$$p_i = (\gamma_i - 1)\rho_i\varepsilon_i - \gamma_i\pi_i. \tag{2.9}$$

For such gas law the fundamental law of thermodynamics (2.2) is satisfied with[1]

$$S_i = C_{vi} \log\left((\varepsilon_i - \pi_i)\tau_i^{\gamma_i - 1}\right) \tag{2.10}$$

and

$$T_i = \frac{\varepsilon_i - \pi_i\tau_i}{C_{vi}}. \tag{2.11}$$

Setting $\pi_i = 0$ allows to consider perfect gases.

[1] Be aware that, in the general case $\pi_i \neq 0$, T_i defined by (2.11) is just an integrand factor to have (2.2), and that it is not the physical temperature. In that sense the designation *isobar-isothermal model* is a misnomer: T must be seen as a parameter of the model rather than the equilibrium temperature. Nevertheless, if $\pi_i = 0$ and if C_{vi} is the correct calorific capacity, (2.11) denotes the physical temperature of a perfect gas.

The previous model (2.5) can be greatly simplified using a splitting between the classical part of the model (assume it corresponds to the equations for τ, u, e) and the non classical part (it corresponds to the equations for α_2, w). Assume that $\rho = \frac{1}{\tau}$ and $T \approx \overline{T}$ are almost constant, and that u is negligible. Then we get (see [4] for further details)

$$\left.\begin{aligned} &\partial_t(\alpha_2) - \partial_x(w(1-\alpha_2)\alpha_2) = 0, \\ &\partial_t(w) + \partial_x\big(k\overline{T}(1-2\alpha_2) - \tfrac{(1-2\alpha_2)}{2} w^2\big) = 0. \end{aligned}\right\} \tag{2.12}$$

Using now non-dimensional variables $a = 2\alpha_2 - 1$ and $b = \frac{w}{\sqrt{2k\overline{T}}}$ and the non-dimensional coordinate $x \leftarrow \sqrt{\frac{2}{k\overline{T}}}x$, we obtain the following system of conservation laws:

$$\left.\begin{aligned} &\partial_t a + \partial_x\big[(a^2-1)b\big] = 0, \\ &\partial_t b + \partial_x\big[(b^2-1)a\big] = 0. \end{aligned}\right\} \tag{2.13}$$

3 Link with the two fluids shallow water model in pipelines

Serguei Gavrilyuk (LMMT, University Aix-Marseille III) has pointed out to us the proximity between (2.13) and the two fluids shallow water model that we recall now. Let us consider a two fluids flow in a pipeline. An illustration is given in Figure 1.

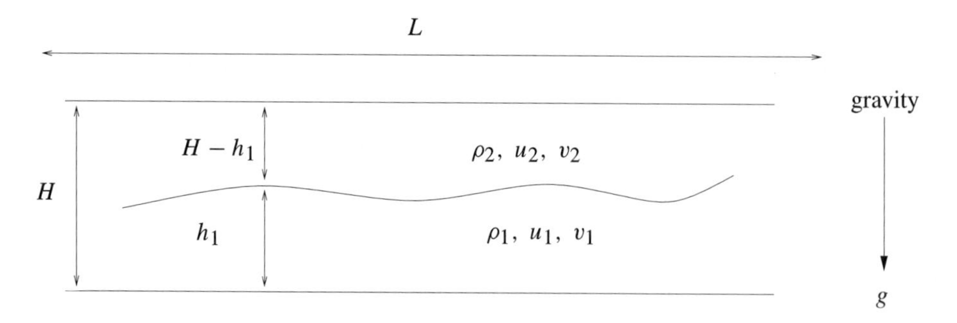

Figure 1. A pipeline.

The flow is stratified in the sense that fluid number 2 is slighter than fluid number 1. It explains why fluid number 2 is above fluid number 1. Then one makes a certain number of assumptions with a scaling with respect to the small parameter ε (here this variable has nothing to do with the internal energy).

Assumption 1. The pipeline is thin. The small parameter is $\frac{H}{L} = \varepsilon$.

Assumption 2. One assumes that the average velocity u is almost constant. Writing the equations in the correspondent moving frame, one gets $h_1 u_1 + h_2 u_2 = O(\varepsilon)$.

Assumption 3. Both fluids are incompressible, that is ρ_1 and ρ_2 are constant in space and time.

Assumption 4. The shallow water approximation is valid, that is $u_1(t,x,y) \approx u_1(t,x)$ and $u_2(t,x,y) \approx u_2(t,x)$.

Assumption 5. One uses the *Boussinesq approximation*, that is,

$$\rho_1 - \rho_2 \text{ is negligible, but } \mu = gH\frac{\rho_1 - \rho_2}{\rho_1} \text{ is not negligible.}$$

Using these assumptions it is possible to derive the two fluids shallow model, where $d_1 = \frac{h_1}{H} \in [0,1]$ is the nondimensional height:

$$\left.\begin{aligned} &\partial_t d_1 + \partial_x\left(d_1(1-d_1)w\right) = 0,\\ &\partial_t w + \partial_x\left(\tfrac{w^2}{2}(1-2d_1)\right) + \mu\partial_x d_1 = 0, \end{aligned}\right\} \tag{3.1}$$

with $w = u_1 - u_2$. Using the scaling

$$2d_1 - 1 = a \in [-1,1], \quad w = \alpha b,\ \alpha = \sqrt{\mu}, \tag{3.2}$$

one gets from (3.1)

$$\left.\begin{aligned} &\partial_t a + \alpha\partial_x\left[(a^2-1)b\right] = 0,\\ &\partial_t b + \alpha\partial_x\left[(b^2-1)a\right] = 0. \end{aligned}\right\} \tag{3.3}$$

So the two fluids shallow water model (3.3) is equivalent to (3.1) up to a redefinition of the time variable. It remains to show how our assumptions give (3.1). For this task we follow the method proposed to us by Serguei Gavrilyuk. We split it into four steps.

Step 1: equations for h_1 and h_2. An elementary volume of fluid number 1 (resp. 2) satisfies

$$\partial_t x(t,X,Y) = u_1 \text{ (resp. } u_2), \quad \partial_t y(t,X,Y) = v_1 \text{ (resp. } v_2).$$

Setting $h_1(t,x)$ the height of fluid one with the relation $h_1(t, x(t,X,Y)) = y(t,X,Y)$ (thus $h_2(t,x) = H - h_1(t,x)$) one gets

$$\partial_t h_1 + u_1\partial_x h_1 = v_1, \quad \partial_t h_2 + u_2\partial_x h_2 = v_2. \tag{3.4}$$

Step 2: incompressibility of the fluid. The incompressibility of both fluids is $\partial_x u_i + \partial_y v_i = 0$. Thus using the wall condition $v_1(y \equiv 0) = 0$ one gets

$$v_1(t,x,y=h_1) = -\int_0^{h_1(t,x)} \partial_x u_1(t,x,y)dy,$$

which we simplify into

$$v_1(t,x,y=h_1) \approx -h_1(t,x)\partial_x u_1(t,x).$$

Using this in (3.4) gives a first conservation law

$$\partial_t h_1 + \partial_x(u_1 h_1) = 0, \tag{3.5}$$

that is, using a non dimensional variable $d_1 = \frac{h_1}{H}$,

$$\partial_t d_1 + \partial_x(u_1 d_1) = 0. \tag{3.6}$$

Step 3: the hydrostatic approximation. Fluids 1 and 2 satisfy

$$\left.\begin{aligned} &\rho_i(\partial_t u_i + u_i\partial_x u_i + v_i\partial_y u_i) + \partial_x p_i = 0,\\ &\rho_i(\partial_t v_i + u_i\partial_x v_i + v_i\partial_y v_i) + \partial_y p_i = -\rho_i g, \end{aligned}\right\} \tag{3.7}$$

with boundary conditions at walls

$$v_1(t, x, y \equiv 0) = v_2(t, x, y \equiv H) = 0.$$

The natural scaling is

$$x \to Lx, \quad y \to Hy, \quad t \to \frac{L}{\sqrt{gH}}t,$$

$$u_i \to \sqrt{gH}u_i, \quad v_i \to \frac{H}{L}\sqrt{gH}v_i, \quad p_i \to \rho_i g H p_i, \quad h_1 \to H d_1.$$

Using $\varepsilon = \frac{H}{L}$ as a small parameter one gets from (3.7)

$$\left.\begin{aligned} &g\varepsilon\rho_i\left(\partial_t u_i + u_i\partial_x u_i + v_i\partial_y u_i\right) + g\varepsilon\rho_i\partial_x p_i = 0,\\ &g\varepsilon^2\rho_i\left(\partial_t v_i + u_i\partial_x v_i + v_i\partial_y v_i\right) + g\rho_i\partial_y p_i = -g\rho_i, \end{aligned}\right\} \tag{3.8}$$

that is

$$\left.\begin{aligned} &\partial_t u_i + u_i\partial_x u_i + v_i\partial_y u_i + \partial_x p_i = 0,\\ &\varepsilon^2\left(\partial_t v_i + u_i\partial_x v_i + v_i\partial_y v_i\right) + \partial_y p_i = -1. \end{aligned}\right\} \tag{3.9}$$

This gives the classical hydrostatic approximation $\partial_y p_i \approx -1$. After integration we obtain

$$\left.\begin{aligned} &p_2(t, x, y) = \tfrac{p^*(t,x)}{\rho_2} - (y - 1),\\ &p_1(t, x, y) = \tfrac{p^*(t,x)}{\rho_1} + \tfrac{\rho_2}{\rho_1} + \lambda d_1(t, x) - y. \end{aligned}\right\} \tag{3.10}$$

Here we have used the continuity of pressure $\rho_1 p_1(t, x, d_1(t, x)) = \rho_2 p_2(t, x, d_1(t, x))$ to integrate the hydrostatic law and the definition of $\lambda = \frac{\rho_1 - \rho_2}{\rho_1}$.

Step 4: equation for $w = u_1 - u_2$. The first equation in (3.9) is rewritten at $y = 0$ (recall that $v_1 = 0$ for $y = 0$ or $y = H$) as

$$\begin{aligned} &\partial_t u_1 + u_1\partial_x u_1 + \partial_x p_1 = 0, \quad y = 0\\ &\partial_t u_2 + u_2\partial_x u_2 + \partial_x p_2 = 0, \quad y = 1. \end{aligned}$$

Since we want to eliminate the unknown pressure $p^*(t, x)$ in the previous two equations, it leaves little choice. One gets by substraction that

$$\rho_1(\partial_t u_1 + u_1\partial_x u_1) - \rho_2(\partial_t u_2 + u_2\partial_x u_2) + \rho_1\lambda\partial_x d_1 = 0.$$

Going back to dimensional variables and unknowns but keeping d_1 instead of h_1, one gets

$$\rho_1(\partial_t u_1 + u_1\partial_x u_1) - \rho_2(\partial_t u_2 + u_2\partial_x u_2) + (gH\rho_1\lambda)\partial_x d_1 = 0.$$

The Boussinesq approximation means that both densities are close to each other, $\rho_1 - \rho_2 = O(\varepsilon)$, but the difference is not small: $0 < \mu = gH\lambda \approx O(1)$. So we write

$$\partial_t w + \frac{1}{2}\partial_x(u_1^2 - u_2^2) + \mu\partial_x d_1 = 0,$$

where $w = u_1 - u_2$. Since $u_{1,2}(t,x,y) \approx u_{1,2}(t,x)$ and $h_1u_1 + h_2u_2 = 0$, we have $u_1 = \left(1 - \frac{h_1}{H}\right)w = (1-d_1)w$ and $u_2 = -\frac{h_1}{H}w = -d_1w$. Thus we get our second conservation law

$$\partial_t w + \frac{1}{2}\partial_x w^2(1-2d_1) + \mu\partial_x h_1 = 0. \tag{3.11}$$

The two fluids shallow water model (3.1) is constituted of equations (3.6) and (3.11).

4 Hyperbolicity of the isobar-isothermal model

This section is devoted to the study of the hyperbolicity of system (2.5). More precisely, we derive sufficient conditions for the jacobian matrix $f'(U)$ to be diagonalizable.

For the sake of legibility we define A and B such that the physical flux of this model in Lagrangian coordinates is given by

$$f(U) = \begin{pmatrix} -u \\ -\frac{w(1-\alpha_2)\alpha_2}{\tau} \\ P \\ \mu_1 - \mu_2 - \frac{1-2\alpha_2}{2}w^2 \\ Pu + \frac{w(1-\alpha_2)\alpha_2}{\tau}\left(\mu_1 - \mu_2 - \frac{1-2\alpha_2}{2}w^2\right) \end{pmatrix} = \begin{pmatrix} -u \\ -A \\ P \\ B \\ Pu + AB \end{pmatrix}.$$

We also define $z = \tau w$. Thus the jacobian matrix $\frac{\partial f}{\partial U}$ is similar to

$$\mathcal{A} = \begin{pmatrix} 0 & 0 & -1 & 0 & 0 \\ -\frac{\partial A}{\partial \tau} & -\frac{\partial A}{\partial \alpha_2} & 0 & -\frac{\partial A}{\partial z} & 0 \\ \frac{\partial P}{\partial \tau} - P\frac{\partial P}{\partial e} & \frac{\partial P}{\partial \alpha_2} - B\frac{\partial P}{\partial e} & \frac{\partial P}{\partial u} + u\frac{\partial P}{\partial e} & \frac{\partial P}{\partial z} + A\frac{\partial P}{\partial e} & \frac{\partial P}{\partial e} \\ \frac{\partial B}{\partial \tau} - P\frac{\partial B}{\partial e} & \frac{\partial B}{\partial \alpha_2} - B\frac{\partial B}{\partial e} & \frac{\partial B}{\partial u} + u\frac{\partial B}{\partial e} & \frac{\partial B}{\partial z} + A\frac{\partial B}{\partial e} & \frac{\partial B}{\partial e} \\ 0 & 0 & 0 & 0 & 0 \end{pmatrix}. \tag{4.1}$$

We immediately see that zero is an eigenvalue. It is therefore sufficient to study the 4×4 matrix composed of the four first lines and rows of (4.1). After some tricky calculations we get for perfect gases laws

$$\mathcal{A} = \begin{pmatrix} 0 & 0 & -1 & 0 \\ 2wE & -D & 0 & -E \\ F & G & 0 & 2wE \\ G & H & 0 & -D \end{pmatrix}$$

with

$$\begin{aligned}
E &= \frac{\alpha_2(1-\alpha_2)}{\tau^2}, \\
D &= \frac{(1-2\alpha_2)w}{\tau}, \\
F &= -\frac{p}{\tau} - \frac{p^2}{\varepsilon} - 3w^2 E, \\
G &= -\frac{p(\mu_1-\mu_2)}{\varepsilon} + \frac{\varepsilon C_{v2} C_{v1}(\gamma_2-\gamma_1)}{\tau\left(\alpha_1 C_{v1} + \alpha_2 C_{v2}\right)^2} + wD, \\
H &= w^2 - \frac{1}{\varepsilon}\left(\mu_1 - \mu_2 + \frac{\varepsilon(C_{v2}-C_{v1})}{\alpha_1 C_{v1} + \alpha_2 C_{v2}}\right)^2 - \frac{2k\varepsilon}{\alpha_1 C_{v1} + \alpha_2 C_{v2}} \\
&\quad - \frac{\varepsilon(C_{v2}(\gamma_2-1) - C_{v1}(\gamma_1-1))^2}{\left(\alpha_1 C_{v1} + \alpha_2 C_{v2}\right)\left(\alpha_1(\gamma_1-1)C_{v1} + \alpha_2(\gamma_2-1)C_{v2}\right)}.
\end{aligned}$$

Let us define

$$\mathcal{B} = \begin{pmatrix} 0 & 0 & 1 & 0 \\ 0 & 0 & 0 & 1 \\ 1 & 0 & 0 & 0 \\ 0 & 1 & 0 & 0 \end{pmatrix}.$$

We notice that $\mathcal{B} = \mathcal{B}^*$ and $\mathcal{B} = \mathcal{B}^{-1}$. Then we define $\mathcal{C} = \mathcal{A}\mathcal{B}$ such that $\mathcal{A} = \mathcal{C}\mathcal{B}$. Here we have

$$\mathcal{C} = \begin{pmatrix} -1 & 0 \\ 0 & \tilde{\mathcal{C}} \end{pmatrix} \quad \text{with } \tilde{\mathcal{C}} = \begin{pmatrix} -E & 2wE & -D \\ 2wE & F & G \\ -D & G & H \end{pmatrix}.$$

If we prove that $\mathcal{C} = \mathcal{C}^* < 0$, then

$$\left(\sqrt{-\mathcal{C}}\right)^{-1} A \left(\sqrt{-\mathcal{C}}\right) = -\left(\sqrt{-\mathcal{C}}\right) B \left(\sqrt{-\mathcal{C}}\right)$$

and thus $\mathcal{A}$ is similar to a symmetric matrix, hence diagonalizable.

A *sufficient* condition for the system (2.5) to be hyperbolic is therefore that $\mathcal{C}$ is symmetric negative definite. We therefore have the following property.

Lemma 4.1. *A sufficient condition for the system* (2.5) *to be hyperbolic is that* $\tilde{\mathcal{C}}$ *is negative definite.*

Now we use the following criterion which gives necessary and sufficient conditions for a matrix to be positive definite. If we write the characteristic polynomial $P(\lambda)$ of a 3×3 matrix $\tilde{\mathfrak{C}}$ in the form $P(\lambda) = -\lambda^3 + \sigma_2\lambda^2 + \sigma_1\lambda + \sigma_0$, we obtain

Lemma 4.2. *A necessary and sufficient condition for the matrix* $\tilde{\mathfrak{C}}$ *to be negative is*

$$\sigma_i < 0, \quad i = 0, 1, 2.$$

Applying this result to our matrix leads to the following result:

Proposition 4.3. *A sufficient condition for the system* (2.5) *to be hyperbolic is that*

$$\left.\begin{aligned} \sigma_2 &= H - E + F < 0, \\ \sigma_1 &= -FH + EH + EF + D^2 + G^2 + 4w^2E^2 < 0, \\ \sigma_0 &= -EFH - FD^2 + EG^2 - 4wEGD - 4w^2E^2H < 0. \end{aligned}\right\} \tag{4.2}$$

The analytical analysis of these conditions seems difficult to handle in the general case. They may, however, be used in a simulation code to flag hyperbolicity defaults.

Notice that in the special case $k >> 1$ and $\alpha_1\alpha_2 \neq 0$, which means that enough mixing entropy has been introduced in the model, it can be shown that a sufficient condition for the system to be hyperbolic is that $\sqrt{\alpha_1\alpha_2}|w| \leq c$, which is a subsonic constraint. This result has already been found in [2].

5 Numerical approximation

The purpose of this section is to propose a numerical method to solve approximately the system and to validate it on classical bifluid test-cases. Then we shall conclude whether this conservative model is physically relevant or not. One of the main difficulties at the discrete level is, briefly, to couple the *classical* part of the system, composed of the equations on τ and u, with the *purely bifluid* part, bringing into play α and τw; the last equation on the total energy is itself a coupling between these two parts. In particular, as we shall deal with discontinuous solutions, we wish the global scheme to be entropy consistent and to degenerate, for a single fluid phase, towards an entropic scheme solving the classical Euler equations. On the other hand, the scheme has to be positive for the mass fractions.

5.1 Numerical scheme for the *classical part*

We first recall the numerical scheme proposed in [2]. This scheme is based upon the canonical structure of the system and is an application of general ideas on Lagrangian system of conservation laws endowed with a natural entropy (see [3]) which allows to easily construct entropic schemes. Such a scheme reads

$$\frac{\Delta x \rho_j^0}{\Delta t}(\boldsymbol{u}_j^{n+1} - \boldsymbol{u}_j^n) + \boldsymbol{f}_{j+\frac{1}{2}}^n - \boldsymbol{f}_{j-\frac{1}{2}}^n = 0, \tag{5.1}$$

where $\boldsymbol{u}_j^n$ is the approximation of $\boldsymbol{u}$ in the jth cell at time step n, and $\boldsymbol{f}_{j+\frac{1}{2}}^n$ is the numerical flux through the right side of the cell, given by

$$\boldsymbol{f}_{j+\frac{1}{2}}^n = \begin{pmatrix} \mathbb{B}_{j+\frac{1}{2}}^+ \Psi_{j+1}^n + \mathbb{B}_{j+\frac{1}{2}}^- \Psi_j^n \\ -\frac{1}{2}(\Psi_{j+1}^n)^t \mathbb{B}_{j+\frac{1}{2}}^+ \Psi_{j+1}^n + (\Psi_j^n)^t \mathbb{B}_{j+\frac{1}{2}}^- \Psi_j^n \end{pmatrix}, \tag{5.2}$$

where $\mathbb{B}_{j+\frac{1}{2}}^+$ and $\mathbb{B}_{j+\frac{1}{2}}^-$ are such that

$$\begin{cases} \mathbb{B}_{j+\frac{1}{2}}^+ + \mathbb{B}_{j+\frac{1}{2}}^- = \mathbb{B}, \\ (\mathbb{B}_{j+\frac{1}{2}}^+)^t = \mathbb{B}_{j+\frac{1}{2}}^+ \geq 0, \\ (\mathbb{B}_{j+\frac{1}{2}}^-)^t = \mathbb{B}_{j+\frac{1}{2}}^- \leq 0. \end{cases} \tag{5.3}$$

In [2], the $\mathbb{B}$ splitting (5.3) used was

$$\mathbb{B}_{j+\frac{1}{2}}^+ = \begin{pmatrix} \frac{1}{\beta_1} & 0 & \frac{1}{2} & 0 \\ 0 & \frac{1}{\beta_2} & 0 & \frac{1}{2} \\ \frac{1}{2} & 0 & \beta_1 & 0 \\ 0 & \frac{1}{2} & 0 & \beta_2 \end{pmatrix} \quad \text{and} \quad \mathbb{B}_{j+\frac{1}{2}}^- = \begin{pmatrix} -\frac{1}{\beta_1} & 0 & \frac{1}{2} & 0 \\ 0 & -\frac{1}{\beta_2} & 0 & \frac{1}{2} \\ \frac{1}{2} & 0 & -\beta_1 & 0 \\ 0 & \frac{1}{2} & 0 & -\beta_2 \end{pmatrix},$$

with

$$\beta_1 = \sqrt{\max(\rho_j c_j^2, \rho_{j+1} c_{j+1}^2) \min(\rho_j^0, \rho_{j+1}^0)},$$
$$\beta_2 = 4\frac{\beta_1}{(\rho_j + \rho_{j+1})^2} \frac{2}{(\alpha_2(1-\alpha_2))_j + (\alpha_2(1-\alpha_2))_{j+1}}.$$

The interest of such a decomposition is that *if the entropy is strictly concave*, then the scheme is entropic under a CFL-like condition $c_j^n \frac{\Delta t}{\Delta m_j} \leq 1$. Although this method has given reasonable results in [2], we emphasize that no results on the concavity of the entropy (2.3) with respect to $\boldsymbol{u}$ (2.8) have been found yet and that the positivity of the mass fractions has not been proven for such a scheme.

However, this strategy seems to be a good choice when the model degenerates towards a single fluid model (classical Euler equations), so that we retain it to design the numerical scheme for the *classical part* of the system, that is to say

$$D_t(\tau) - D_m(u) = 0,$$
$$D_t(u) + D_m(p + \rho\alpha_1\alpha_2 w^2) = 0,$$

will be solved with the scheme

$$\frac{\Delta m_j}{\Delta t}\begin{pmatrix} \tau_j^{n+1} - \tau_j^n \\ u_j^{n+1} - u_j^n \end{pmatrix} + \left[(\tilde{\mathbb{B}}_{j+\frac{1}{2}}^+ \tilde{\Psi}_{j+1}^n + \tilde{\mathbb{B}}_{j+\frac{1}{2}}^- \tilde{\Psi}_j^n) - (\tilde{\mathbb{B}}_{j-\frac{1}{2}}^+ \tilde{\Psi}_j^n + \tilde{\mathbb{B}}_{j-\frac{1}{2}}^- \tilde{\Psi}_{j-1}^n) \right] = 0, \tag{5.4}$$

where

$$\tilde{\mathbb{B}}^+_{j+\frac{1}{2}} = \begin{pmatrix} \frac{1}{2\beta} & \frac{1}{2} \\ \frac{1}{2} & \frac{\beta}{2} \end{pmatrix}, \quad \tilde{\mathbb{B}}^-_{j+\frac{1}{2}} = \begin{pmatrix} -\frac{1}{2\beta} & \frac{1}{2} \\ \frac{1}{2} & -\frac{\beta}{2} \end{pmatrix}, \quad \tilde{\Psi}^n_j = \begin{pmatrix} p + \rho\alpha_1\alpha_2 w^2 \\ -u \end{pmatrix}^n_j$$

and $\beta = \sqrt{\max(\rho_j c_j^2, \rho_{j+1} c_{j+1}^2)\min(\rho_j^0, \rho_{j+1}^0)}$. With simpler notations, this scheme reads

$$\frac{\Delta m_j}{\Delta t}\left(\tau_j^{n+1} - \tau_j^n\right) + \left(f^\tau_{j+\frac{1}{2}} - f^\tau_{j-\frac{1}{2}}\right) = 0,$$
$$\frac{\Delta m_j}{\Delta t}\left(u_j^{n+1} - u_j^n\right) + \left(f^u_{j+\frac{1}{2}} - f^u_{j-\frac{1}{2}}\right) = 0,$$

with

$$f^\tau_{j+\frac{1}{2}} = -\frac{u^n_{j+1} + u^n_j}{2} + \frac{1}{2\beta}\left(\left(p + \rho\alpha_1\alpha_2 w^2\right)^n_{j+1} - \left(p + \rho\alpha_1\alpha_2 w^2\right)^n_j\right), \tag{5.5}$$

and

$$f^u_{j+\frac{1}{2}} = -\frac{\left(p + \rho\alpha_1\alpha_2 w^2\right)^n_{j+1} + \left(p + \rho\alpha_1\alpha_2 w^2\right)^n_j}{2} + \frac{\beta}{2}\left(u^n_{j+1} - u^n_j\right). \tag{5.6}$$

Remark 5.1. When $\alpha_1\alpha_2 w^2$ is zero, the "classical" acoustic Riemann solver of Godunov is retrieved.

5.2 Study of the 2 × 2 bifluid sub-model (2.13)

5.2.1 The 2 x 2 submodel. As was said in the introduction of this section, the bifluid model studied here is composed of two subsystems: the standard Euler system for compressible gas dynamics and a purely bifluid part with the two unknowns α_2 and τw. We focus here on this second subsystem, which has already been studied in [4] and which has been recalled at the end of Section 2, namely

$$\left.\begin{aligned} \partial_t a + \partial_x[(a^2 - 1)b] = 0, \\ \partial_t b + \partial_x[(b^2 - 1)a] = 0. \end{aligned}\right\} \tag{5.7}$$

The reader will find a detailed analysis of this system in [4]. Especially, it is shown that the domain of real physical eigenvalues is compact in $\mathbb{R}^2$

$$\overline{\mathcal{D}} = \{-1 \le a \le 1 \text{ and } -1 \le b \le 1\},$$

and that the numerical solution is stable provided that the numerical dissipation is large enough: Lax–Friedrichs schemes are used in [4].

5.2.2 Numerical scheme. The aim of this paragraph is to present another numerical scheme which is also stable. To do this we define

$$\left.\begin{aligned} a^{\pm} &= a \pm 1, \\ b^{\pm} &= b \pm 1, \end{aligned}\right\} \tag{5.8}$$

and the fluxes

$$\left.\begin{aligned} f^a &= a^+ a^- b = (a^2 - 1)b, \\ f^b &= b^+ b^- a = (b^2 - 1)a. \end{aligned}\right\} \tag{5.9}$$

The scheme writes

$$\left.\begin{aligned} a_i^{n+1} &= a_i^n - \tfrac{\Delta t}{\Delta x}\left(f^a_{i+\frac{1}{2}} - f^a_{i-\frac{1}{2}}\right), \\ b_i^{n+1} &= b_i^n - \tfrac{\Delta t}{\Delta x}\left(f^b_{i+\frac{1}{2}} - f^b_{i-\frac{1}{2}}\right). \end{aligned}\right\} \tag{5.10}$$

It remains to define the numerical fluxes $f^{a,b}_{i+\frac{1}{2}}$. To do this, we define a mean value of a and b at the interface $x_{i+\frac{1}{2}}$ by

$$\left.\begin{aligned} a^*_{i+\frac{1}{2}} &= \tfrac{a_{i+1}+a_i}{2}, \\ b^*_{i+\frac{1}{2}} &= \tfrac{b_{i+1}+b_i}{2}, \end{aligned}\right\} \tag{5.11}$$

and write these fluxes

$$\left.\begin{aligned} f^a_{i+\frac{1}{2}} &= a^+_{i+\frac{1}{2}} a^-_{i+\frac{1}{2}} b^*_{i+\frac{1}{2}}, \\ f^b_{i+\frac{1}{2}} &= b^+_{i+\frac{1}{2}} b^-_{i+\frac{1}{2}} a^*_{i+\frac{1}{2}}. \end{aligned}\right\} \tag{5.12}$$

We now notice that a^+ and b^+ belong to $[0, 2]$ and are therefore positive, while a^- and b^- belong to $[-2, 0]$ so that they are negative. A way to ensure stability of the scheme is to upwind $a^{\pm}_{i+\frac{1}{2}}$ and $b^{\pm}_{i+\frac{1}{2}}$ according to the sign of $b^*_{i+\frac{1}{2}}$ and $a^*_{i+\frac{1}{2}}$ respectively.

If $b^*_{i+\frac{1}{2}}$ is positive, we have

$$a^+_{i+\frac{1}{2}} b^*_{i+\frac{1}{2}} \geq 0,$$

$$a^-_{i+\frac{1}{2}} b^*_{i+\frac{1}{2}} \leq 0.$$

It therefore seems natural to upwind $a^-_{i+\frac{1}{2}}$ on the left and $a^+_{i+\frac{1}{2}}$ on the right. On the contrary, if $b^*_{i+\frac{1}{2}}$ is negative, we have

$$a^+_{i+\frac{1}{2}} b^*_{i+\frac{1}{2}} \leq 0,$$

$$a^-_{i+\frac{1}{2}} b^*_{i+\frac{1}{2}} \geq 0,$$

and it seems natural to upwind $a^-_{i+\frac{1}{2}}$ on the right and $a^+_{i+\frac{1}{2}}$ on the left.

The numerical scheme is therefore given by (5.10)–(5.12) and

$$a^+_{i+\frac{1}{2}} = \begin{cases} a_{i+1} \text{ if } b^*_{i+\frac{1}{2}} \geq 0, \\ a_i \text{ otherwise,} \end{cases} \qquad b^+_{i+\frac{1}{2}} = \begin{cases} b_{i+1} \text{ if } a^*_{i+\frac{1}{2}} \geq 0, \\ b_i \text{ otherwise,} \end{cases}$$
$$\text{and} \tag{5.13}$$
$$a^-_{i+\frac{1}{2}} = \begin{cases} a_i \text{ if } b^*_{i+\frac{1}{2}} \geq 0, \\ a_{i+1} \text{ otherwise,} \end{cases} \qquad b^-_{i+\frac{1}{2}} = \begin{cases} b_i \text{ if } a^*_{i+\frac{1}{2}} \geq 0, \\ b_{i+1} \text{ otherwise.} \end{cases}$$

We now state the following result whose proof will be found in Appendix A.

Proposition 5.2. *Under the CFL condition*

$$\max_i \left(|a^*_{i+\frac{1}{2}}| + |a^*_{i-\frac{1}{2}}|, |b^*_{i+\frac{1}{2}}| + |b^*_{i-\frac{1}{2}}|\right) \frac{\Delta t}{\Delta x} \leq \frac{1}{2}, \tag{5.14}$$

the scheme (5.10)–(5.13) *satisfies the following property:*

$$(a^n_i, b^n_i) \in \overline{\mathcal{D}} \text{ for all } j \in \mathbb{Z} \implies (a^{n+1}_i, b^{n+1}_i) \in \overline{\mathcal{D}} \text{ for all } j \in \mathbb{Z}.$$

5.2.3 Numerical results. We present some numerical results obtained with this scheme to illustrate the behavior of solutions of the 2×2 model. The two test-cases are taken from [4] where a detailed analysis of the solutions is described.

A Riemann problem with linear degeneracy. The initial conditions are

$$a = 0.3, \quad b = 0.9 \quad \text{if } x \in [0, 0.5],$$

$$a = 0.9, \quad b = 0.1 \quad \text{if } x \in [0.5, 1].$$

The computation is done with 10000 cells and the final time is 0.3. The solution is composed of a first classical shock plus a shock attached to a rarefaction fan. The attached shock is located at $a = b$ which is the locus of local linear degeneracy of the non linear wave (see [4] for further details).

A Riemann problem starting from the boundary. We take for initial conditions

$$a = -0.7, \quad b = 1 \quad \text{if } x \in [0, 0.5],$$

$$a = 0.4, \quad b = 0.8 \quad \text{if } x \in [0.5, 1].$$

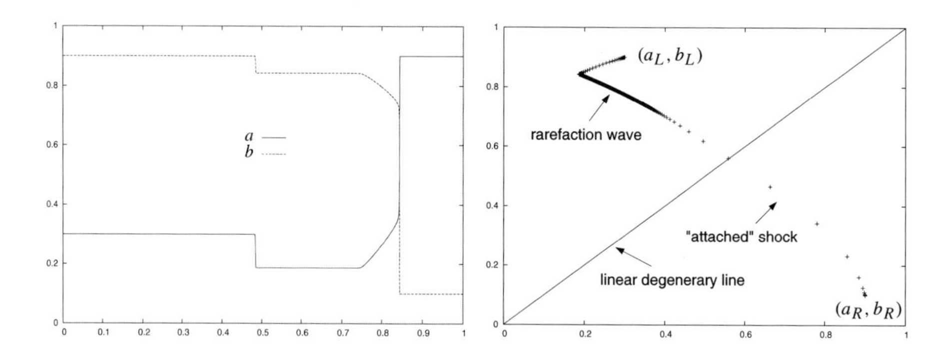

Figure 2. A Riemann problem with linear degeneracy (figure on the right shows the solution (a, b) in the phase plane).

The computation is done with 10000 cells and the final time is 0.2. As stated in Proposition 5.2, all points are inside the square $[-1, 1]^2$.

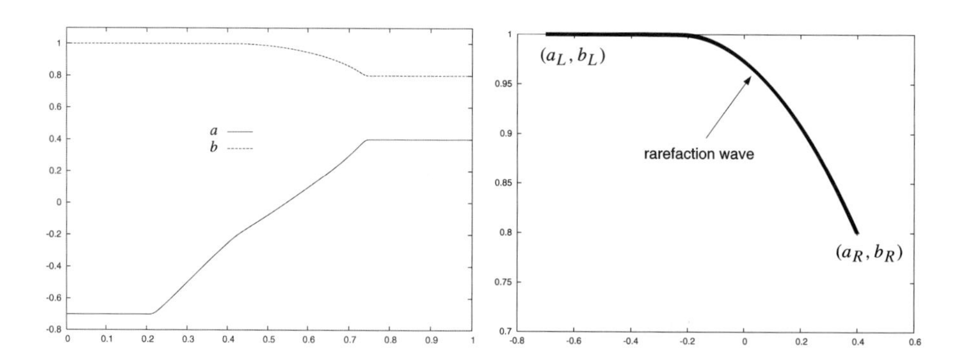

Figure 3. A Riemann problem starting from the boundary (figure on the right shows the solution (a, b) in the phase plane).

5.3 Numerical scheme for the *purely bifluid part*

5.3.1 A numerical scheme for the mass fractions equations. Now we apply the scheme described in the preceding paragraph to the equations of evolution of the mass fractions, namely

$$\left.\begin{aligned} D_t(\alpha_1) + D_m(\rho w \alpha_1 \alpha_2) = 0, \\ D_t(\alpha_2) - D_m(\rho w \alpha_1 \alpha_2) = 0. \end{aligned}\right\} \tag{5.15}$$

This scheme must also satisfy the stability conditions

$$(\alpha_i)_j \in [0,1],\ i \in \{1,2\}, \quad \text{and} \quad (\alpha_1)_j + (\alpha_2)_j = 1 \quad \text{for all } j \in \mathbb{Z}. \tag{5.16}$$

The scheme we propose for (5.15) can be written in the following form:

$$\left.\begin{aligned} &\frac{(\alpha_1)_j^{n+1} - (\alpha_1)_j^n}{\Delta t} + \frac{1}{\Delta m_j}\left(f^{\alpha_1}_{j+\frac12} - f^{\alpha_1}_{j-\frac12}\right) = 0, \\ &\frac{(\alpha_2)_j^{n+1} - (\alpha_2)_j^n}{\Delta t} + \frac{1}{\Delta m_j}\left(f^{\alpha_2}_{j+\frac12} - f^{\alpha_2}_{j-\frac12}\right) = 0, \end{aligned}\right\} \tag{5.17}$$

with

$$f^{\alpha_1}_{j+\frac12} = \rho_{j+\frac12} w_{j+\frac12} (\alpha_1)_{j+\frac12} (\alpha_2)_{j+\frac12} = -f^{\alpha_2}_{j+\frac12}. \tag{5.18}$$

We denote by

$$v_{j+\frac12} = (\rho w)_{j+\frac12}. \tag{5.19}$$

Following the ideas of paragraph 5.2.2, we upwind α_1 and α_2 according to the sign of $v_{j+\frac12}$:

$$(\alpha_1)_{j+\frac12} = \begin{cases} (\alpha_1)_j & \text{if } v_{j+\frac12} \geq 0, \\ (\alpha_1)_{j+1} & \text{if } v_{j+\frac12} < 0, \end{cases} \tag{5.20}$$

and

$$(\alpha_2)_{j+\frac12} = \begin{cases} (\alpha_2)_{j+1} & \text{if } v_{j+\frac12} \geq 0, \\ (\alpha_2)_j & \text{if } v_{j+\frac12} < 0. \end{cases} \tag{5.21}$$

Therefore, we have

Proposition 5.3. *Under the CFL condition*

$$\max_{j\in\mathbb{Z}} \left(|v_{j+\frac12}| + |v_{j-\frac12}|\right) \frac{\Delta t}{\Delta m_j} \leq 1, \tag{5.22}$$

the scheme (5.17)–(5.21) *satisfies:*

i) $(\alpha_i)_j^n \in [0,1]$ *for all* $j \in \mathbb{Z}$, $i = 1,2 \Rightarrow (\alpha_i)_j^{n+1} \in [0,1]$ *for all* $j \in \mathbb{Z}$, $i = 1,2$.

ii) $(\alpha_1)_j^n + (\alpha_2)_j^n = 1$ *for all* $j \in \mathbb{Z} \Rightarrow (\alpha_1)_j^{n+1} + (\alpha_2)_j^{n+1} = 1$ *for all* $j \in \mathbb{Z}$.

The proof of point i) is the same as the one for Proposition 5.2 (see Appendix A), and the proof of point ii) is immediate since the numerical fluxes for α_1 and α_2 are of opposite sign.

5.3.2 A numerical scheme for the τw equation. We cannot directly apply the same ideas on the (τw) equation because of the part of the chemical potentials which were not taken into account in the 2×2 model (see the assumptions made at the end of Section 2) and a more detailed analysis needs to be achieved in order to do so.

In a first step we have decided to discretize the equation

$$D_t(\tau w) + D_m\big(\mu_1 - \mu_2 - \frac{1 - 2\alpha_2}{2} w^2\big) = 0 \tag{5.23}$$

with a Lax–Friedrichs scheme which writes

$$\frac{(\tau w)_j^{n+1} - (\tau w)_j^n}{\Delta t} + \frac{1}{\Delta m_j}\big(f_{j+\frac{1}{2}}^{\tau w} - f_{j-\frac{1}{2}}^{\tau w}\big) = 0, \tag{5.24}$$

where

$$f_{j+\frac{1}{2}}^{\tau w} = \frac{f_{j+1}^{\tau w} + f_j^{\tau w}}{2} - \frac{1}{2}\frac{\Delta m_j}{\Delta t}\left((\tau w)_{j+1} - (\tau w)_j\right), \tag{5.25}$$

with $f^{\tau w} = \mu_1 - \mu_2 - \frac{1-2\alpha_2}{2} w^2$.

5.4 Numerical scheme for the total energy equation

Now it remains to discretize the total energy equation

$$D_t e + D_m f^e = 0, \tag{5.26}$$

with

$$f^e = \big(p + \rho\alpha_1\alpha_2 w^2\big)u + \rho\alpha_1\alpha_2 w\big(\mu_1 - \mu_2 - \frac{1 - 2\alpha_2}{2} w^2\big). \tag{5.27}$$

Noticing that $f^e = -\,(f^\tau.f^u + f^{\alpha_2}.f^{\tau w})$, we discretize (5.26) the following way

$$\frac{e_j^{n+1} - e_j^n}{\Delta t} + \frac{1}{\Delta m_j}\big(f_{j+\frac{1}{2}}^e - f_{j-\frac{1}{2}}^e\big) = 0, \tag{5.28}$$

where $f_{j+\frac{1}{2}}^e = -\big(f_{j+\frac{1}{2}}^\tau . f_{j+\frac{1}{2}}^u + f_{j+\frac{1}{2}}^{\alpha_2} . f_{j+\frac{1}{2}}^{\tau w}\big)$, using respectively the numerical fluxes (5.5), (5.6), (5.18) and (5.25) for $f_{j+\frac{1}{2}}^\tau$, $f_{j+\frac{1}{2}}^u$, $f_{j+\frac{1}{2}}^{\alpha_2}$ and $f_{j+\frac{1}{2}}^{\tau w}$.

6 Numerical results

Here we give some numerical results to illustrate the interest of the scheme. In most of them the numerical strategy consists in solving the equations on a fixed grid, first by the Lagrangian step previously described, and then by a projection back onto the initial grid (*Lagrangian + remapping scheme*). We refer the reader for instance to [7] for the general procedure of projection, and to [6] for more advanced non dissipative algorithms (not used here).

When no further precisions will be given, the simulations are run with 1000 cells.

6.1 Weak shock

We first consider a weak shock in a homogeneous medium. Both fluids are described with a perfect gas law $p = (\gamma - 1)\rho\varepsilon$ and $\varepsilon_i = C_{vi}T$ with

$$\left.\begin{aligned} \gamma_1 = 1.4, \quad C_{v1} = 1, \\ \gamma_1 = 1.8, \quad C_{v2} = 1. \end{aligned}\right\} \tag{6.1}$$

Some mixing entropy is added choosing $k = 0.1$. The initial discontinuous state is defined by

$$\left.\begin{aligned} \rho_L = 1, \quad p_L = 0.12, \quad u_L = 0, \quad c_{1,L} = 0.5, \\ \rho_R = 1, \quad p_R = 0.1, \quad u_R = 0, \quad c_{1,R} = 0.5. \end{aligned}\right\} \tag{6.2}$$

Results at time $T = 0.5$ are shown in Figure 4. There are not any simple analytical reference solutions, but the results are in accordance with those in [2]. Furthermore this scheme retrieves the fact that behind the shock propagating to the right, the velocities satisfy $u_2 \approx 2u_1$, which has already been predicted in [2] by an analysis in the asymptotic regime of a weak shock. Eventually note that the pike in the mass fraction around the contact discontinuity (after $x = 0.5$) results from the projection step that needs the product between the two discontinuous variables ρ and α_2.

6.2 Strong shock

We now consider a strong shock in an inhomogeneous medium. We still suppose that both fluids are perfect gases with

$$\left.\begin{aligned} \gamma_1 = 1.4, \quad C_{v1} = 1, \\ \gamma_1 = 1.8, \quad C_{v2} = 0.6. \end{aligned}\right\} \tag{6.3}$$

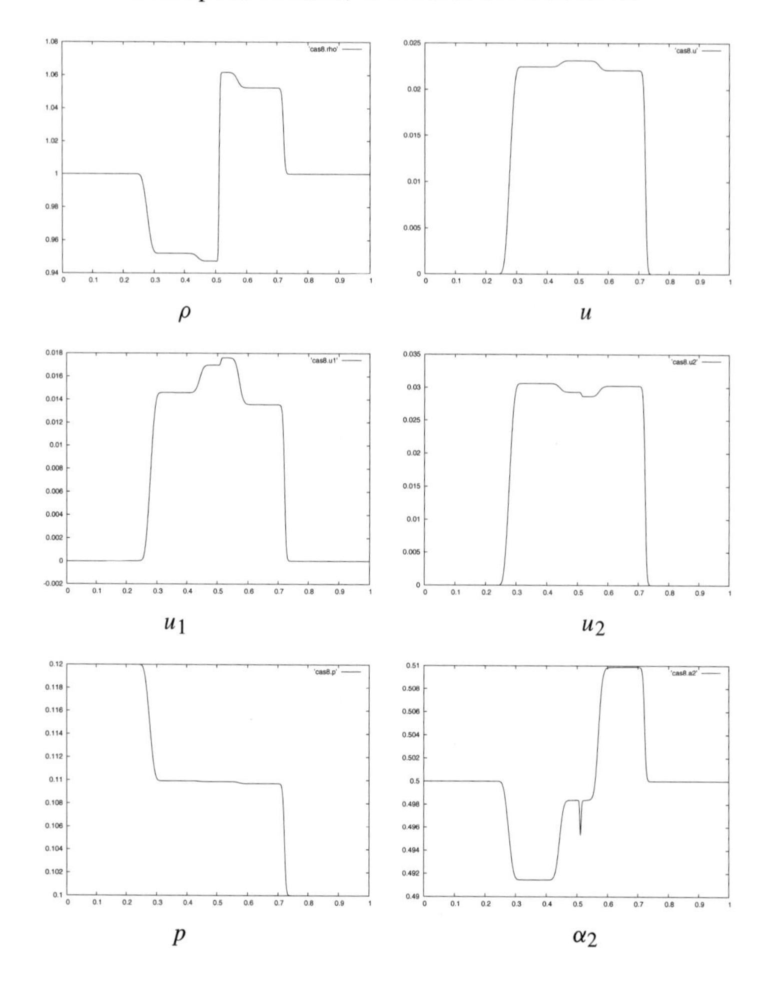

Figure 4. Weak shock, $T = 0.5$, Lagrange + projection.

We set $k = 0.4$ and initiate the simulation like the classical Sod shock tube problem:

$$\left.\begin{array}{llll} \rho_L = 1, & p_L = 1, & u_L = 0, & c_{1,L} = 0.5, \\ \rho_R = 0.125, & p_R = 0.1, & u_R = 0, & c_{1,R} = 0.5. \end{array}\right\} \tag{6.4}$$

The results at time $T = 0.14$ in Figure 5 show the capture of the right waves: a rarefaction fan propagating to the left, a contact discontinuity near $x = 0.6$ and two shocks propagating to the right. Differences with the Sod shock profiles are essentially visible for $x \in [0.73, 0.78]$; this corresponds to a large ratio between u_1 and u_2.

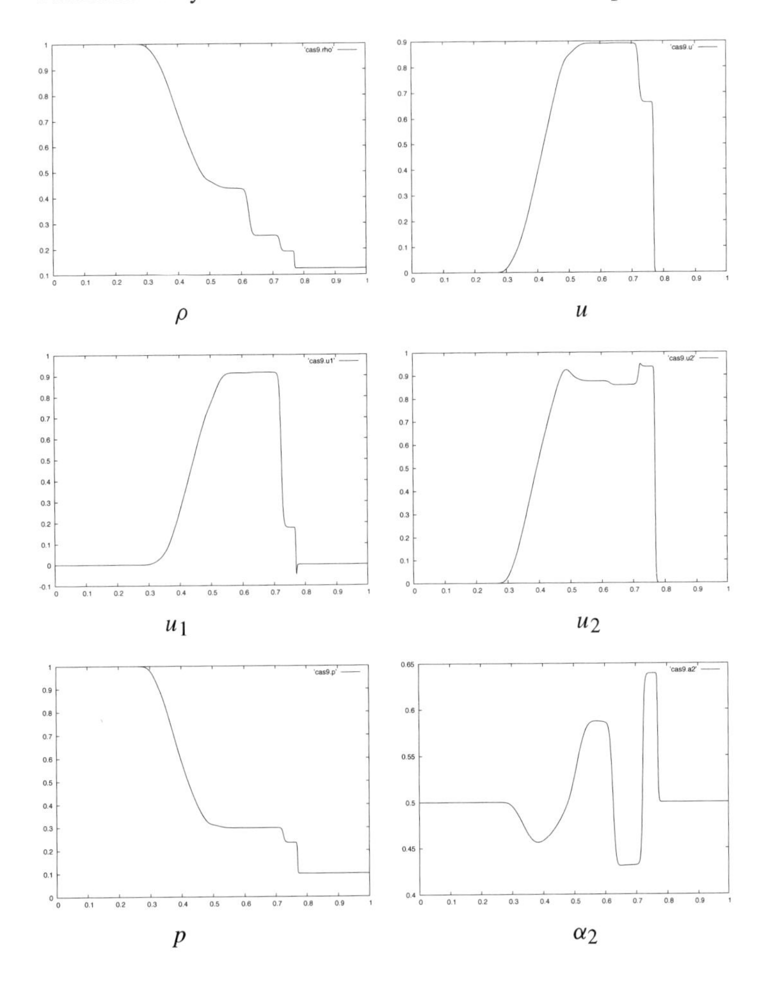

Figure 5. Strong shock, $T = 0.14$, Lagrange + projection.

6.3 Strong shock with drag force

This test has the same parameter than the previous one but we add in our system a *drag force* in the equation for w. More precisely, we solve the following system:

$$\left.\begin{aligned}
&D_t(\tau) - D_m(u) = 0,\\
&D_t(\alpha_2) - D_m\big(\tfrac{w(1-\alpha_2)\alpha_2}{\tau}\big) = 0,\\
&D_t(u) + D_m(P) = 0,\\
&D_t(\tau w) + D_m\big(\mu_1 - \mu_2 - \tfrac{1-2\alpha_2}{2} w^2\big) = -\tfrac{1}{\kappa}\tau w,\\
&D_t(e) + D_m\Big(Pu + \tfrac{1}{\tau} w(1-\alpha_2)\alpha_2\Big(\mu_1 - \mu_2 - \tfrac{1-2\alpha_2}{2} w^2\Big)\Big) = 0.
\end{aligned}\right\} \tag{6.5}$$

This new term is expected to relax the difference of velocities to zero. We present in Figure 6 the numerical solution for $\kappa = 0.01$. Compared to the previous case, the slower shock has almost disappeared and both velocities are almost equal everywhere.

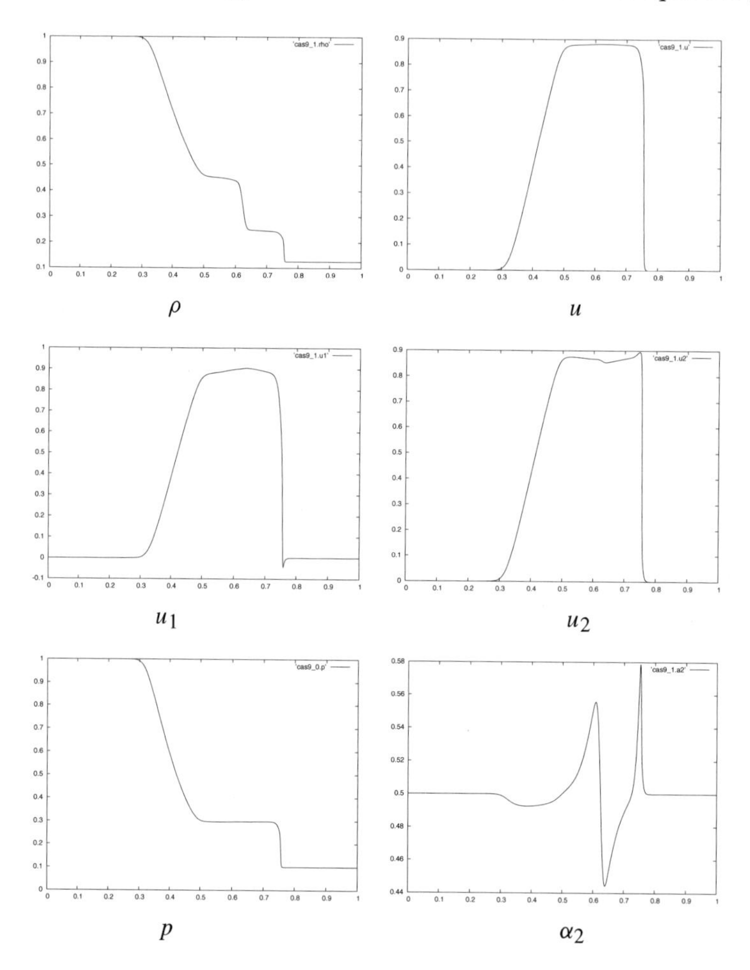

Figure 6. Strong shock with drag force, $\kappa = 0.01$, $T = 0.14$, Lagrange + projection.

6.4 Strong shock with infinite drag force

In this test we solve the system with an infinite drag force for the difference $w = u_1 - u_2$: formally, we take $\kappa = 0$ in (6.5).

We easily obtain that $w = 0$, and therefore that $u_1 = u_2$. If we start from the initial state of Subsection 6.2, we also have that $\alpha_1 = \alpha_2 = 0.5$. In fact it is an easy matter to check that this test case is equivalent to the monofluid classical Sod shock tube problem with $\gamma = \frac{\alpha_1 C_{v1}\gamma_1 + \alpha_2 C_{v2}\gamma_2}{\alpha_1 C_{v1} + \alpha_2 C_{v2}}$. Numerically, the scheme indeed captures this solution, without oscillations in the mass fraction as in [2].

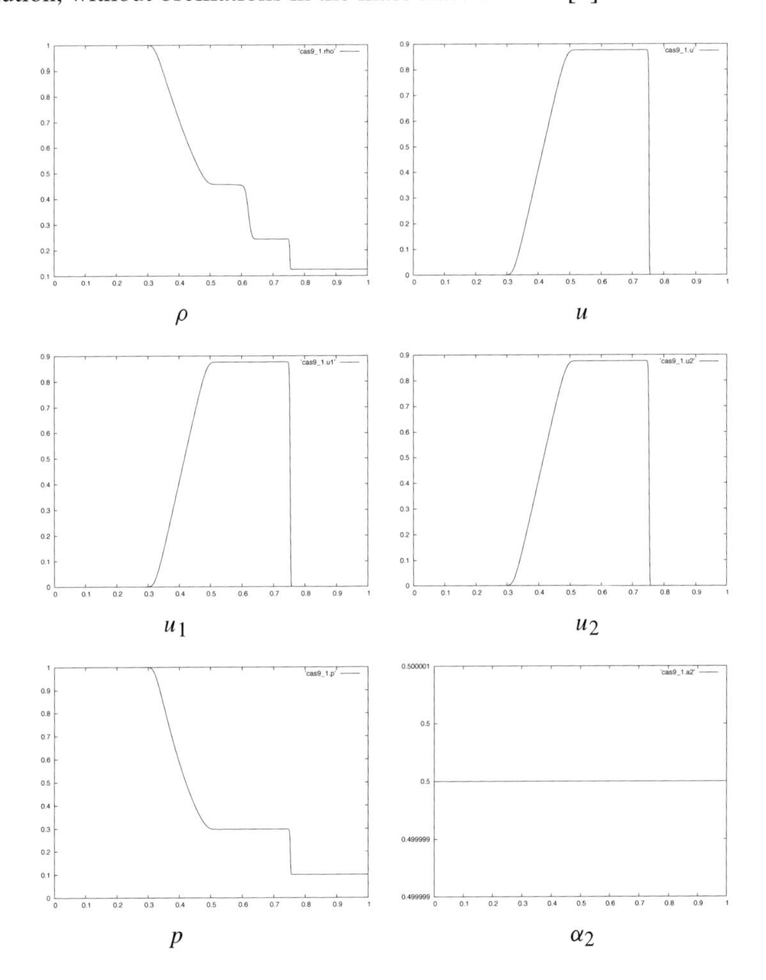

Figure 7. Strong shock with infinite drag force, $\kappa \to 0$, $T = 0.14$, Lagrange + projection.

6.5 A water-air Riemann problem

We now consider the case where one fluid (fluid number 2, the water) is no more described by a perfect gas law but by a stiffened-gas equation of state (2.9)–(2.11) with

$$\left.\begin{array}{lll} \gamma_1 = 1.4, & C_{v1} = 1, & \pi_1 = 0, \\ \gamma_2 = 4.4, & C_{v2} = 0.1, & \pi_2 = 6\ 10^6. \end{array}\right\} \qquad (6.6)$$

This test case is given in [9]. It supposes that at initial time the water is at the left of the tube and the air at the right with

$$\left.\begin{array}{llll}\rho_L = 1000, & p_L = 10^9, & u_L = 0, & \alpha_{1,L} = 0,\\ \rho_R = 50, & p_R = 10^5, & u_R = 0, & \alpha_{1,R} = 1.\end{array}\right\} \tag{6.7}$$

In this test problem (and in the following), in order to recover all the quantities of the system from α_2, τ, u, w and e, we need to solve a second order equation for the pressure p of the form

$$(\alpha_1 C_{v1} + \alpha_2 C_{v2}) p^2 + \eta_1 p + \eta_2 = 0, \tag{6.8}$$

where η_1 and η_2 are two coefficients depending on the conservative quantities; this calculation is given in Appendix B.

The results in Figure 8 are in accordance with those of [9] and [2]; in particular the interface between air and water is respected, no mixing appears. Furthermore the positivity of mass fraction (what here corresponds to α_i exactly equal to 0 or 1) has been obtained without artificial technique.

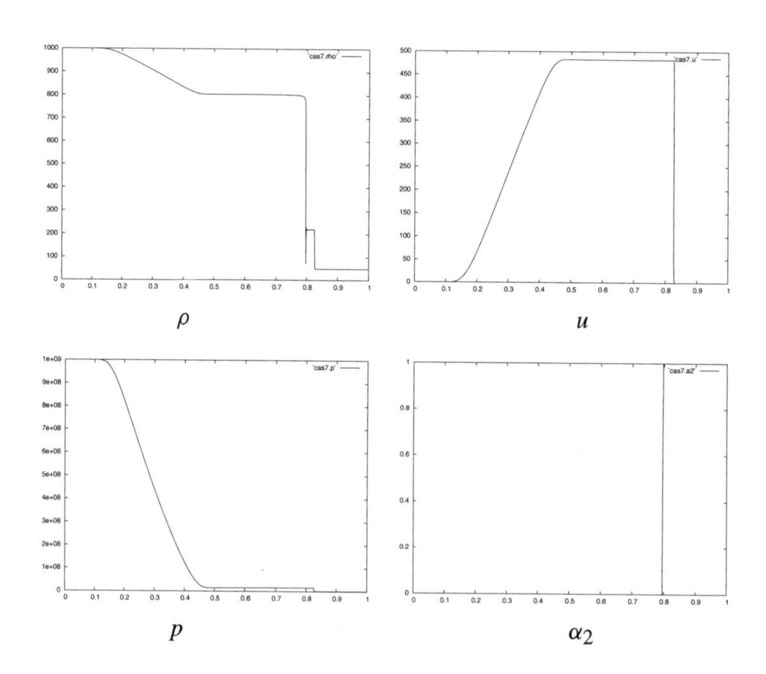

Figure 8. Water-air tube shock problem, $T = 0.0002$, Lagrange.

6.6 The Ransom faucet problem

This test is more classical than the previous one and still considers the case of water-air problem. More precisely, we consider a vertical tube where a vertical jet of water is falling under the action of the gravity. In the tube, the water (indexed by 2) is always surrounded by the air (indexed by 1).

For the initial state, we consider that the water fills a uniform column and is surrounded by a stagnant air. The coefficients are

$$\left.\begin{array}{llll} \rho_1 = 1.286, & p_1 = 10^5, & u_1 = 0, & \alpha_1 \approx 0.000321, \\ \rho_2 = 1000, & p_2 = 10^5, & u_2 = 10, & \alpha_2 \approx .999679. \end{array}\right\} \tag{6.9}$$

The mass fractions have been chosen in order to get an exact gas volume fraction $\beta_1 = 0.2$. The boundary conditions are specified velocities of $u_1 = 0$ for the air and $u_2 = 10$ for the water at the inlet and a pressure of 10^5 at the outlet.

First simulations on coarse grids (see Figure 9 with 100, 200, 400 and 800 cells) capture the same solution than other approaches (see [9] for example concerning the mass fraction). There must remain a problem in the left boundary condition for the

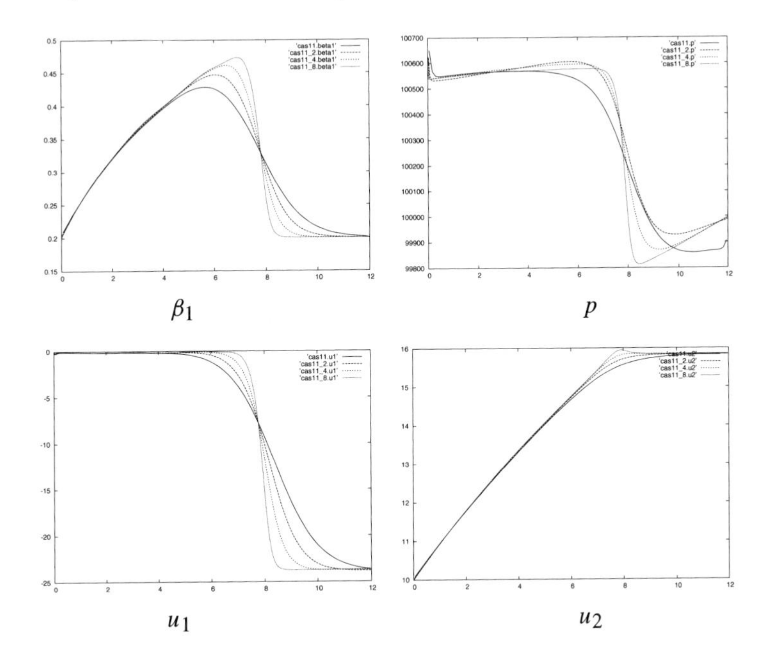

Figure 9. Ransom problem, $T = 0.6$, $\{100 * 2^i\}_{0 \le i \le 3}$ cells, Lagrange + projection.

pressure, but the rest of the profile is right, without oscillations. When refining the mesh (see Figure 10 with 1600 cells), an undershoot appears at the foot of mass fraction, pressure and velocity of the water, pointing out the instability of the scheme.

In fact, more than a numerical problem, this phenomenon is related in all likelihood to the lack of hyperbolicity of models involving pressure equality between phases (see [10]). The scheme thus cannot be stable, since the model develops imaginary wave velocities.

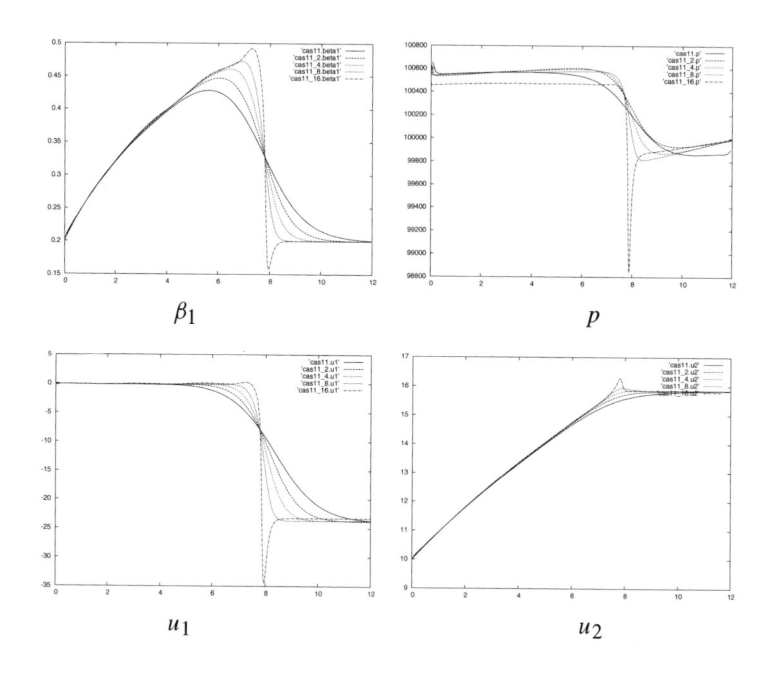

Figure 10. Ransom problem, $T = 0.6$, $\{100 * 2^i\}_{0 \leq i \leq 4}$ cells, Lagrange + projection.

7 Conclusion

This study shows that it is possible to compute some basic two-phase benchmarks (such as the Ransom faucet problem) with a fully conservative and positive *Lagrange + Projection* numerical scheme.

The model studied here has been derived by choosing isobar and isothermal closures, but we emphasize that the temperature of the model has to be understood as an *adhoc* closure and has *a priori* no physical interpretation. Nevertheless, a drawback of this model is that it explicitly needs the entropy in the definition of the fluxes (through the chemical potentials), which is only possible for simple equations of state, like stiffened gas laws. Other closures, as for instance

$$\frac{D_t\tau_1}{\tau_1} = \frac{D_t\tau_2}{\tau_2},$$
$$T_1 D_t S_1 = T_2 D_t S_2,$$

are an alternative, since it can be shown that the chemical potentials are replaced by enthalpies in the conservative equation of differential velocity. Furthermore they are more realistic from the physical point of view. Numerical methods based on the one presented here are being examined for such extensions.

A Proof of Proposition 5.2

Notice that the first equation of (5.7) can be rewritten

$$\partial_t(a+1) + \partial_x f^a = 0.$$

We use this form to show that under the CFL condition (5.14), $a+1$ is positive. Suppose that $(a^n, b^n) \in \overline{\mathcal{D}}$ so that $|a_i^n - 1| \leq 2$ for all i. Four cases are to be investigated.

1) $b^*_{i+\frac{1}{2}} \geq 0$ and $b^*_{i-\frac{1}{2}} \geq 0$.

The scheme writes

$$\begin{aligned}(a+1)_i^{n+1} &= (a+1)_i^n\left(1 + \frac{\Delta t}{\Delta x} b^*_{i-\frac{1}{2}}(a-1)^n_{i-1}\right)\\ &\quad - \frac{\Delta t}{\Delta x} b^*_{i+\frac{1}{2}}(a-1)_i^n (a+1)^n_{i+1},\\ &\geq 0 \quad \text{if } 1 + \frac{\Delta t}{\Delta x} b^*_{i-\frac{1}{2}}(a-1)^n_{i-1} \geq 0.\end{aligned}$$

This condition is satisfied if

$$\frac{\Delta t}{\Delta x}|b^*_{i-\frac{1}{2}}| \leq \frac{1}{2}.$$

2) $b^*_{i+\frac{1}{2}} \geq 0$ and $b^*_{i-\frac{1}{2}} \leq 0$.

The scheme writes

$$\begin{aligned}(a+1)_i^{n+1} &= (a+1)_i^n - \frac{\Delta t}{\Delta x} b^*_{i+\frac{1}{2}}(a-1)_i^n (a+1)^n_{i+1}\\ &\quad + \frac{\Delta t}{\Delta x} b^*_{i-\frac{1}{2}}(a-1)_i^n (a+1)^n_{i-1},\\ &\geq 0.\end{aligned}$$

3) $b^*_{i+\frac{1}{2}} \le 0$ and $b^*_{i-\frac{1}{2}} \le 0$.

The scheme writes

$$(a+1)_i^{n+1} = (a+1)_i^n \left(1 - \frac{\Delta t}{\Delta x} b^*_{i+\frac{1}{2}} (a-1)_{i+1}^n\right) + \frac{\Delta t}{\Delta x} b^*_{i-\frac{1}{2}} (a-1)_i^n (a+1)_{i-1}^n,$$

$$\ge 0 \quad \text{if } 1 - \frac{\Delta t}{\Delta x} b^*_{i+\frac{1}{2}} (a-1)_{i+1}^n \ge 0.$$

This condition is satisfied if

$$\frac{\Delta t}{\Delta x} |b^*_{i+\frac{1}{2}}| \le \frac{1}{2}.$$

4) $b^*_{i+\frac{1}{2}} \le 0$ and $b^*_{i-\frac{1}{2}} \ge 0$.

The scheme writes

$$(a+1)_i^{n+1} = (a+1)_i^n \left(1 - \frac{\Delta t}{\Delta x} b^*_{i+\frac{1}{2}} (a-1)_{i+1}^n + \frac{\Delta t}{\Delta x} b^*_{i-\frac{1}{2}} (a-1)_{i-1}^n\right),$$

$$\ge 0 \quad \text{if } 1 - \frac{\Delta t}{\Delta x} b^*_{i+\frac{1}{2}} (a-1)_{i+1}^n + \frac{\Delta t}{\Delta x} b^*_{i-\frac{1}{2}} (a-1)_{i-1}^n \ge 0.$$

This condition is satisfied if

$$\frac{\Delta t}{\Delta x} \left(|b^*_{i-\frac{1}{2}}| + |b^*_{i+\frac{1}{2}}|\right) \le \frac{1}{2}.$$

These conditions are satisfied under the most restrictive one, that is under the CFL condition (5.14).

Notice that the first equation of (5.7) could also be rewritten

$$\partial_t (a-1) + \partial_x f^a = 0.$$

The same kind of proof applied on this equation would also show that under the CFL condition (5.14), $a-1$ is negative, which shows that for all $i \in \mathbb{Z}$, $a_i^{n+1} \in \overline{\mathcal{D}}$. Since the 2×2 system (5.7) is symmetric in a and b, the same result holds on b, which ends the proof.

B Determination of the pressure from the conservative variables for a stiffened-gas law

We want to compute the pressure with respect to the conservative variables τ, α_2, u, w and e for stiffened gas laws. Since both phases are in pressure and temperature equilibrium[2], we have

$$p = (\gamma_1 - 1)\frac{\varepsilon_1}{\tau_1} - \gamma_1\pi_1 = (\gamma_2 - 1)\frac{\varepsilon_2}{\tau_2} - \gamma_2\pi_2$$

and

$$T = \frac{\varepsilon_1 - \pi_1\tau_1}{C_{v1}} = \frac{\varepsilon_2 - \pi_2\tau_2}{C_{v2}},$$

which implies that

$$p = (\gamma_1 - 1)\frac{C_{v1}T}{\tau_1} - \pi_1 = (\gamma_2 - 1)\frac{C_{v2}T}{\tau_2} - \pi_2.$$

Hence we obtain $\tau_i, i = 1, 2$, from T and p:

$$\tau_i = (\gamma_i - 1)\frac{C_{vi}T}{p + \pi_i}.$$

Therefore, by definition of τ we deduce that

$$\alpha_1(\gamma_1 - 1)\frac{C_{v1}T}{p + \pi_1} + \alpha_2(\gamma_2 - 1)\frac{C_{v2}T}{p + \pi_2} = \tau. \tag{2.1}$$

On the other hand, we have that

$$\varepsilon_i = \frac{p + \gamma_i\pi_i}{\gamma_i - 1}\tau_i = \frac{p + \gamma_i\pi_i}{p + \pi_i}C_{vi}T$$

and by using the definition of ε we get

$$\alpha_1\frac{p + \gamma_1\pi_1}{p + \pi_1}C_{v1}T + \alpha_2\frac{p + \gamma_2\pi_2}{p + \pi_2}C_{v2}T = \varepsilon. \tag{2.2}$$

The equations (2.1) and (2.2) yield

$$\frac{\tau}{\varepsilon} = \frac{\alpha_1(\gamma_1 - 1)\frac{C_{v1}T}{p+\pi_1} + \alpha_2(\gamma_2 - 1)\frac{C_{v2}T}{p+\pi_2}}{\alpha_1\frac{p+\gamma_1\pi_1}{p+\pi_1}C_{v1}T + \alpha_2\frac{p+\gamma_2\pi_2}{p+\pi_2}C_{v2}T}$$

and thus

$$\frac{\tau}{\varepsilon} = \frac{\alpha_1(\gamma_1 - 1)C_{v1}(p + \pi_2) + \alpha_2(\gamma_2 - 1)C_{v2}(p + \pi_1)}{\alpha_1(p + \gamma_1\pi_1)(p + \pi_2)C_{v1} + \alpha_2(p + \gamma_2\pi_2)(p + \pi_1)C_{v2}}.$$

Consequently, the pressure p is a solution of the second order equation

$$(\alpha_1C_{v1} + \alpha_2C_{v2})p^2 + \eta_1p + \eta_2 = 0 \tag{2.3}$$

[2] By misnomer, see footnote 1, page 181.

with

$$\eta_1 = \big[\alpha_1 C_{v1}(\gamma_1\pi_1 + \pi_2) + \alpha_2 C_{v2}(\gamma_2\pi_2 + \pi_1) \\ - \frac{\varepsilon}{\tau}\big(\alpha_1 C_{v1}(\gamma_1 - 1) + \alpha_2 C_{v2}(\gamma_2 - 1)\big)\big]$$

and

$$\eta_2 = (\alpha_1 C_{v1}\gamma_1 + \alpha_2 C_{v2}\gamma_2)\,\pi_1\pi_2 - \frac{\varepsilon}{\tau}\,(\alpha_1 C_{v1}(\gamma_1 - 1)\pi_2 + \alpha_2 C_{v2}(\gamma_2 - 1)\pi_1)\,.$$

We notice that if $\pi_2 = 0$, which is the case of a water-air problem, we have that

$$\eta_2 = -\frac{\varepsilon}{\tau}\alpha_2 C_{v2}(\gamma_2 - 1)\pi_1 \leqslant 0,$$

and so there is only one positive solution p of (2.3).

References

[1] F. Coquel, K. El. Amine, E. Godlewski, B. Perthame and P. Rascle, A numerical method using upwind schemes for the resolution of two-phase flows, *Int. J. Multiphase Flow* 12 (6) (1986), 861–889.

[2] B. Després, *Conservation laws for compressible multiphase flows with shocks*, Publications du Laboratoire d'Analyse Numérique, Université Paris 6, France, 2001.

[3] B. Després, Lagrangian systems of conservation laws, *Numer. Math.* 89 (2001), 99–134.

[4] B. Després, F. Lagoutière and D. Ramos, Stability of a thermodynamically coherent multi-phase model, *Math. Mod. Meth. App. Sci* 13 (10) (2003), 1463–1487.

[5] F. De Vuyst, Stable and accurate hybrid finite volume methods based on pure convexity arguments for hyperbolic systems of conservation law, preprint.

[6] F. Lagoutière, Modélisation mathématique et résolution numérique de problèmes de fluides compressibles à plusieurs constituants, Doctoral dissertation, Thèse de l'Université Pierre et Marie Curie - Paris VI (2000).

[7] E. Godlewski and P. A. Raviart, *Numerical approximation of hyperbolic systems of conservation laws*, Appl. Math. Sci. 118, Springer-Verlag, New York 1996.

[8] L. Sainsaulieu, Finite volume approximations of two phase fluid flows based on approximate Roe-type Riemann solver, *J. Comput. Phys.* 121 (1995), 1–28.

[9] R. Saurel and R. Abgrall, A multiphase Godunov method for compressible multifluid and multiphase flows, *J. Comput. Phys.* 150 (1999), 425–467.

[10] Nicolas Seguin, Modélisation et simulation numérique des écoulements diphasiques, Doctoral dissertation, Thèse de l'Université de Provence (2002).

[11] H. B. Stewart and B. Wendroff, Two phase flows: models and methods, *J. Comput. Phys.* 56 (1984), 363–409.

[12] E. I. Romensky, Hyperbolic systems of thermodynamically compatible conseration laws in continuum mechanics, *Math. Comp. Modelling* 28 (10) (1998), 115–130.

[13] E. I. Romensky, Thermodynamics and hyperbolic systems of balance laws in continuum mechanics, in *Godunov methods: theory and applications* (E. F. Toro, ed.), Kluwer/Plenum, New York 2001, 745–761.

DINMOD: A diffuse interface model for two-phase flows modelling

F. Caro, F. Coquel, D. Jamet and S. Kokh

DEFA, ONERA Châtillon, 29 avenue de la Division Leclerc
92322 Châtillon Cedex, France
email: `florian.caro@onera.fr`

LJLL, Université Paris VI, 175 rue du Chevaleret, 75013 Paris, France
email: `coquel@ann.jussieu.fr`

DEN/DTP/SMTH/LDTA, CEA Grenoble, 17 rue des Martyrs
38054 Grenoble Cedex 9, France
email: `didier.jamet@cea.fr`

DEN/DM2S/SFME/LETR, CEA Saclay, 91191 Gif-sur-Yvette Cedex, France
email: `samuel.kokh@cea.fr`

Abstract. We present an isothermal model for liquid-vapor phase transition. We use a Hamilton principle to derive the conservative part of the system and we propose additional stiff phase change source terms that are compatible with the second thermodynamic principle. These terms are considered as relaxation terms and define an equilibrium manifold for the system. We formally study the equilibrium return process. While the original system is hyperbolic, the equilibrium limit system is only weakly hyperbolic. Finally, we propose a numerical scheme based on the relaxed system.

1 Introduction

With the rise of computing power the simulation of small scales phenomena is becoming an important challenge for the study of phase transitions. This aims at the description of physical processes involved in a wide range of industrial applications such as the boiling crisis prediction.

We adopt here a compressible diffuse interface model: the interface is defined thanks to the introduction of an abstract order parameter z. This parameter takes the value 1 (*resp.* 0) in phase 1 (*resp.* 2) denoting here the liquid (*resp.* vapor) phase. Therefore, the interface evolution is totally described by the function $(x, t) \mapsto z(x, t)$.

For a given interface position, the properties of the fluids are prescribed through the following lines: for points belonging to the interface, *i.e.* for (x, t) such that $0 < z(x, t) < 1$, fluids governing equations are coupled thanks to a non-classical thermodynamical description which allows mass transfer between phases. In pure fluids areas, each phase is subjected to its own classical thermodynamical Equation Of State (EOS).

We restrict our study to the case of isothermal flows and isothermal equilibrium between phases. In order to derive a thermodynamically consistent model including mass transfer we shall make use of the following lines: prescribe a free energy for the diffuse interface system (potential energy), retrieve the motion equations thanks to a Hamilton principle. The mathematical entropy production (free energy plus kinetic energy) is then constrained to be non-negative, which provides us with an entropy inequality. It also leads us to a dissipation compatible model for both mass transfer and evolution process for z.

The resulting system issued from our modelling task can be analyzed as a non-homogeneous hyperbolic system endowed with an entropy inequality that embraces the stiff source terms actions. Instead of a direct numerical discretization work we pursue the model study by looking into the equilibrium states manifolds connected to the source terms. We determine two such manifolds, the first one is related to a mechanical process, which we shall refer to as the “M-equilibrium”. The second one relates to a thermodynamical process occurring after the M-equilibrium. This shall be referred to as the “MT-equilibrium”. We shall see that the return to the MT-equilibrium process can be interpreted thanks to energy based arguments. This sheds some light on the MT-equilibrium system structures. Unfortunately this system fails to fit the classical hyperbolic well-posedness criteria. Nevertheless, the relaxed systems will allow us to design two numerical schemes: when tested against two simple simulations, these schemes display numerical stability and a coherent behaviour for the MT-equilibrium states when reaching space convergence.

2 System modelling

2.1 General modelling assumptions

The starting point of our roadmap is the postulate of a Lagrangian associated with the system. We introduce a few notations. Let ρ_α be the density of the phase $\alpha = 1, 2$. We define the system density ρ by

$$\rho = z\rho_1 + (1 - z)\rho_2.$$

Let $y_\alpha = \rho_\alpha z_\alpha / \rho$ be the phase $\alpha = 1, 2$ concentration, where $z_1 = z$, $z_2 = 1 - z$ and $y = y_1$. Fluids are supposed to have the same velocity $\boldsymbol{u}$ and we denote by $\boldsymbol{j} = \rho \boldsymbol{u}$ the fluid momentum. We first make the following general assumption concerning the system frcc energy f: we suppose that

$$f = f(\rho, y, z, \nabla z).$$

Let us note that we allow dependences on ∇z until applications in Section 3 by analogy with microstructure driven continuum mechanics ([2], [11]) such as capillarity effects in second gradient fluids theory.

Finally, we define the lagrangian density L for the two-phase system using the classical expression

$$L(\rho, \boldsymbol{j}, y, z, \nabla z) = \frac{1}{2}\rho |\boldsymbol{u}|^2 - \rho f = \frac{|\boldsymbol{j}|^2}{2\rho} - \rho f. \tag{2.1}$$

2.2 Lagrangian action minimization

We consider the motion of a volume of fluid Ω between two instants t_1 and $t_2 > t_1$. According to (2.1), this volume is subjected to transformations which are totally described by $(x, t) \mapsto (\rho, \boldsymbol{j}, y, z, \nabla z)(x, t)$. If we denote by X the lagrangian material reference coordinates and by x the eulerian coordinates, there exists a map χ such that $x = \chi(X, t)$.

Let $(x, t, \nu) \mapsto (\hat{\rho}, \hat{\boldsymbol{j}}, \hat{y}, \hat{z}, \widehat{\nabla z})(x, t, \nu)$ be a perturbation of given functions $(x, t) \mapsto (\rho, \boldsymbol{j}, y, z, \nabla z)(x, t)$ parametrized by a real coefficient ν. We add here some classical assumptions: the perturbated motion verifies mass conservation, for any physical quantity a, if $\hat{a}$ is its corresponding perturbed state, we suppose that $\hat{a}(x, t, \nu = 0) = a(x, t)$ and that $\hat{a}(x, t, \nu) = 0$ for $x \in \partial\Omega$.

Let us define the virtual displacement $\boldsymbol{\xi}$ associated to the perturbation

$$\boldsymbol{\xi}(X, t) = \left(\frac{\partial \hat{x}}{\partial \nu}\right)_{X,t} (X, t, \nu = 0).$$

For any physical quantity a we define the operator δ as follows:

$$\delta a(x, t) = \left(\frac{\partial \hat{a}}{\partial \nu}\right)_{\hat{x},t} (x, t, \nu = 0).$$

We shall now connect $\delta\rho$ and $\delta\boldsymbol{j}$ to $\boldsymbol{\xi}$ thanks to the mass conservation ansatz. This hypothesis reads

$$\left(\frac{\partial \hat{\rho}}{\partial \nu}\right)_{\hat{x},t} + \sum_i \frac{\partial}{\partial \hat{x}_i}\left[\hat{\rho}\left(\frac{\partial \hat{x}_i}{\partial \nu}\right)_{X,t}\right] = 0,$$

which immediately gives

$$\delta\rho = -\mathrm{div}(\rho\boldsymbol{\xi}). \tag{2.2}$$

We now turn to the expression of $\delta\boldsymbol{u}$. We have

$$\left(\frac{\partial\hat{\boldsymbol{u}}}{\partial\nu}\right)_{X,t}=\frac{\partial}{\partial\nu}\left(\hat{\boldsymbol{u}}\left[\hat{\chi}(X,t,\nu),t,\nu\right]\right)=\left(\frac{\partial\hat{\boldsymbol{u}}}{\partial\nu}\right)_{\hat{x},t}+\left(\frac{\partial\hat{\boldsymbol{u}}}{\partial\hat{x}}\right)_{\nu,t}\left(\frac{\partial\hat{\chi}}{\partial\nu}\right)_{X,t}. \tag{2.3}$$

The definition of $\hat{\boldsymbol{u}}$ reads $\hat{\boldsymbol{u}}(X,t,\nu)=(\partial\hat{\chi}/\partial t)_{X,\nu}$. Differentiating this expression according to t and using the definition of $\boldsymbol{\xi}$ gives

$$\left(\frac{\partial\hat{\boldsymbol{u}}}{\partial\nu}\right)_{X,t}(X,t,\nu=0)=\left(\frac{\partial\boldsymbol{\xi}}{\partial t}\right)_{X}(X,t)=\frac{\mathrm{d}\boldsymbol{\xi}}{\mathrm{d}t}(x,t),$$

where $\mathrm{d}/\mathrm{d}t=\partial/\partial t+\boldsymbol{u}\cdot\nabla$. Hence, by setting $\nu=0$ in (2.3), using the above relation we find

$$\delta\boldsymbol{u}=\frac{\mathrm{d}\boldsymbol{\xi}}{\mathrm{d}t}-\frac{\partial\boldsymbol{u}}{\partial x}\boldsymbol{\xi}.$$

As $\delta\boldsymbol{j}=\rho\delta\boldsymbol{u}+\boldsymbol{u}\delta\rho$ and by relation (2.2), we have

$$\delta\boldsymbol{j}=\frac{\partial}{\partial t}(\rho\boldsymbol{\xi})+\operatorname{div}(\boldsymbol{\xi}\otimes\boldsymbol{j}-\boldsymbol{j}\otimes\boldsymbol{\xi}), \tag{2.4}$$

where if $a=(a_i)_i$ and $b=(b_i)_i$ are two vectors, $\operatorname{div}(a\otimes b)$ is a vector defined as

$$\operatorname{div}(a\otimes b)_i=\sum_j\frac{\partial(a_ib_j)}{\partial x_j}.$$

In the following we will write

$$R=\frac{\partial L}{\partial\rho}. \tag{2.5}$$

We now define the Hamilton action associated to Ω between t_1 and t_2:

$$\mathcal{A}(\nu)=\int_{t_1}^{t_2}\int_{\Omega}L(\hat{\rho},\hat{\boldsymbol{j}},\hat{y},\hat{z},\widehat{\nabla z})\,\mathrm{d}x\,\mathrm{d}t.$$

According to the classical least action principle, $(\rho,\boldsymbol{j},y,z,\nabla z)$ satisfies relations that extremize $\mathcal{A}$ over the set of selected perturbations. Moreover, supposing that the virtual displacements of $\boldsymbol{\xi}$, δz and δy are regular functions, by differentiation according to ν we find that

$$\delta\mathcal{A}=\int_{\Omega\times[t_1,t_2]}\left\{\frac{\partial L}{\partial\boldsymbol{j}}\delta\boldsymbol{j}+\frac{\partial L}{\partial\rho}\delta\rho+\frac{\partial L}{\partial y}\delta y+\frac{\partial L}{\partial z}\delta z+\frac{\partial L}{\partial(\nabla z)}\delta(\nabla z)\right\}\mathrm{d}x\,\mathrm{d}t. \tag{2.6}$$

As differentiation according to ν at the constant $\hat{x}$ commutes with differentiation according to $\hat{x}$, we have the straightforward commutation property

$$\delta(\nabla z)=\nabla(\delta z). \tag{2.7}$$

Thus, integrating by parts in (2.6) together with (2.2), (2.4) and (2.7) yields

$$\delta \mathcal{A} = \int_{\Omega\times[t_1,t_2]} \left\{ -\rho \frac{\partial \boldsymbol{u}}{\partial t} - (\nabla \times \boldsymbol{u}) \times \boldsymbol{j} + \rho \nabla R \right\} \cdot \boldsymbol{\xi} \, \mathrm{d}x \, \mathrm{d}t$$
$$+ \int_{\Omega\times[t_1,t_2]} \left\{ \frac{\partial L}{\partial z} - \operatorname{div}\left(\frac{\partial L}{\partial (\nabla z)} \right) \right\} \delta z \, \mathrm{d}x \, \mathrm{d}t + \int_{\Omega\times[t_1,t_2]} \left\{ \frac{\partial L}{\partial y} \right\} \delta y \, \mathrm{d}x \, \mathrm{d}t.$$

Since the above relation is true for arbitrary $\boldsymbol{\xi}$, δz and δy, the minimization of the Hamilton action $\mathcal{A}$ is equivalent to

$$\left.\begin{aligned} \rho \frac{\partial \boldsymbol{u}}{\partial t} + (\nabla \times \boldsymbol{u}) \times \boldsymbol{j} - \rho \nabla R = 0, \\ \frac{\partial L}{\partial z} - \operatorname{div}\left(\frac{\partial L}{\partial (\nabla z)} \right) = 0, \\ \frac{\partial L}{\partial y} = 0, \end{aligned}\right\} \tag{2.8}$$

which provides the following equation for the impulsion:

$$\begin{aligned} \frac{\partial (\rho \boldsymbol{u})}{\partial t} + \operatorname{div}(\rho \boldsymbol{u} \otimes \boldsymbol{u}) - \operatorname{div}\big[\big(\rho |\boldsymbol{u}|^2 + \rho R\big) \, \mathrm{Id} \big] \\ - \operatorname{div}\left(\nabla z \otimes \frac{\partial L}{\partial (\nabla z)} \right) + \operatorname{div}(L \, \mathrm{Id}) = 0. \end{aligned} \tag{2.9}$$

The motion equations have been derived using a general free energy for the system. We shall detail in the next section a specific free energy that will be used for modelling work.

2.3 Free energy choice

In the following we will denote by f_α, g_α and P_α the free energy, the free enthalpy (or Gibbs enthalpy) and the pressure of the phase α. In the case of an isothermal process, these quantities only depend on ρ_α and are connected by

$$\left.\begin{aligned} P_\alpha = \rho_\alpha^2 \frac{\mathrm{d} f_\alpha}{\mathrm{d}\rho_\alpha}, \\ \frac{\mathrm{d} P_\alpha}{\mathrm{d}\rho_\alpha} = \rho_\alpha \frac{\mathrm{d} g_\alpha}{\mathrm{d}\rho_\alpha}, \end{aligned}\right\} \tag{2.10}$$

which also read equivalently in terms of the volumic free energy $F_\alpha = \rho_\alpha f_\alpha$ as

$$\left.\begin{aligned} g_\alpha = \frac{\mathrm{d} F_\alpha}{\mathrm{d}\rho_\alpha}, \\ P_\alpha = \rho_\alpha \frac{\mathrm{d} F_\alpha}{\mathrm{d}\rho_\alpha} - F_\alpha. \end{aligned}\right\} \tag{2.11}$$

We now postulate the following expression for f:

$$\rho f = \sum_{\alpha} \rho_\alpha z_\alpha f_\alpha + \mathcal{H}(z) + \frac{\mu_0}{2}|\nabla z|^2, \tag{2.12}$$

where $\mu_0 \geq 0$ and $\mathcal{H}$ is a double-well Cahn–Hilliard type potential.

The definition (2.5) of R gives

$$R = -\frac{1}{2}\frac{|\boldsymbol{j}|^2}{\rho^2} - f - \rho \sum_{\alpha} \frac{y_\alpha^2}{z_\alpha}\frac{\mathrm{d} f_\alpha}{\mathrm{d}\rho_\alpha}.$$

Together with relation (2.10) this reads

$$\rho R = -\frac{1}{2}\rho|\boldsymbol{u}|^2 - \rho f - \sum_{\alpha} z_\alpha P_\alpha. \tag{2.13}$$

Relations (2.9), (2.12) and (2.13) provide the following equation for the impulsion:

$$\frac{\partial(\rho \boldsymbol{u})}{\partial t} + \operatorname{div}(\rho \boldsymbol{u} \otimes \boldsymbol{u}) + \operatorname{div}(P\mathrm{Id}) = -\mu_0 \operatorname{div}(\nabla z \otimes \nabla z), \tag{2.14}$$

where P is a generalized pressure given by

$$P = \sum_{\alpha} z_\alpha P_\alpha - \mathcal{H}(z) - \frac{\mu_0}{2}|\nabla z|^2. \tag{2.15}$$

2.4 Dissipative structures

Euler equations for isothermal flows are equipped with the natural entropy

$$\eta = \rho\left(f + \frac{|\boldsymbol{u}|^2}{2}\right).$$

In order to prescribe the flux mass between phase 1 and phase 2, we will use the second thermodynamical principle which claims that the dissipation of η has to be non-negative.

The evolution of η reads

$$\rho \frac{\mathrm{d}}{\mathrm{d}t}\left(f + \frac{|\boldsymbol{u}|^2}{2}\right) = -\operatorname{div}(\boldsymbol{q}) - \operatorname{div}[(P\mathrm{Id} + \mu_0 \nabla z \otimes \nabla z)\boldsymbol{u}] - \Delta_\eta, \tag{2.16}$$

where Δ_η is the entropy dissipation and q is a heat flux.

Multiplying the impulsion equation (2.9) by $\boldsymbol{u}$ gives the kinetic energy equation

$$\rho \frac{\mathrm{d}}{\mathrm{d}t}\left(\frac{|\boldsymbol{u}|^2}{2}\right) = -\boldsymbol{u} \cdot \operatorname{div}(P\mathrm{Id} + \mu_0 \nabla z \otimes \nabla z).$$

We compute the left hand side of (2.16) using (2.10) and (2.12):

$$\rho \frac{\mathrm{d}}{\mathrm{d}t}\left(f + \frac{|\boldsymbol{u}|^2}{2}\right) = \operatorname{div}\left(\mu_0 \nabla z \frac{\mathrm{d}z}{\mathrm{d}t}\right) - \operatorname{div}(P\mathrm{Id} + \mu_0 \nabla z \otimes \nabla z \boldsymbol{u})$$
$$+ \rho \frac{\mathrm{d}y}{\mathrm{d}t}(g_1 - g_2) + \frac{\mathrm{d}z}{\mathrm{d}t}\left(P_2 - P_1 + \frac{\mathrm{d}\mathcal{H}}{\mathrm{d}z} - \operatorname{div}(\mu_0 \nabla z)\right).$$

The above relation and the relation (2.16) provide

$$\boldsymbol{q} = -\mu_0 \nabla z \frac{\mathrm{d}z}{\mathrm{d}t}$$

and

$$\Delta_\eta = \rho \frac{\mathrm{d}y}{\mathrm{d}t}(g_2 - g_1) + \frac{\mathrm{d}z}{\mathrm{d}t}\left(P_1 - P_2 - \frac{\mathrm{d}\mathcal{H}}{\mathrm{d}z} + \operatorname{div}(\mu_0 \nabla z)\right). \tag{2.17}$$

Let us remark that no ansatz has been made so far concerning the relations satisfied by $\mathrm{d}y/\mathrm{d}t$ and $\mathrm{d}z/\mathrm{d}t$. Indeed, such assumptions are still to be prescribed. We see that relation (2.17) leads us to a very simple choice for $\mathrm{d}y/\mathrm{d}t$ and $\mathrm{d}z/\mathrm{d}t$ that ensures Δ_η positivity, namely

$$\left.\begin{aligned} \rho \frac{\mathrm{d}y}{\mathrm{d}t} &= \lambda(g_2 - g_1), \\ \frac{\mathrm{d}z}{\mathrm{d}t} &= \kappa\left(P_1 - P_2 - \frac{\mathrm{d}\mathcal{H}}{\mathrm{d}z} + \mu_0 \Delta z\right), \end{aligned}\right\} \tag{2.18}$$

where λ and κ are non-negative modelization functions which can be interpreted as relaxation parameters.

3 System properties

We assume now that $\mathcal{H} \equiv 0$ and $\mu_0 = 0$. For regular solutions, the model reads

$$\left.\begin{aligned} \partial_t(\rho_1 z_1) + \operatorname{div}(\rho_1 z_1 \boldsymbol{u}) &= \lambda(g_2 - g_1), \\ \partial_t(\rho_2 z_2) + \operatorname{div}(\rho_2 z_2 \boldsymbol{u}) &= -\lambda(g_2 - g_1), \\ \partial_t(\rho \boldsymbol{u}) + \operatorname{div}(\rho \boldsymbol{u} \otimes \boldsymbol{u} + P\mathrm{Id}) &= 0, \\ \partial_t z + \boldsymbol{u} \cdot \nabla z &= \kappa(P_1 - P_2), \end{aligned}\right\} \tag{3.1}$$

with P defined as

$$P = \sum_\alpha z_\alpha P_\alpha.$$

We will consider in the following the study of the sole convective terms

$$\left.\begin{aligned} \partial_t(\rho_1 z_1) + \mathrm{div}(\rho_1 z_1 \boldsymbol{u}) &= 0,\\ \partial_t(\rho_2 z_2) + \mathrm{div}(\rho_2 z_2 \boldsymbol{u}) &= 0,\\ \partial_t(\rho \boldsymbol{u}) + \mathrm{div}(\rho \boldsymbol{u} \otimes \boldsymbol{u} + P\mathrm{Id}) &= 0,\\ \partial_t z + \boldsymbol{u}\cdot\nabla z &= 0. \end{aligned}\right\} \tag{3.2}$$

3.1 Hyperbolicity of the homogeneous system

Considering smooth solutions of (3.2) and asuuming $\boldsymbol{V} = (\rho_1 z_1, \rho_2 z_2, \rho u, z)^\top$, (3.2) reads in 1-D

$$\partial_t \boldsymbol{V} + A(\boldsymbol{V})\partial_x \boldsymbol{V} = 0,$$

$$A(\boldsymbol{V}) = \begin{pmatrix} u y_2 & -u y_1 & y_1 & 0\\ -u y_2 & u y_1 & y_2 & 0\\ c_1^2 - u^2 & c_2^2 - u^2 & 2u & M\\ 0 & 0 & 0 & u \end{pmatrix},$$

where

$$c_\alpha^2 = \frac{\mathrm{d}P_\alpha}{\mathrm{d}\rho_\alpha}, \qquad M = \frac{\partial P}{\partial z}.$$

Denoting by c the mixture sound velocity

$$c^2 = \sum_\alpha y_\alpha c_\alpha^2, \tag{3.3}$$

A possesses the following eigenvalues:

$$\lambda_1 = u - c, \quad \lambda_2 = \lambda_3 = u, \quad \lambda_4 = u + c,$$

with the corresponding left eigenvectors l_k and right eigenvectors r_k

$$l_1 = \frac{1}{2c^2}\begin{pmatrix} c_1^2 + uc\\ c_2^2 + uc\\ -c\\ M \end{pmatrix}, \quad l_2 = \frac{1}{c^2}\begin{pmatrix} -y_1 c_1^2 + c^2\\ -y_1 c_2^2\\ 0\\ -y_1 M \end{pmatrix},$$

$$l_3 = \frac{1}{c^2}\begin{pmatrix} -y_2 c_1^2\\ c^2 - y_2 c_2^2\\ 0\\ -y_2 M \end{pmatrix}, \quad l_4 = \frac{1}{2c^2}\begin{pmatrix} c_1^2 - uc\\ c_2^2 - uc\\ c\\ M \end{pmatrix},$$

$$r_1 = \begin{pmatrix} y_1 \\ y_2 \\ u - c \\ 0 \end{pmatrix}, \quad r_2 = \begin{pmatrix} 1 \\ 0 \\ u \\ -c_1^2/M \end{pmatrix},$$

$$r_3 = \begin{pmatrix} 0 \\ 1 \\ u \\ -c_2^2/M \end{pmatrix}, \quad r_4 = \begin{pmatrix} y_1 \\ y_2 \\ u + c \\ 0 \end{pmatrix}.$$

This shows that the system (3.2) is hyperbolic. Nevertheless, the entropy of the system is not strictly convex. Indeed, there is a loss of convexity in the z direction, thus we do not know if the system (3.2) is symmetrizable.

3.2 Limit system $\kappa = +\infty$, $\lambda < +\infty$ (M-equilibrium)

3.2.1 Equilibrium existence and unicity. In the first place, we remark that λ^{-1} and κ^{-1} are two characteristic time scales for the source terms that drive the system back to equilibrium. The $(P_2 - P_1)$ term corresponds to a mechanical equilibrium, while $(g_2 - g_1)$ corresponds to a thermodynamical equilibrium. We suppose here that the mechanical equilibrium occurs faster than the thermodynamical one. Therefore we are now interested in the (formal) behavior of (3.1) when $\kappa \to \infty$ while λ remains finite (M-equilibrium).

The assumption $\kappa = +\infty$ in the system (3.1) is formally equivalent to computing $z \in [0, 1]$ such that

$$P_1\left(\frac{m_1}{z}\right) = P_2\left(\frac{m_2}{1-z}\right)$$

for given fixed values of $m_1 = \rho_1 z_1$ and $m_2 = \rho_2 z_2$ (see [3]).

Proposition 3.1. *We suppose that the functions $P_\alpha \colon \rho_\alpha \mapsto P_\alpha(\rho_\alpha)$ are $\mathcal{C}^1(\mathbb{R}^+)$, strictly increasing on $\mathbb{R}^+$ and tend to $+\infty$ when $\rho_\alpha \to +\infty$. Then there exists a unique solution $z^*(m_1, m_2) \in [0, 1]$ of the equation*

$$P_1\left(\frac{m_1}{z}\right) = P_2\left(\frac{m_2}{1-z}\right). \tag{3.4}$$

Moreover,

- *$m_1 \mapsto z^*(m_1, m_2)$ is increasing on $\mathbb{R}^+$,*
- *$m_2 \mapsto z^*(m_1, m_2)$ is non-increasing on $\mathbb{R}^+$,*
- *$(m_1, m_2) \mapsto z^*(m_1, m_2)$ is regular as the pressure P_α, $\alpha = 1, 2$.*

Proof. Let us define

$$\phi\colon [0,1] \longrightarrow \mathbb{R}$$

$$z \longmapsto P_1\left(\frac{m_1}{z}\right) - P_2\left(\frac{m_2}{1-z}\right).$$

Then $\phi \to +\infty$ when $z \to 0$ and $\phi \to -\infty$ when $z \to 1$. Moreover,

$$\frac{\mathrm{d}\phi}{\mathrm{d}z} = -\frac{m_1}{z^2}\frac{\mathrm{d}P_1}{\mathrm{d}\rho_1}\left(\frac{m_1}{z}\right) - \frac{m_2}{(1-z)^2}\frac{\mathrm{d}P_2}{\mathrm{d}\rho_2}\left(\frac{m_2}{1-z}\right) < 0,$$

which provides existence and uniqueness for $z^* \in [0,1]$. The regularity of z^* follows from the implicit function theorem.

For the monotonicity of z^*, using the equation (3.4) we have

$$\frac{\partial z^*}{\partial m_1} = \frac{c_1^2/z^*}{\frac{c_1^2 m_1}{(z^*)^2} + \frac{c_2^2 m_2}{(1-z^*)^2}} \geq 0$$

and

$$\frac{\partial z^*}{\partial m_2} = -\frac{c_2^2/(1-z^*)}{\frac{c_1^2 m_1}{(z^*)^2} + \frac{c_2^2 m_2}{(1-z^*)^2}} \leq 0,$$

which ends the proof. □

Let us briefly remark that it is possible to obtain an analytical expression for z^* for some given specific EOS. For example, if we consider as Chanteperdrix, Villedieu and Vila in [3] the case of two stiffened gases, then P_α reads

$$P_\alpha = P^0 + c_\alpha^2(\rho_\alpha - \rho_\alpha^0),$$

where c_α, ρ_α^0 and P^0 are constants. Let us define q, $\tilde{q}$ by

$$q = \rho_2^0 c_2^2 - \rho_1^0 c_1^2, \quad \tilde{q} = m_2 c_2^2 - m_1 c_1^2.$$

Then z^* satisfies

$$z^* = \frac{\beta}{1+\beta}, \quad \beta = \frac{q - \tilde{q} + \sqrt{(q-\tilde{q})^2 + 4 m_1 m_2 c_1^2 c_2^2}}{2 m_2 c_2^2}.$$

We now turn back to the study of $\kappa = +\infty$ limit system in the general case. Formally, the system reads

$$\left.\begin{aligned} \partial_t(\rho_1 z_1) + \mathrm{div}(\rho_1 z_1 \boldsymbol{u}) &= \lambda(g_2 - g_1),\\ \partial_t(\rho_2 z_2) + \mathrm{div}(\rho_2 z_2 \boldsymbol{u}) &= -\lambda(g_2 - g_1),\\ \partial_t(\rho \boldsymbol{u}) + \mathrm{div}(\rho \boldsymbol{u} \otimes \boldsymbol{u} + P\mathrm{Id}) &= 0, \end{aligned}\right\} \tag{3.5}$$

with

$$P = \sum_{\alpha} z_{\alpha}^{*} P_{\alpha} = P_1 = P_2.$$

Let us show that (3.5) is strictly hyperbolic. Let $\boldsymbol{W} = (\rho_1 z_1, \rho_2 z_2, u)^{\top}$. For smooth solutions, system (3.5) is equivalent to

$$\partial_t \boldsymbol{W} + B(\boldsymbol{W})\partial_x \boldsymbol{W} = \mathcal{S}$$

with

$$B(\boldsymbol{W}) = \begin{pmatrix} u & 0 & \rho_1 z_1 \\ 0 & u & \rho_2 z_2 \\ 1/\rho\partial_{m_1} P & 1/\rho\partial_{m_2} P & u \end{pmatrix}, \quad \mathcal{S} = \begin{pmatrix} \lambda(g_2 - g_1) \\ -\lambda(g_2 - g_1) \\ 0 \end{pmatrix}.$$

B possesses the three following distinct eigenvalues:

$$\lambda_1 = u - c^*, \quad \lambda_2 = u, \quad \lambda_3 = u + c^*$$

with c^* such that

$$\frac{1}{\rho(c^*)^2} = \sum_{\alpha} \frac{(z_{\alpha}^*)^2}{m_{\alpha} c_{\alpha}^2},$$

which proves the strict hyperbolicity of the pressure equilibrium system (3.5).

We can also show that there is no 3×3 symmetric matrix $B_0(\boldsymbol{W})$ such that $B_0(\boldsymbol{W})B(\boldsymbol{W})$ is symmetric and then the system (3.5) is not symmetrizable. As in part 3.1, this comes from convexity loss of the entropy.

3.3 $\kappa = +\infty, \lambda = +\infty$ limit system (MT-equilibrium)

3.3.1 Conditional equilibrium existence and unicity. The condition $\kappa = +\infty$, $\lambda = +\infty$ (MT-equilibrium) is formally equivalent to determine m_1, m_2 and z, for a given ρ, such that

$$\left.\begin{aligned} P_1\left(\frac{m_1}{z}\right) &= P_2\left(\frac{m_2}{1-z}\right), \\ g_1\left(\frac{m_1}{z}\right) &= g_2\left(\frac{m_2}{1-z}\right), \\ \rho &= m_1 + m_2. \end{aligned}\right\} \tag{3.6}$$

Solving system (3.6) is equivalent to find the solutions ρ_1, ρ_2 and z of the following system:

$$\left.\begin{aligned} P_1(\rho_1) &= P_2(\rho_2), \\ g_1(\rho_1) &= g_2(\rho_2), \\ \rho &= z\rho_1 + (1-z)\rho_2, \end{aligned}\right\} \tag{3.7}$$

with the constraint

$$\rho_\alpha \geq 0 \quad \text{and} \quad z \in [0,1]. \tag{3.8}$$

We propose a sufficient solvability condition for (3.7)–(3.8) based on a geometrical argument.

Proposition 3.2. *Let $F_\alpha = \rho_\alpha f_\alpha$ be the volumic free energy. If there exists a unique $\rho^* \in (0, +\infty)$ such that*

$$F_1(\rho^*) = F_2(\rho^*),$$

then the system

$$\left.\begin{aligned} P_1(\rho_1) &= P_2(\rho_2), \\ g_1(\rho_1) &= g_2(\rho_2) \end{aligned}\right\} \tag{3.9}$$

has a unique solution $(\rho_1^*, \rho_2^*) \in (0, +\infty)^2$.

Let $\mathfrak{C}_\alpha$ be the curves $\rho_\alpha \mapsto F_\alpha(\rho_\alpha)$ in the (ρ, F) plane. Suppose $(\rho_1^*, \rho_2^*, z^*)$ is a solution of (3.7). Then consider D_α the tangent line to $\mathfrak{C}_\alpha$ at ρ_α^*. Then (3.7) imposes D_1 and D_2 to have the same slope and the same F-intercept (see Figure 1). In this case z^* is given by

$$\left.\begin{aligned} z^* &= 1 && \text{if } \rho < \rho_1^*, \\ z^* &= \frac{\rho - \rho_2^*}{\rho_1^* - \rho_2^*} && \text{if } \rho_1^* < \rho < \rho_2^*, \\ z^* &= 0 && \text{if } \rho_2^* < \rho. \end{aligned}\right\} \tag{3.10}$$

Using this geometrical interpretation, we can easily see, as depicted in Figure 2, that it is possible to draw two curves $\mathfrak{C}_1$ and $\mathfrak{C}_2$ such that (3.7) has no solution. Consequently, it is not possible to ensure existence nor uniqueness for a general choice of EOS. Nevertheless, given two specific EOS, the above lines allow us to check whether both existence and uniqueness of the equilibrium states stand. Moreover, the same lines provide (at least geometrically) the equilibrium states ρ_1^* and ρ_2^* when they exist. This means we have a tool to check the EOS compatibility as regards the MT-equilibrium return process.

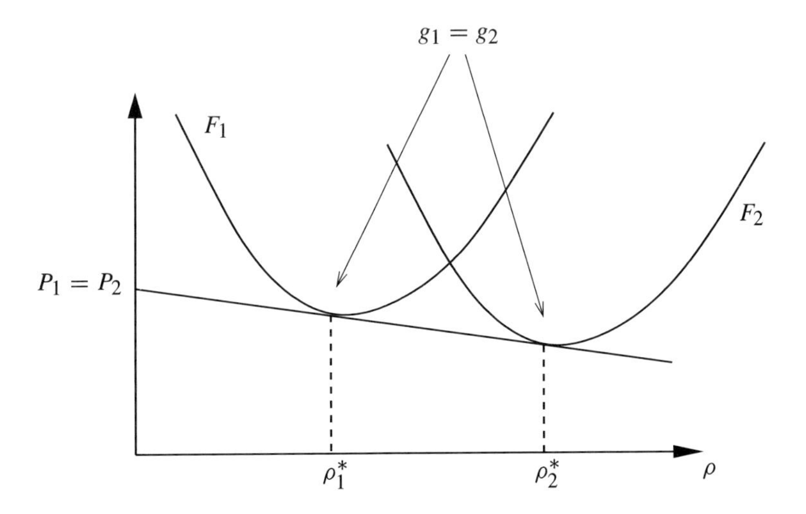

Figure 1. A geometric condition for the existence of an equilibrium.

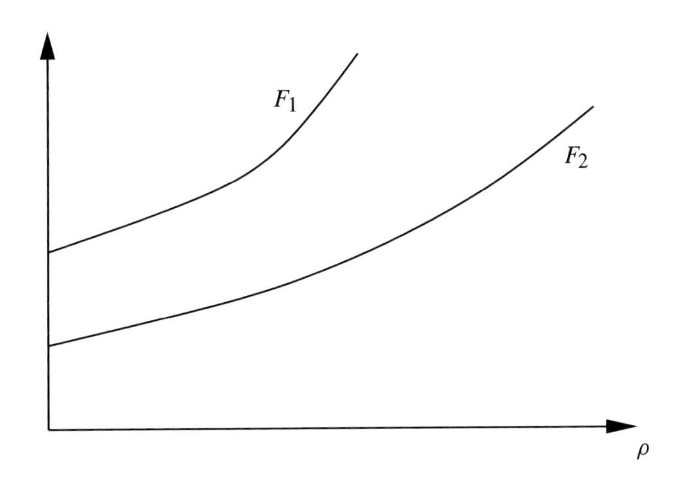

Figure 2. An example of F_α where there is no equilibrium points.

In the following, we suppose that there exists a unique solution (ρ_1^*, ρ_2^*) of (3.9) and without loss of generality we suppose $\rho_1^* < \rho_2^*$. The MT-equilibrium system reads

$$\left.\begin{aligned} \partial_t \rho + \operatorname{div}(\rho \boldsymbol{u}) = 0 \\ \partial_t(\rho \boldsymbol{u}) + \operatorname{div}(\boldsymbol{u} \otimes \boldsymbol{u} + P\mathrm{Id}) = 0 \end{aligned}\right\} \tag{3.11}$$

with

$$P = P(\rho).$$

Before going any further, we shall look into the behaviour of the MT-equilibrium pressure law $\rho \mapsto P(\rho)$. If we consider the volumic free energy F_α graph in the $F - \rho$ plane, then the MT-equilibrium free energy can be interpreted as the convex envelope of the function $\rho \mapsto \min\{F_1(\rho), F_2(\rho)\}$ (solid line in Figure 3). The pressure behaviour follows from $P = \rho \mathrm{d}F/\mathrm{d}\rho - F$. As depicted in Figure 3, we see that the pressure (and consequently the system flux) is $\mathcal{C}^1$ for $\rho \in \mathbb{R}^+$ except for $\rho = \rho_1^*$ and $\rho = \rho_2^*$ where it is only $\mathcal{C}^0$. Let us note that the result system corresponds to a particular case of the system derived by S. Jaouen [9] using a convexification principle.

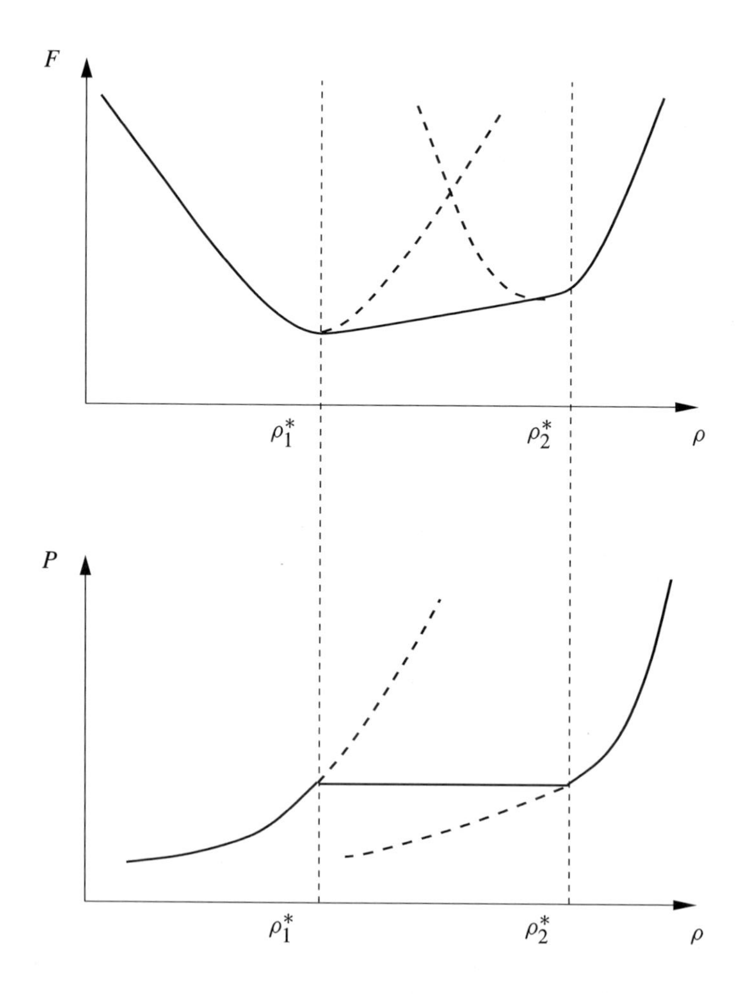

Figure 3. Volumic free energy F and pressure at the equilibrium.

For $\rho \notin \{\rho_1^*, \rho_2^*\}$ and $\rho \in \mathbb{R}^+$, we can compute the jacobian matrix A of the system flux (3.11). In primitives variables $\boldsymbol{U} = (\rho, u)^\top$, we have

$$A(\boldsymbol{U}) = \begin{pmatrix} u & \rho \\ 1/\rho(\partial P/\partial \rho) & u \end{pmatrix}$$

with eigenvalues

$$\lambda_1 = u - c, \quad \lambda_2 = u + c, \quad c = \frac{\mathrm{d}P}{\mathrm{d}\rho}.$$

For $\rho \notin (\rho_1^*, \rho_2^*)$, c is strictly positive therefore the system is strictly hyperbolic. In the region $\rho_1^* < \rho < \rho_2^*$, we have $c = 0$, thus the characteristic polynomial associated with $A(\boldsymbol{U})$ possesses a single root u. Consequently, the system is not hyperbolic.

3.3.2 First-order correction for the system (3.5). Let us write the system (3.5) in lagrangian coordinates

$$\left.\begin{aligned} \partial_t y &= \lambda(g_2 - g_1), \\ \partial_t \tau - \partial_X u &= 0, \\ \partial_t u + \partial_X P &= 0, \end{aligned}\right\} \tag{3.12}$$

with P defined as

$$P = \sum_\alpha z_\alpha^* P_\alpha\left(\frac{y_\alpha}{\tau z_\alpha^*}\right) = P_1\left(\frac{y_1}{\tau z_1^*}\right) = P_2\left(\frac{y_2}{\tau z_2^*}\right).$$

We now want to check whether the solutions of the relaxed system (3.12) formally possess another attractor with good stability properties – besides the solutions of the MT-equilibrium system (3.11). We remark that (3.12) is of the form of the system studied in [4]:

$$\left.\begin{aligned} \partial_t \boldsymbol{V} + \partial_x \mathcal{F}(a, \boldsymbol{V}) &= 0, \\ \partial_t a + \partial_x \mathcal{G}(a, \boldsymbol{V}) &= \frac{1}{\epsilon}\mathcal{S}(a, \boldsymbol{V}), \end{aligned}\right\} \tag{3.13}$$

with $\boldsymbol{V}(x, t) \in \mathbb{R}^N$ and $a(t, x) \in \mathbb{R}$. In [4], the equation

$$\mathcal{S}(a, \boldsymbol{V}) = 0$$

is supposed to have a unique solution $a^*(\boldsymbol{V})$ for all $\boldsymbol{V} \in \mathbb{R}^N$ with $a \mapsto \mathcal{S}(a, \boldsymbol{V})$ strictly monotonic. System (3.13) is formally shown to be equivalent for $\epsilon \to 0$ to the first-order system

$$\begin{aligned} \partial_t \boldsymbol{V} + \partial_x \mathcal{F}^*(\boldsymbol{V}) = \epsilon \partial_x \Bigg\{ \frac{\partial \mathcal{F}}{\partial a}\left(\frac{\partial \mathcal{S}}{\partial a}\right)^{-1} \Bigg[&\frac{\partial a^*}{\partial \boldsymbol{V}}\left(\frac{\partial \mathcal{F}}{\partial a}\frac{\partial a^*}{\partial \boldsymbol{V}}\partial_x \boldsymbol{V} + \frac{\partial \mathcal{F}}{\partial \boldsymbol{V}}\partial_x \boldsymbol{V}\right) \\ &- \left(\frac{\partial \mathcal{G}}{\partial a}\frac{\partial a^*}{\partial \boldsymbol{V}} + \frac{\partial \mathcal{G}}{\partial \boldsymbol{V}}\right)\partial_x \boldsymbol{V}\Bigg]\Bigg\}. \end{aligned} \tag{3.14}$$

We apply formula (3.14) with $\mathcal{G} = 0$, $V = (\tau, u)^\top$, $\mathcal{F} = (-u, P)^\top$, $a = y$, $\mathcal{S} = g_2 - g_1$ and $\epsilon = \lambda^{-1}$. Using the notations

$$M = \frac{1}{\tau}\left(\frac{1-y^*}{1-z^*}c_2^2 - \frac{y^*}{z^*}c_1^2\right), \quad K = \sum_\alpha \frac{c_\alpha^2 y_\alpha^*}{(z_\alpha^*)^2},$$

then

$$\frac{\partial z^*}{\partial \tau} = \frac{1}{\tau}\frac{M}{K}, \quad \frac{\partial z^*}{\partial y} = \frac{1}{K}\sum_\alpha \frac{c_\alpha^2}{z_\alpha^*},$$

and

$$\begin{aligned}\frac{\partial \mathcal{S}}{\partial y} &= -\sum_\alpha \frac{c_\alpha^2}{y_\alpha^*} + \frac{1}{K}\left(\sum_\alpha \frac{c_\alpha^2}{z_\alpha^*}\right)^2 \\ &= -\frac{c_1^2 c_2^2}{Kz^*(1-z^*)y^*(1-yç*)}\left(z^*(1-y^*) - y^*(1-z^*)\right)^2 < 0.\end{aligned}$$

The first-order correction for the system (3.5) reads

$$\left.\begin{aligned}&\partial_t \tau - \partial_X u = 0,\\ &\partial_t u + \partial_X P = \frac{1}{\lambda}\partial_X\left(\frac{\partial P}{\partial y}\frac{\partial y^*}{\partial \tau}\left(-\frac{\partial \mathcal{S}}{\partial y}\right)\partial_X u\right),\end{aligned}\right\} \tag{3.15}$$

with

$$\frac{\partial P}{\partial y} = \frac{c_1^2 - c_2^2}{\tau} + \frac{M}{\tau K}\left(\sum_\alpha \frac{c_\alpha^2}{z_\alpha^*}\right), \quad \frac{\partial y^*}{\partial \tau} = \frac{\dfrac{c_1^2 - c_2^2}{\tau} + \dfrac{M}{\tau K}\left(\sum_\alpha \dfrac{c_\alpha^2}{z_\alpha^*}\right)}{\sum_\alpha \dfrac{c_\alpha^2}{y_\alpha^*} - \dfrac{1}{K}\left(\sum_\alpha \dfrac{c_\alpha^2}{z_\alpha^*}\right)^2} = \frac{\dfrac{\partial P}{\partial y}}{-\dfrac{\partial h}{\partial y}}.$$

Hence (3.15) is equivalent to the following system

$$\begin{aligned}&\partial_t \tau - \partial_X u = 0\\ &\partial_t u + \partial_X P = \frac{1}{\lambda}\partial_X\left(\left(\frac{\partial P/\partial y}{\partial \mathcal{S}/\partial y}\right)^2 \partial_X u\right),\end{aligned}$$

which proves that the λ^{-1} first-order correction of the system (3.5) is dissipative.

4 Numerical treatment

From now on, we suppose both fluids to be perfect gases. We recall the expressions of P_α, f_α and g_α

$$\begin{aligned} P_\alpha &= b_\alpha \rho_\alpha, \\ f_\alpha &= a_\alpha + b_\alpha\big(\log(\rho_\alpha) - 1\big), \\ g_\alpha &= a_\alpha + b_\alpha \log(\rho_\alpha), \\ b_\alpha &= (\gamma_\alpha - 1) C_{v\alpha} T, \end{aligned}$$

where a_α is constant.

4.1 Numerical scheme based on the M-equilibrium system ($\kappa = +\infty$, $\lambda < +\infty$)

In the sequel, this solver will be referred to as "scheme 1". In the case of a two-perfect gas flow, we have a simple expression for pressure equilibrium z^*

$$z^* = \frac{b_1 m_1}{b_1 m_1 + b_2 m_2}. \tag{4.1}$$

The adopted numerical method is a time splitting method wich consists of two steps:

1. Solve the system (3.2)

$$\begin{aligned} \partial_t(\rho_1 z_1) + \operatorname{div}(\rho_1 z_1 \boldsymbol{u}) &= 0, \\ \partial_t(\rho_2 z_2) + \operatorname{div}(\rho_2 z_2 \boldsymbol{u}) &= 0, \\ \partial_t(\rho \boldsymbol{u}) + \operatorname{div}(\rho \boldsymbol{u} \otimes \boldsymbol{u} + P\mathrm{Id}) &= 0, \\ \partial_t z + \boldsymbol{u} \cdot \nabla z &= 0, \end{aligned}$$

   with a classical hyperbolic solver. This step computes from a state

$$\boldsymbol{V}_i^n := \big((\rho_1 z_1)_i^n, (\rho_2 z_2)_i^n, (\rho \boldsymbol{u})_i^n, z_i^n\big)^\top$$

   an intermediate state $\boldsymbol{V}_i^{n+1/2}$.

2. Solve the differential system

$$\left.\begin{aligned} \partial_t m_1 &= \lambda(g_2 - g_1), \\ \partial_t \rho &= 0, \\ \partial_t(\rho u) &= 0, \\ \partial_t z &= \kappa(P_1 - P_2), \end{aligned}\right\} \tag{4.2}$$

over $[0, \Delta t]$ with $V_i^{n+1/2}$ as initial condition for large values of λ and κ.

In (4.2), we use $\lambda := \tilde{\lambda} m_1(\rho - m_1)$ where $\tilde{\lambda} > 0$ is a constant; we freeze $(g_2 - g_1)$ to its time $t^{n+1/2}$ value and integrate exactly over $[0, \Delta t]$

$$\partial_t m_1 = \tilde{\lambda} m_1(\rho_i^{n+1/2} - m_1)(g_2 - g_1)_i^{n+1/2},$$
$$m_1(t = 0) = (m_1)_i^{n+1/2},$$

which gives $(m_1)_i^{n+1} := m_1(\Delta t; (m_1)_i^{n+1/2})$. Using $\partial_t \rho = 0,\ \partial_t u = 0$, we define

$$(m_2)_i^{n+1} = \rho_i^{n+1/2} - (m_1)_i^{n+1},$$
$$\rho_i^{n+1} = \rho_i^{n+1/2},$$
$$u_i^{n+1} = u_i^{n+1/2}.$$

In order to define z_i^{n+1}, the parameter κ is formally set to $+\infty$ which implies that z_i^{n+1} is given by

$$z_i^{n+1} := z^*((m_1)_i^{n+1}, (m_2)_i^{n+1}) = \frac{b_1 (m_1)_i^{n+1}}{b_1 (m_1)_i^{n+1} + b_2 (m_2)_i^{n+1}}$$

4.2 Numerical scheme based on the MT-equilibrium system ($\kappa = \lambda = +\infty$)

In the sequel, this solver will be referred to as “scheme 2”. For the MT-equilibrium system, we have to compute ρ_1^* and ρ_2^* such that $P_1 = P_2$ and $g_1 = g_2$. Using the perfect gas EOS, a simple calculation gives

$$\rho_1^* = \exp\left(-\frac{a_1 - a_2}{b_1 - b_2}\right)\left(\frac{b_1}{b_2}\right)^{\frac{b_2}{b_1 - b_2}},$$

$$\rho_2^* = \exp\left(-\frac{a_1 - a_2}{b_1 - b_2}\right)\left(\frac{b_1}{b_2}\right)^{\frac{b_1}{b_1 - b_2}},$$

and

$$z^* = \begin{cases} 1 & \text{if } \rho < \rho_1^*, \\ \dfrac{\rho - \rho_2^*}{\rho_1^* - \rho_2^*} & \text{if } \rho_1^* < \rho < \rho_2^*, \\ 0 & \text{if } \rho_2^* < \rho. \end{cases}$$

As previously, the adopted numerical method is a time splitting method which consists of two steps:

1. Solve the system (3.2) with a classical hyperbolic solver. This step computes from a state
$$\boldsymbol{V}_i^n = \left((\rho_1 z_1)_i^n, (\rho_2 z_2)_i^n, (\rho \boldsymbol{u})_i^n, z_i^n\right)^\top$$
an intermediate state $\boldsymbol{V}_i^{n+1/2}$.

2. Set formally $\kappa = +\infty$ and $\lambda = +\infty$ by projecting $\boldsymbol{V}_i^{n+1/2}$ on the equilibrium manifold. This projection gives z_i^{n+1} by
$$\left.\begin{aligned} z_i^{n+1} &= 1 && \text{if } \rho_i^{n+1/2} < \rho_1^*, \\ z_i^{n+1} &= \frac{\rho_i^{n+1/2} - \rho_2^*}{\rho_1^* - \rho_2^*} && \text{if } \rho_1^* < \rho_i^{n+1/2} < \rho_2^*, \\ z_i^{n+1} &= 0 && \text{if } \rho_2^* < \rho_i^{n+1/2}, \end{aligned}\right\} \tag{4.3}$$
and $(m_\alpha)_i^{n+1/2}$, $\alpha = 1, 2$, by
$$\left.\begin{aligned} &(m_1)_i^{n+1} = \rho_i^{n+1/2}, \ (m_2)_i^{n+1} = 0 && \text{if } \rho_i^{n+1/2} < \rho_1^*, \\ &(m_1)_i^{n+1} = \rho_1^* z_i^{n+1}, \ (m_2)_i^{n+1} = \rho_2^*(1 - z_i^{n+1}) && \text{if } \rho_1^* < \rho_i^{n+1/2} < \rho_2^*, \\ &(m_1)_i^{n+1} = 0, \quad (m_2)_i^{n+1} = \rho_i^{n+1/2} && \text{if } \rho_2^* < \rho_i^{n+1/2}, \end{aligned}\right\} \tag{4.4}$$
and we set $\rho_i^{n+1} = \rho_i^{n+1/2}$, $u_i^{n+1} = u_i^{n+1/2}$.

4.3 Numerical results

4.3.1 Pure interface advection. We consider a one dimensional advection problem: a 1 m long domain contains liquid and vapor initially separated by an interface located at $x = 0.5$ m. The characteristic parameters of the fluids are given in Table 1. The

Table 1. Fluids parameters for the advection test.

	left	right
C_v (J.kg^{-1}.K^{-1})	5 000	25
$\gamma = C_p/C_v$	3	5
a (kJ.kg^{-1})	8 407	47 038
b (kJ.kg^{-1}.K^{-1})	10^3	10^3
T (K)	100	100

velocity of both fluids is initially $u = 100$ m.s^{-1}, and the densities are chosen on each side of the interface so that we have a constant pressure in the whole domain, namely $\rho = \rho_1^* \simeq 2.25\ 10^{-4}$ kg.m^{-3} at the left of the interface and $\rho = \rho_2^* \simeq 225.10^{-4}$ kg.m^{-3} at the right (pressure equilibrium).

Boundary conditions are constant states kept on both side of the domain which is discretized over 500 cells.

Figure 4 displays the solution obtained with the scheme 2 (the solid line) and the solution obtained with a Rusanov scheme applied to the system (3.2) (dots): neglecting the source terms in favor of the convective part.

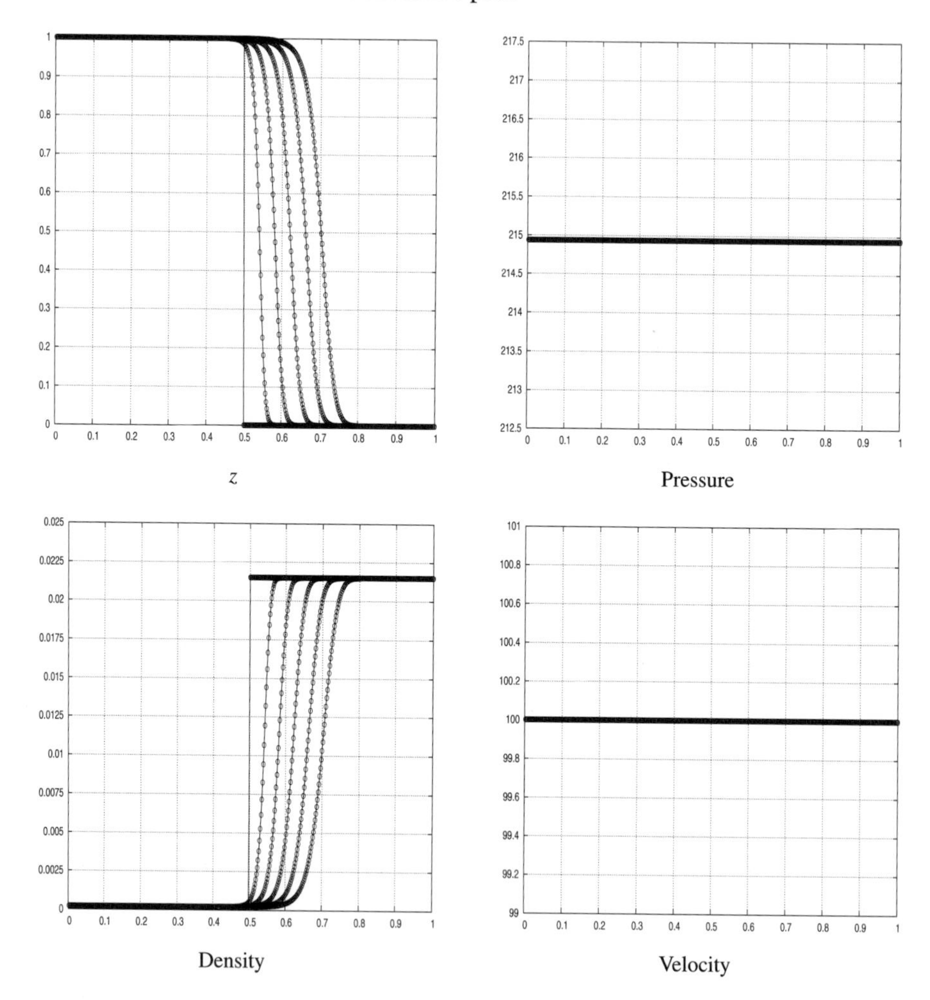

Figure 4. Time evolution of the MT-equilibrium solution computed with scheme 2 and the solution of the pure convective system (3.2) computed with a Rusanov scheme; t varies from $t = 0$ s to $T_{\max} = 0.002$ s.

Although this test is a simple Riemann problem which does not involve phase change but purely kinetic effects, we emphasize the fact that the MT-equilibrium system we deal with does not match the classical hyperbolic well-posedness criteria. However we can check that the scheme 2 seems to capture the same advected profile as the Rusanov scheme applied to (3.2). Indeed, despite the source terms effects, we have the same propagation speed without any pressure nor velocity perturbations.

4.3.2 A one dimensional isothermal phase change problem. We now turn to the case of a one dimensional phase change test. As previously, we consider a 1 m domain containing liquid and vapor initially separated by an interface located at $x = 0.5$ m (the characteristic parameters of the fluids are given in Table 2). In the whole domain the fluid is supposed to be initially at rest, and the densities are once again chosen so that pressure is constant: $\rho = \rho_1^* \simeq 6.10^{-4}$ kg.m^{-3} at the left of the interface and $\rho = \rho_2^* \simeq 33.10^{-4}$ kg.m^{-3} at the right.

Table 2. Fluids parameters for the piston test.

	left	right
C_v (J.kg^{-1}.K^{-1})	1 816	1 040
$\gamma = C_p/C_v$	2.35	1.43
a (kJ.kg^{-1})	1 504	−57.3
b (kJ.kg^{-1}.K^{-1})	245.16	245.16
T (K)	100	100

We now suppose that the left boundary is a piston which moves towards left at constant speed. This motion will generate an acoustic wave travelling from left to right that will reach the interface location, perturbing its thermodynamical equilbrium and triggering the system source terms. A phase change process occurs.

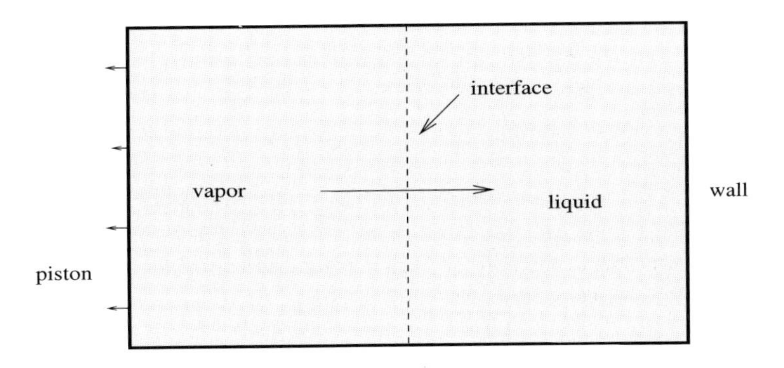

Figure 5. Sketch of the one-dimensional piston isothermal phase-change problem.

The piston effect is realized by imposing a left boundary condition $u_p = -100\,\mathrm{m.s}^{-1}$ with a null mass flux, while the right boundary condition is kept constant.

We use here the scheme 2 with different space steps. Figures 6, 7 and 8 display the profiles of z, pressure P, velocity u, density ρ and partial densities m_α for $\alpha = 1, 2$, at instants (from the top to the bottom) 3.33 ms, 6.66 ms and 10 ms. The solid lines represent the approximate solution with 9000 cells, the $\times$ symbol lines represent the solution with 1000 cells and the ∇ symbol lines represent the solution with 100 cells. This allows us to check the convergence behaviour of the solution when the space step goes to zero. We note that no instability occurs.

We now turn to the convergence of the solution obtained with the scheme 1 to the one obtained with scheme 2. Figures 9 and 10 show the profiles computed with both schemes for z, ρ, P and u at $t = 10$ ms in the case of the piston test. We use a λ range from 1 up to 10^4 (from the top to the bottom). The solid lines represent the scheme 1 solution while the dots represent the scheme 2 one.

We take the scheme 2 solution $\boldsymbol{V}_h^{\mathrm{ref}}$ as the reference solution. Let $\boldsymbol{V}_h^{\lambda}$ be the solution obtained with the off-equilibrium scheme 1. Figure 11 displays the function $\lambda \mapsto \ln\left(\|\boldsymbol{V}_h^{\mathrm{ref}} - \boldsymbol{V}_h^{\lambda}\|_{L^1}\right)$. We numerically check that for $\lambda \to +\infty$

$$\|\boldsymbol{V}_h^{\lambda} - \boldsymbol{V}_h^{\mathrm{ref}}\|_{L^1} = \mathcal{O}\left(1/\lambda\right)$$

5 Conclusion and perspectives

We have presented an isothermal model for two-phase flow problems. The core method used to derive the system is general and it ensures a thermodynamical consistency. The additional source terms we obtain are interpreted as relaxation coefficients related to equilibrium manifolds states. Although the equilibrium system is not hyperbolic, we are able to design two numerical solvers based on the relaxed system. The first one uses the mechanical equilibrium and a source term integration step while the second one uses both mechanical and thermodynamical equilibriums. The solvers have been tested against two simulations: good behaviour for the space and time convergence is observed. The schemes preserve properties like density positivity or maximum principle for z. The convergence of the source term integration method to the mechanical and thermodynamical equilibrium solution has been checked.

A straightforward sequel to this study would be an extensive test of the obtained models with various EOS along with comparisons to basic physical experiments. It is possible to extend the present work by looking into systems involving more complex equilibrium manifolds, *e.g.* by adding extra-potential functions to the system energy. The $+\infty$ relaxation parameters limit in the continuous systems are still to be carefully studied within a proper mathematical framework. The same modelling lines can be used when working in an anisothermal context. However the derived systems involve

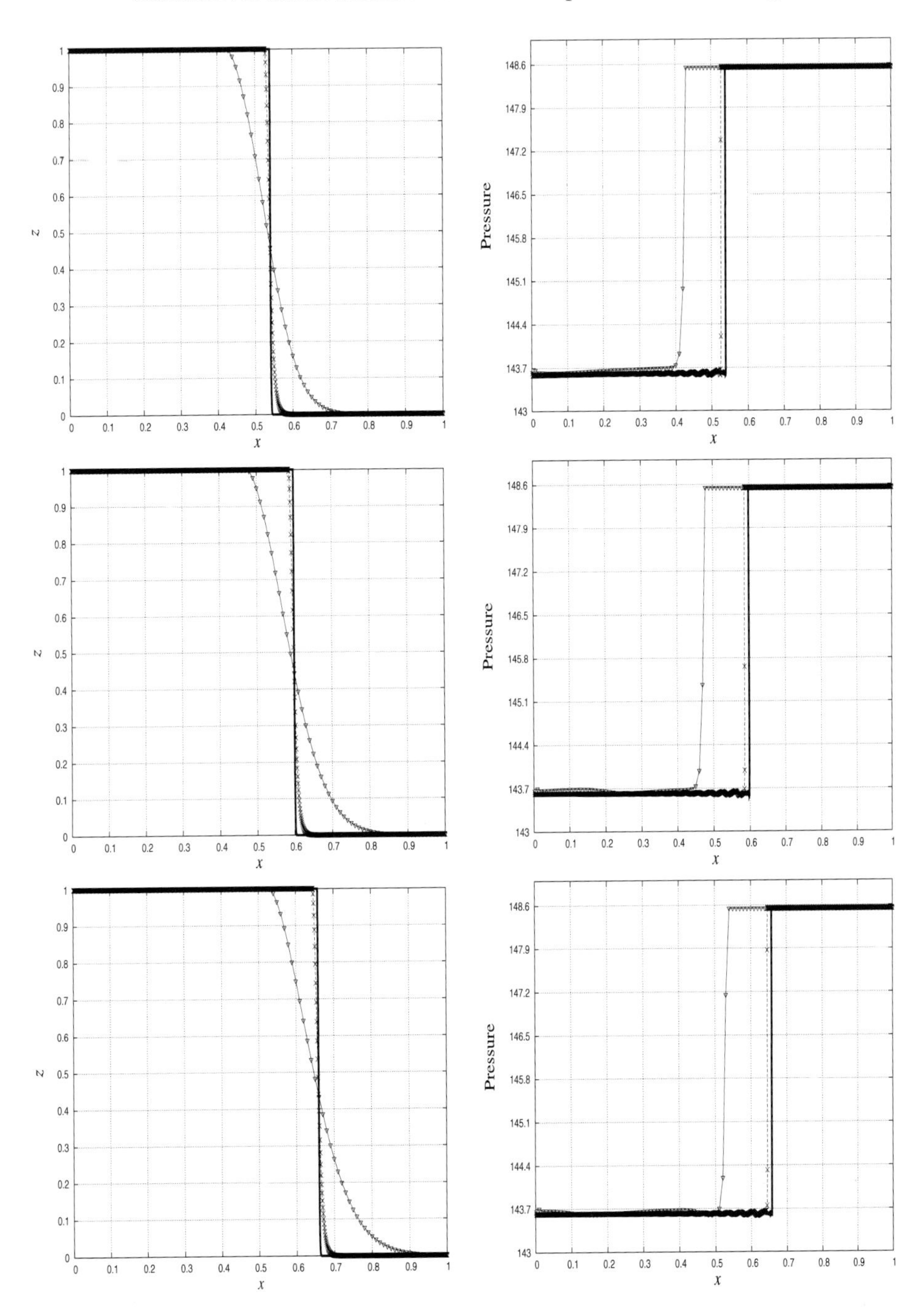

Figure 6. Profile of z and P with scheme 2.

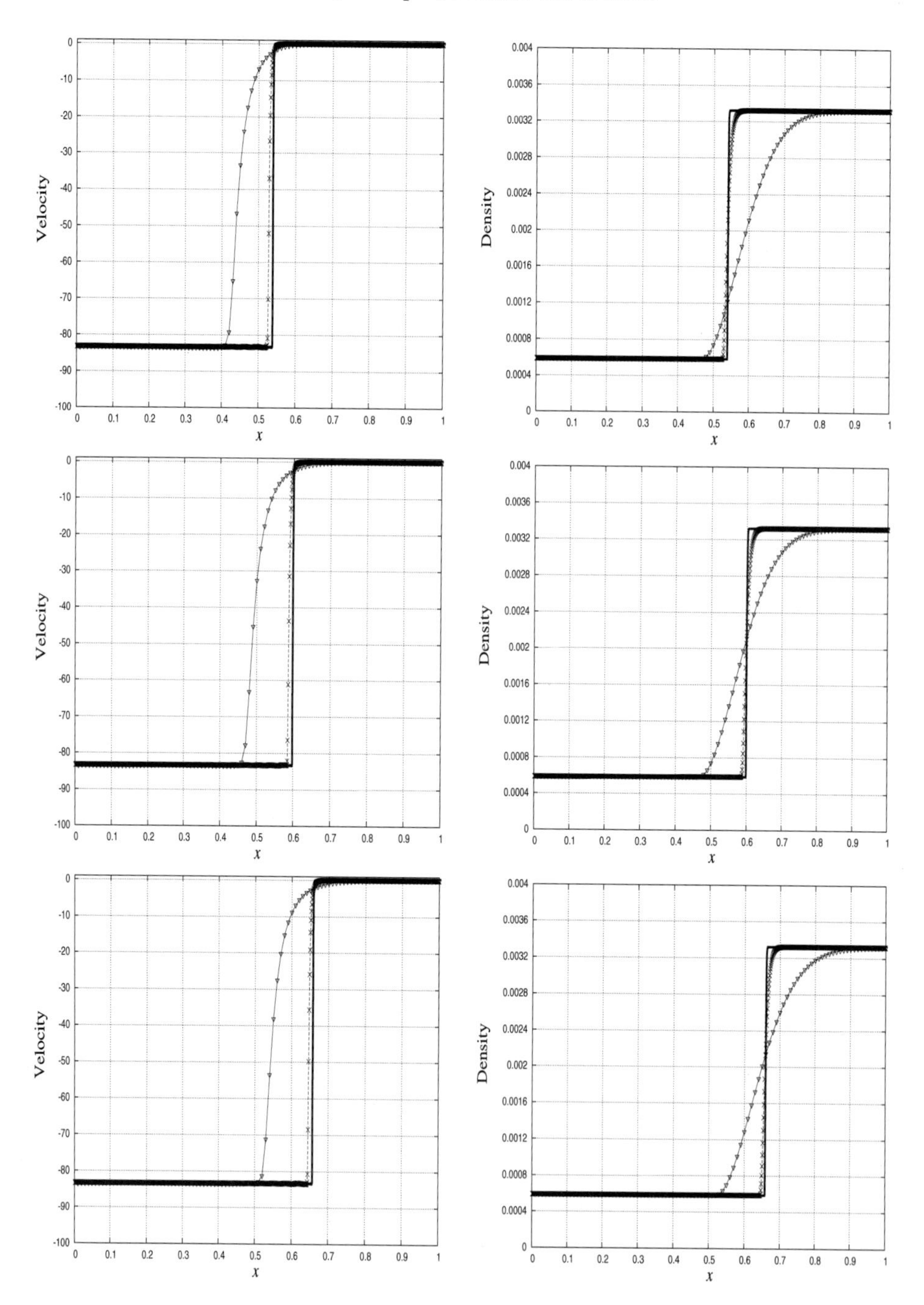

Figure 7. Profile of u and ρ with scheme 2.

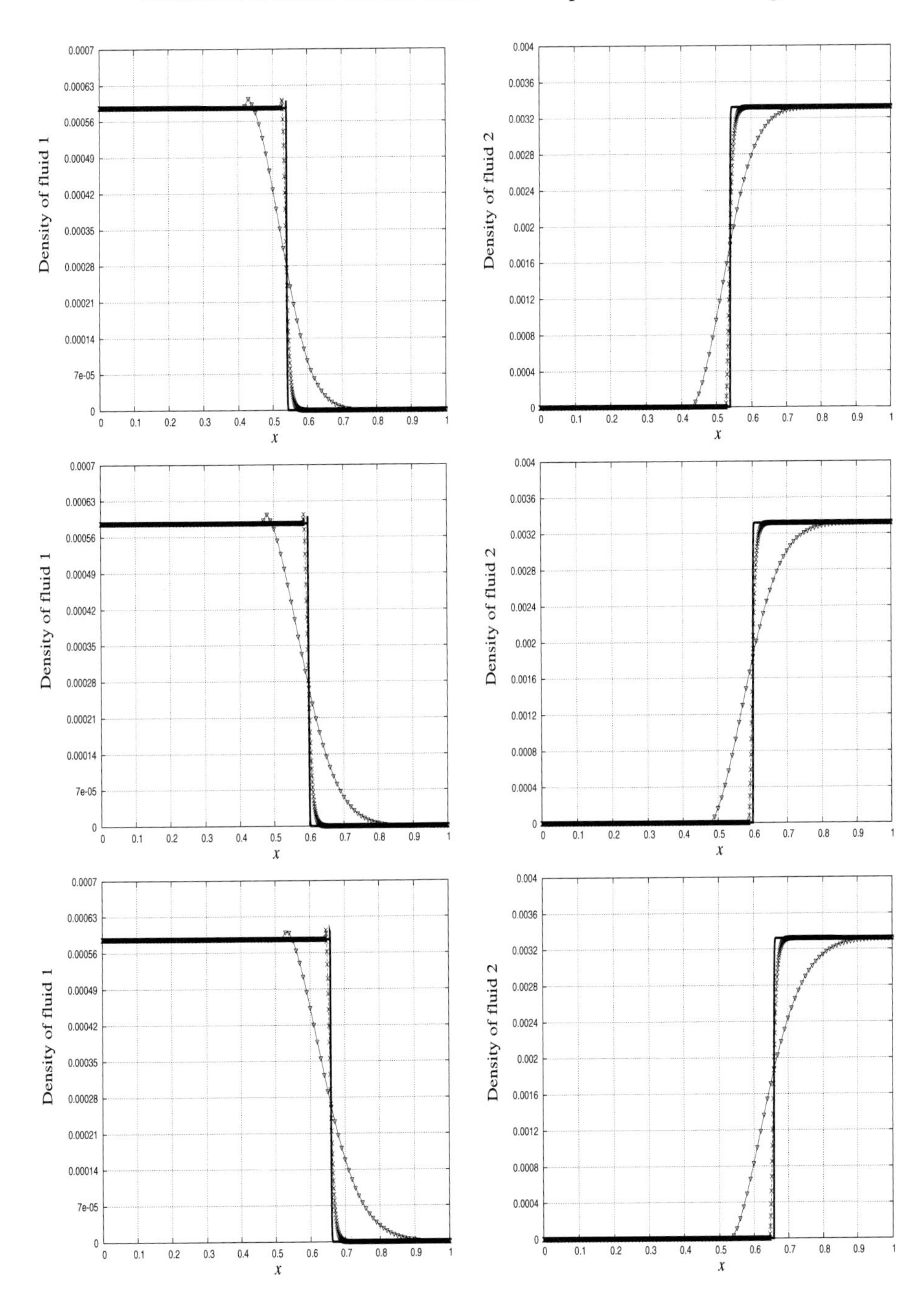

Figure 8. Profile of m_1 and m_2 with scheme 2.

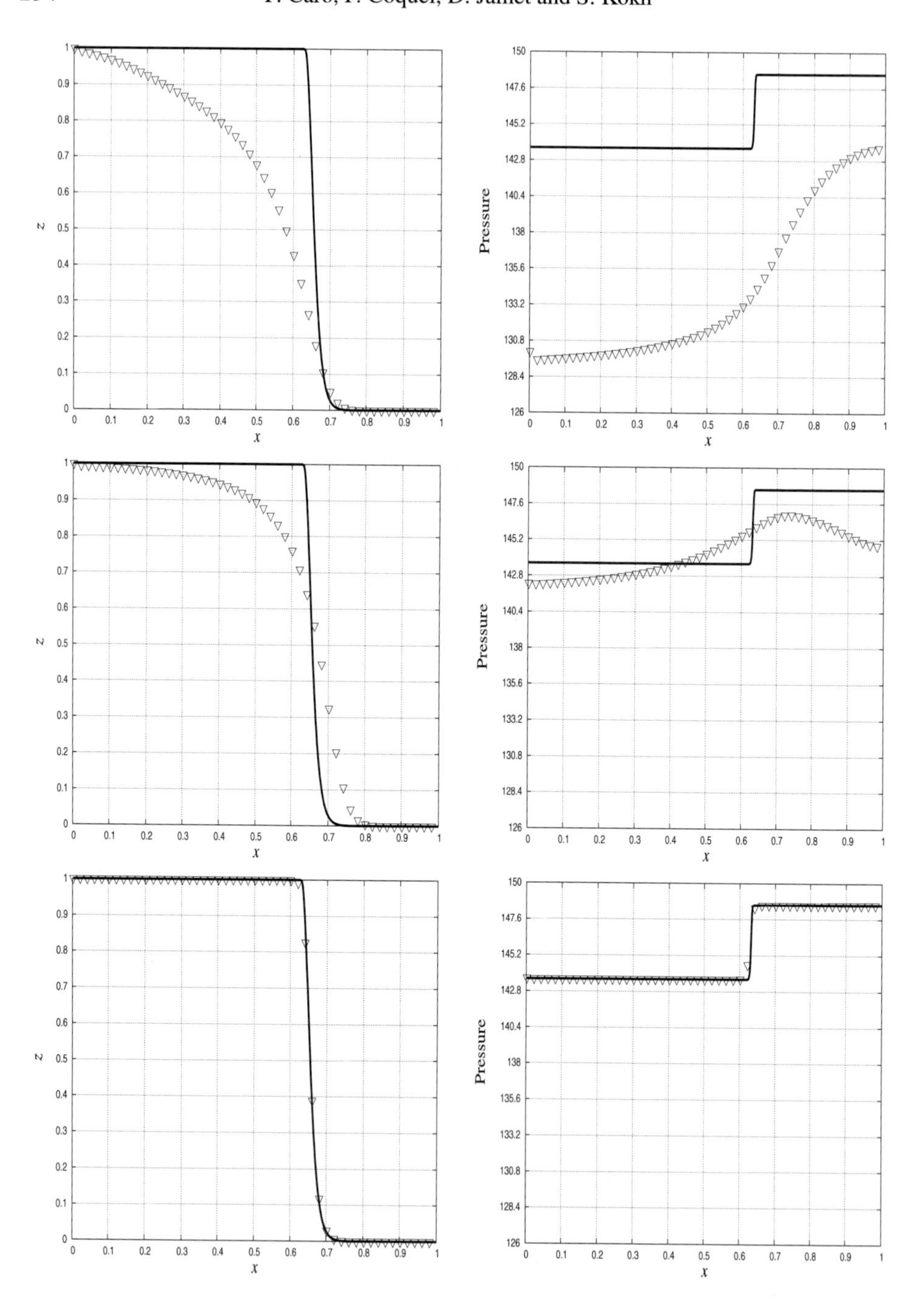

Figure 9. Profiles comparison between scheme 1 and 2 for z and P at time $t = 0.01$ s, $\lambda = 1$ (top), $\lambda = 10^2$ (center) and $\lambda = 10^4$ (bottom).

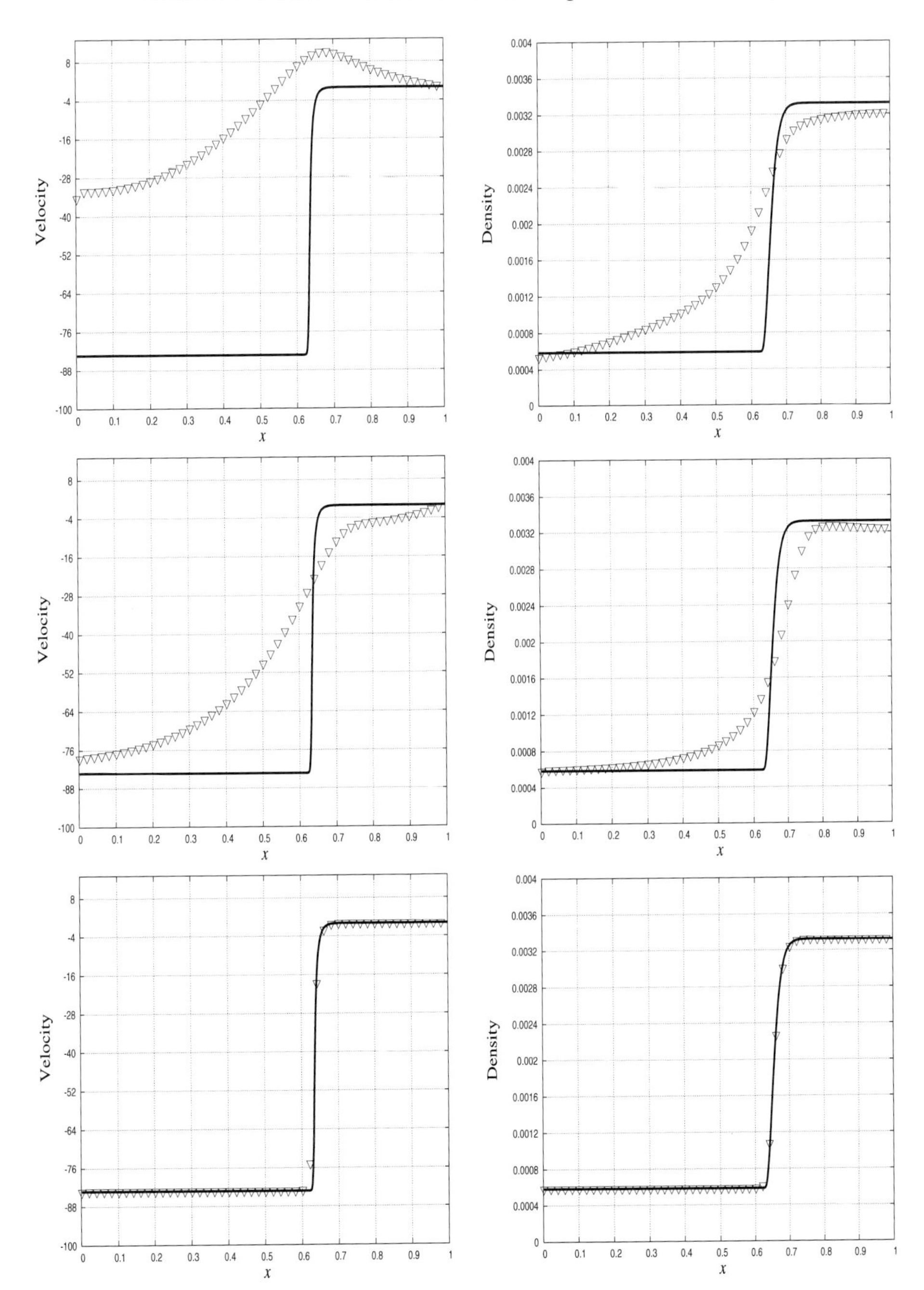

Figure 10. Profiles comparison between scheme 1 and 2 for u and ρ at time $t = 0.01$ s, $\lambda = 1$ (top), $\lambda = 10^2$ (center) and $\lambda = 10^4$ (bottom).

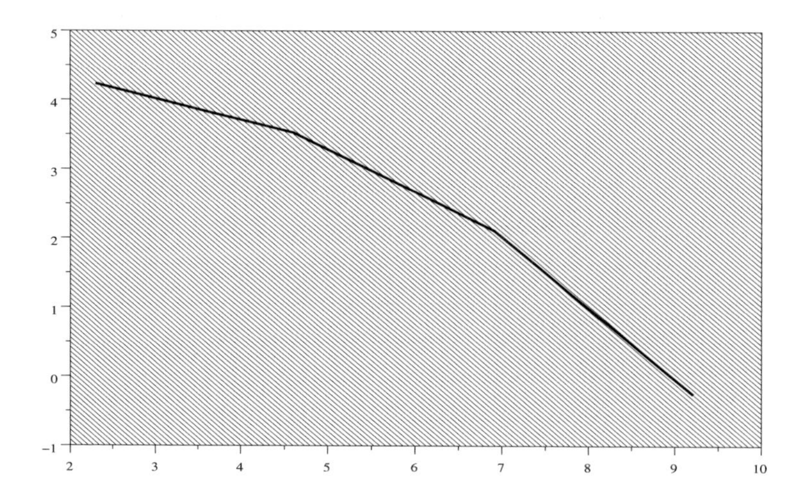

Figure 11. $\lambda \mapsto \ln\left(\|V_h^{\text{ref}} - V_h^{\lambda}\|_{L^1}\right)$ graph.

a much higher complex structure, and therefore more sophisticated theoretical tools seem required for their study.

References

[1] G. Allaire, S. Clerc, and S. Kokh, A five-equation model for the simulation of interfaces between compressible fluids, *J. Comput. Phys.*, 181 (2002), 577–616.

[2] A. Bedford, *Hamilton's principle in continuum mechanics*, Research Notes in Mathematics, Pitman (Advanced Publishing Program), Boston, MA, 1985.

[3] C. Chanteperdrix, P. Villedieu, and J. P. Vila, A compressible model for separated two-phase flows computations, *Proc. of ASME FEDSM'02*, 2002.

[4] Gui-Quiang Chen, C. David Levermore, and Tai-Ping Liu, Hyperbolic conservation laws with stiff relaxation terms and entropy, *Comm. Pure Appl. Math.* 47 (1994), 787–830.

[5] S. Gavrilyuk and H. Gouin, A new form of governing equations of fluids arising from hamilton's principle, *J. Engrg. Sci.* 37 (1999), 1495–1520.

[6] S. Gavrilyuk and R. Saurel, Mathematical and numerical modeling of two-phase compressible flows with micro-inertia, *J. Comput. Phys.* 175 (2002), 326–360.

[7] H. Gouin and S. L. Gavrilyuk, Hamilton's principle and Rankine-Hugoniot conditions for general motions of mixtures, *Meccanica* 34 (1999), 39–47.

[8] D. Jamet, O. Lebaigue, N. Coutris, and J. M. Delhaye, The second gradient method for the direct numerical simulation of liquid-vapor flows with phase change, *J. Comput. Phys.* 169 (2001), 624–651.

[9] S. Jaouen, *Etude mathématique et numérique de stabilité pour des modèles hydrodynamiques avec transition de phase*, PhD thesis, Univ. Paris 6, 2001.

[10] R. Saurel and R. Abgall, A simple method for compressible multifluid flows, *SIAM J. Sci. Comput.* 21 (1999), 1115–1145 (electronic).

[11] L. Truskinovsky, Kinks versus shocks, in *Shock Induced Transitions and Phase Structures in General Media* (R. Fosdick et al., eds.), IMA Vol. Math. Appl. 52, Springer-Verlag, Berlin 1991, 185–229.

Sharp and diffuse interface methods for phase transition problems in liquid-vapour flows

F. Coquel[1], *D. Diehl*[2], *C. Merkle*[2] *and C. Rohde*[2]

[1]*Laboratoire Jacques-Louis Lions, Université Pierre et Marie Curie*
4, place Jussieu, 75252 Cedex 05, France
email: coquel@ann.jussieu.fr

[2]*Mathematisches Institut, Universität Freiburg*
Hermann-Herder-Str. 10, 79104 Freiburg, Germany
email: dennis@mathematik.uni-freiburg.de
christian@mathematik.uni-freiburg.de
chris@mathematik.uni-freiburg.de

Abstract. We consider the dynamics of a compressible fluid exhibiting phase transitions between a liquid and a vapour phase. As the basic mathematical model we use the Euler equations for a sharp interface approach and local and global versions of the Navier–Stokes–Korteweg equations for the diffuse interface approach. The mathematical models are discussed and we introduce discretization methods for both approaches. Finally numerical simulations in one and two space dimensions are presented.

1 Introduction

For the numerical simulation of homogeneous compressible flow problems like the dynamics of a perfect gas reliable and efficient methods have been developed in the last two decades (cf. the monographs [GR96], [Krö97], [Lev02]). On the other hand methods on a comparable level for flows *with phase transitions* are not available up to now. Difficulties start even on the level of modelling by time-dependent partial differential equations.

In this contribution we consider an isotropic compressible fluid at a fixed temperature $T_* > 0$ such that the fluid can occur in a liquid and a vapour state. As the natural order parameter to identify the phases we choose the density $\rho \in (0, b)$, $b > 0$. We assume a free energy function $W \in C^2((0, b))$ to be defined by

$$W(\rho) = \rho w(\tau), \quad \tau = \frac{1}{\rho},$$

where $w = w(\tau)$ is the given internal energy and $\tau \in (1/b, \infty)$ is the specific volume. The pressure $p = p(\rho)$ is then defined from the standard thermomechanical relation

$$p(\rho) = \rho W'(\rho) - W(\rho) = -w'\left(\frac{1}{\rho}\right). \tag{1.1}$$

In the case where phase transitions occur W has a shape as in Figure 1 and we can define phases as follows. The fluid phase with density ρ is called

$$\begin{array}{lll} \text{vapour} & \text{if} & \rho \in (0, \alpha_1), \\ \text{spinodal} & \text{if} & \rho \in (\alpha_1, \alpha_2), \\ \text{liquid} & \text{if} & \rho \in (\alpha_2, b). \end{array}$$

The numbers $\alpha_1, \alpha_2 \in (0, b)$ are given as in Figure 1. The numbers $\beta_1 < \beta_2$ from Figure 1 are the unique numbers such that the chord connecting $(\beta_1, W(\beta_1))$ and $(\beta_2, W(\beta_2))$ has the same slope as the tangents of W in β_1 and β_2. They are called Maxwell-states.

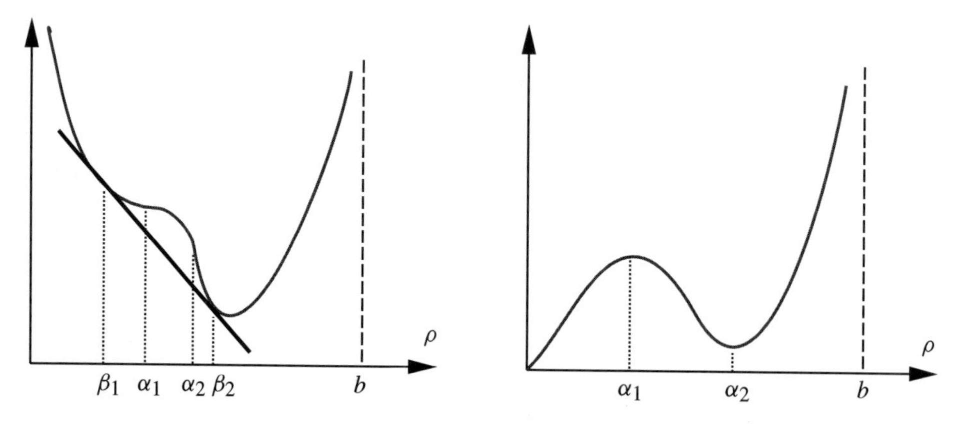

Figure 1. The lefthand picture shows the graph of a typical free energy density W in the presence of a liquid and vapour phase for the fluid under consideration. The density values that belong to convex branches of W are liquid or vapour states, the others are spinodal. The righthand picture shows a (Van-der-Waals) pressure function that is related to the energy W and the internal energy w by (1.1).

To capture the dynamics of the fluid correctly we want to use as our basic model an extension of the compressible isothermal Navier–Stokes equations. More specifically we consider the Navier–Stokes–Korteweg system (NSK-system). It contains an additional term that takes into account capillarity effects close to phase boundaries.

In Section 2 we derive two versions of the NSK-system to model capillarity. Starting from the description of the equilibrium situation there is one version with a classical third order derivative term and an another new one with a non-local integral term. Together with the sharp-interface model that is obtained by neglecting the viscous and capillarity terms in the NSK-systems we then have three different (but of course related) mathematical models to describe dynamical phase transitions. The section concludes with basic analytical statements on all three models.

Sections 3 and 4 are concerned with the discretization of initial boundary value problems for the NSK-systems and the sharp-interface approach, respectively.

In Section 3 we concentrate on the reliable simulation of the evolution towards equilibrium at $t = \infty$ in NSK-systems. It turns out that a direct discretization of the NSK-system leads to poor convergence rates. An improvement by means of a relaxation procedure is suggested and tested for several problems.

Finally in Section 4 we present a scheme for the sharp-interface case. Here it is in particular important to approximate the speed of a dynamical phase boundary correctly. Note that in the framework of our sharp-interface model phase interfaces are shock waves, however usually not of standard Lax but of undercompressive type. Based on the concept of the kinetic relation we extend a numerical method which originally has been developed for solid-solid phase transitions in elastodynamics.

In all cases we show what kinds of typical difficulties occur for fluid flow undergoing phase transitions and what prevents us from having such strong numerical tools as for homogeneous fluid flow.

2 Mathematical models for liquid-vapour phase transitions: equilibrium and dynamics

2.1 Equilibrium phase transitions

Let $\Omega \subset \mathbb{R}^2$ be an open and bounded set. For prescribed total mass we search for the density distributions $\rho : \Omega \to (0, b)$ that describe the equilibria of a liquid-vapour mixture in Ω. To find the solutions we consider three different approaches that determine the density via a minimization problem.

2.1.1 The sharp interface approach. For $m > 0$ we introduce an admissibility set for the density by

$$\mathbb{A}^0 = \Big\{\rho \in L^1(\Omega) \,\Big|\, W(\rho) \in L^1(\Omega), \int_\Omega \rho(x)\, dx = m\Big\}.$$

To find a static equilibrium we search for minimizers $\rho \in \mathbb{A}^0$ of the functional

$$F^0 = F^0[\rho] := \int_\Omega W(\rho(\boldsymbol{x}))\, d\boldsymbol{x}. \tag{2.1}$$

In view of the mass conservation constraint elementary variational calculus shows that a non-constant but piecewise constant minimizer of F^0 can only take the Maxwell states β_1 and β_2 (if it exists).

We consider an example. Let Ω be the box $(-0.5, 0.5)^2$ with volume $|\Omega| = 1$ and $m \in (\beta_1, (\beta_1 + \beta_2)/2)$. It is easy to check that the function

$$\rho(\boldsymbol{x}) = \begin{cases} \beta_1 & \boldsymbol{x} \in B_r(0) \\ \beta_2 & \boldsymbol{x} \in \Omega \setminus B_r(0), \end{cases} \qquad r^2 = \frac{\beta_2 - m}{\pi(\beta_2 - \beta_1)}$$

is a global minimizer of F^0. This is a physically relevant solution since it stands for a vapour bubble surrounded by liquid material. In the same way we can construct a liquid drop with $m \in ((\beta_1 + \beta_2)/2), \beta_2)$.

However there are many other global minimizers in $\mathbb{A}^0$ taking only the values β_1 and β_2 (cf. Figure 2). Mathematically speaking we need selection criteria to identify relevant minimizers of F^0. The physically relevant minimizers away from the boundary $\partial\Omega$ in an isotropic fluid are those where the length of the phase interface is minimal.

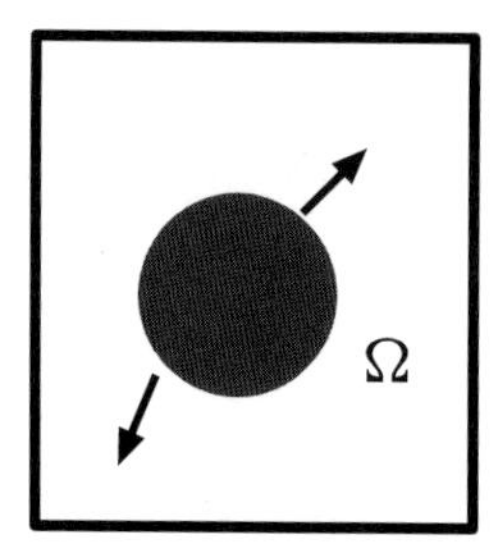

Figure 2. Possible minimizers of F^0. The regions of black color stand for $\rho \equiv \beta_1$ while otherwise we have $\rho \equiv \beta_2$. As long as the density configuration satisfies the mass side-condition all piecewise constant functions taking only values β_1, β_2 are minimizers of F^0.

2.1.2 The local diffuse interface approach. Van-der-Waals seemed to be the first who realized the nonuniqueness problem for the sharp interface approach and who suggested a modified approach ([vdW94]). The idea for his selection principle is to penalize the occurrence of phase interfaces appropriately by derivative terms.
Let the parameter $\varepsilon > 0$ be given and define the admissibility set $\mathbb{A}^{\text{local}}$ by

$$\mathbb{A}^{\text{local}} = H^1(\Omega) \cap \mathbb{A}^0.$$

For $\gamma > 0$ we look for a function $\rho^\varepsilon \in \mathbb{A}^{\text{local}}$ which is a minimizer of the modified functional

$$F^\varepsilon_{\text{local}} = F^\varepsilon_{\text{local}}[\rho^\varepsilon] = \int_\Omega \left(W(\rho^\varepsilon(\boldsymbol{x})) + \frac{\varepsilon^2\gamma}{2} \left|\nabla\rho^\varepsilon(\boldsymbol{x})\right|^2 \right) d\boldsymbol{x}. \tag{2.2}$$

Under appropriate growth conditions on W it can be shown that there are minimizers of $F^\varepsilon_{\text{local}}$. If a sequence of minimizers $\{\rho^\varepsilon\} \subset \mathbb{A}^{\text{local}}$ of (2.2) converges in $L^1(\Omega)$ then it is known that the limit selects a physically relevant solution, i.e., a solution which takes only the values β_1, β_2 and minimizes the length of the curve constituting the phase interface (cf. [Mod87] for the precise result in terms of BV-functions).

Finally let us write down the Euler–Lagrange equation for $F^{\varepsilon}_{\text{local}}$. Taking into account the mass constraint there is a constant $c \subset \mathbb{R}$ such that a minimizer ρ^{ε} of (2.2) satisfies almost everywhere in Ω the elliptic equation

$$D^{\varepsilon}_{\text{local}}[\rho^{\varepsilon}] := \varepsilon^2 \gamma \Delta \rho^{\varepsilon} - W'(\rho^{\varepsilon}) = c. \tag{2.3}$$

2.1.3 The global diffuse interface approach. The local diffuse interface approach requires more regular solutions than in the original sharp interface approach. We present an alternative that avoids spatial derivatives. Define the admissibility set $\mathbb{A}^{\text{global}}$ by

$$\mathbb{A}^{\text{global}} = \mathbb{A}^0 \cap L^2(\Omega).$$

An even and nonnegative function $\phi \in C^1(\mathbb{R}^2)$ is called interaction potential if ϕ satisfies $\int_{\mathbb{R}^2} \phi(\boldsymbol{x})\, d\boldsymbol{x} = 1$. For a given interaction potential ϕ we define the scaled interaction potential

$$\phi_{\varepsilon}(\boldsymbol{x}) := \frac{1}{\varepsilon^2} \phi\Big(\frac{\boldsymbol{x}}{\varepsilon}\Big) \quad (\boldsymbol{x} \in \mathbb{R}^2). \tag{2.4}$$

Now, for $\gamma > 0$, we search for functions $\rho^{\varepsilon} \in \mathbb{A}^{\text{global}}$ that minimize the modified functional

$$F^{\varepsilon}_{\text{global}} = F^{\varepsilon}_{\text{global}}[\rho^{\varepsilon}] = \int_{\Omega} \Big(W(\rho^{\varepsilon}(\boldsymbol{x})) + \frac{\gamma}{4} \int_{\Omega} \phi_{\varepsilon}(\boldsymbol{x} - \boldsymbol{y})(\rho^{\varepsilon}(\boldsymbol{y}) - \rho^{\varepsilon}(\boldsymbol{x}))^2\, d\boldsymbol{y} \Big)\, d\boldsymbol{x}. \tag{2.5}$$

The nonlocal term penalizes rapid variations in the density field on an ε-scale if the interaction potential has its support (essentially) in a ball around zero with $\mathcal{O}(\varepsilon)$-radius. The set $\mathbb{A}^{\text{global}}$ contains functions with jumps, i.e., sharp interfaces. A result similar to that of Modica has been established that shows that in the limit $\varepsilon \to 0$ the minimizers of $F^{\varepsilon}_{\text{global}}$ depict a physically relevant minimizer of F^0 ([AB98]).

We conclude with the Euler–Lagrange equation for $F^{\varepsilon}_{\text{global}}$, that is the integral equation

$$D^{\varepsilon}_{\text{global}}[\rho^{\varepsilon}] := \gamma(\phi_{\varepsilon} * \rho^{\varepsilon} - \rho^{\varepsilon}) - W'(\rho^{\varepsilon}) = c. \tag{2.6}$$

2.2 The Navier–Stokes–Korteweg systems

2.2.1 The derivation. We turn to the description of time-dependent liquid-vapour flow in two space dimensions. We search for three evolution equations that govern the dynamics of the three unknowns

$$\rho = \rho(\boldsymbol{x}, t),\ \boldsymbol{v} = (v_1, v_2)^T = (v_1(\boldsymbol{x}, t), v_2(\boldsymbol{x}, t))^T.$$

Here $\boldsymbol{v}$ denotes the velocity which vanishes for the equilibrium case in Section 2.1 and $(\boldsymbol{x}, t)$ is from the set $\Omega_T := \mathbb{R}^2 \times (0, T)$ for $T > 0$. Herivel and Lin have shown that the equations of motion for ideal fluids can be obtained by a variant of Hamilton's principle (cf. [Her55] and the review article [Ser59] of Serrin).

Assuming sufficient regularity for $\rho, \boldsymbol{v}$ this principle can also be applied here and leads to the following equations of motion in Ω_T:

$$\boldsymbol{v}_t + \boldsymbol{v} \cdot \nabla \boldsymbol{v} = \nabla D^{\varepsilon}[\rho]. \tag{2.7}$$

The term D^{ε} is given by either $D^{\varepsilon}_{\text{local}}$ from (2.3) or $D^{\varepsilon}_{\text{global}}$ from (2.6), depending whether we start from the local functional $F^{\varepsilon}_{\text{local}}$ or the nonlocal $F^{\varepsilon}_{\text{global}}$. If we use the relation $p'(\rho) = \rho W''(\rho)$ for the thermodynamic pressure $p = p(\rho)$ (cf. Figure 1) and add the law of mass conservation we finally arrive at the *local/global isentropic Navier–Stokes–Korteweg system*

$$\begin{aligned} \rho_t + \quad & \operatorname{div}(\rho \boldsymbol{v}) && = 0, \\ (\rho \boldsymbol{v})_t + \; & \operatorname{div}(\rho \boldsymbol{v} \otimes \boldsymbol{v} + p\mathbf{I}) && = \varepsilon \alpha \Delta \boldsymbol{v} + \gamma \rho \nabla \begin{cases} \varepsilon^2 \Delta \rho \\ \phi_\varepsilon * \rho - \rho, \end{cases} \end{aligned} \tag{2.8}$$

valid in Ω_T. Additionally we supplemented the ε-scaled viscosity term $\varepsilon\alpha\Delta\boldsymbol{v}$ with viscosity parameter $\alpha \geq 0$. The NSK-system extends the classical Navier–Stokes system by the capillarity or Korteweg terms. We are interested in small values of the parameter ε. The choice of scaling in ε between viscous and capillarity term in (2.8) is artificial up to now. However – as we see below – it is the correct one. Note that we skipped the upper index ε for the functions $\rho, \boldsymbol{v}$ for the sake of simplicity.

We have now three different models to describe phase transitions in a compressible medium:

Sharp interface model:	(2.8) with $\alpha = \gamma = 0$.
Local diffuse interface model:	(2.8) with $\alpha, \gamma > 0$ and the term $\varepsilon^2 \Delta \rho$.
Global diffuse interface model:	(2.8) with $\alpha, \gamma > 0$ and the term $\phi_\varepsilon * \rho - \rho$.

The sharp interface approach results of course in the Euler equations. Each of these models can be taken as a starting point for numerical discretization if an adequate wellposedness theory is available. For more background on the NSK-system we refer to [AMW98], [FS02] (local case) and [Roh04] (global case).

In the next section we discuss some analytical results for the three models as far as it is important for the construction of numerical methods or helps to understand the difficulties in numerics.

2.2.2 Energy estimates for the Navier–Stokes–Korteweg systems. Classical solutions of (2.8) with $\alpha, \gamma > 0$ show up with basic energy inequalities that we present here. The importance of energy inequalities for the numerical discretization will become clear in Section 3. We will show that standard discretizations do not satisfy (a discrete version of) the energy inequality. This failure is connected with a bad behaviour close to static equilibria. A modified discretization will cure this problem.

Proposition 2.1. *Let $(\rho, \boldsymbol{v})$ be a classical solution of the local diffuse interface model such that ρ and its spatial derivatives up to order three, v_i, $i = 1, 2$ and all its spatial derivatives vanish for $|\boldsymbol{x}| \to \infty$ and $t \in (0, T)$.*
Then we have for $t \in (0, T)$

$$\frac{d}{dt}\Big(F^{\varepsilon}_{\text{local}}[\rho(.,t)] + \int_{\mathbb{R}^2} \frac{1}{2}\rho(.,t)|\boldsymbol{v}(.,t)|^2\Big) + \alpha\varepsilon \int_{\mathbb{R}^2} |\nabla v_1(.,t)|^2 + |\nabla v_2(.,t)|^2 = 0,$$

provided all integrals exist. The domain of integration for $F^{\varepsilon}_{\text{local}}$ is $\Omega = \mathbb{R}^2$.

Proof. We proceed as in the case of the compressible Navier–Stokes equations. Let

$$\eta = \eta(\rho, \boldsymbol{m}) = \frac{|\boldsymbol{m}|^2}{2\rho} + W(\rho), \qquad \boldsymbol{m} = \rho \boldsymbol{v}.$$

We multiply each equation in the system (2.8) with the corresponding component of the gradient of η, that are $-\frac{1}{2}|\boldsymbol{v}|^2 + W'(\rho)$, v_1, v_2. We add up all equations and integrate with respect to space. Using the decay estimates to eliminate the fluxes and performing several partial integrations for the viscosity terms we obtain the energy inequality. The only non-standard term is the energy term $\frac{\gamma\varepsilon^2}{2}|\nabla\rho|^2$ which comes from the product $\gamma\varepsilon^2\rho\boldsymbol{v}\cdot\nabla\Delta\rho$ of $\boldsymbol{v}$ with the capillarity term and is derived as follows.

$$\begin{aligned}
\gamma\varepsilon^2 \int_{\mathbb{R}^2} \boldsymbol{v}(\boldsymbol{x},t)\rho(\boldsymbol{x},t)\cdot\nabla\Delta\rho(\boldsymbol{x},t)\,d\boldsymbol{x} &= -\gamma\varepsilon^2 \int_{\mathbb{R}^2} \operatorname{div}(\boldsymbol{v}(\boldsymbol{x},t)\rho(\boldsymbol{x},t))\Delta\rho(\boldsymbol{x},t)\,d\boldsymbol{x} \\
&= \gamma\varepsilon^2 \int_{\mathbb{R}^2} \rho_t(\boldsymbol{x},t)\Delta\rho(\boldsymbol{x},t)\,d\boldsymbol{x} \\
&= -\frac{d}{dt}\int_{\mathbb{R}^2} \frac{\gamma\varepsilon^2}{2}|\nabla\rho(\boldsymbol{x},t)|^2\,d\boldsymbol{x}.
\end{aligned}$$

We have used the decay estimates for ρ and the equation of mass conservation. □

Proposition 2.2. *Let $(\rho, \boldsymbol{v})$ be a classical solution of the global diffuse interface model such that ρ, v_i, $i = 1, 2$ and its first order spatial derivatives decay for $|\boldsymbol{x}| \to \infty$ and $t \in (0, T)$.*

Then we have for $t \in (0, T)$

$$\frac{d}{dt}\Big(F^{\varepsilon}_{\text{global}}[\rho(.,t)] + \int_{\mathbb{R}^2} \frac{1}{2}\rho(.,t)|\boldsymbol{v}(.,t)|^2\Big) + \alpha\varepsilon \int_{\mathbb{R}^2} |\nabla v_1(.,t)|^2 + |\nabla v_2(.,t)|^2 = 0,$$

provided all integrals exist. The domain of integration for $F^{\varepsilon}_{\text{global}}$ is $\Omega = \mathbb{R}^2$.

Proof. We can apply the same energy technique for (2.8) as in Proposition 2.1. The only non-standard term is now the nonlocal contribution in $F^{\varepsilon}_{\text{global}}$ which arises in the following manner.

$$\gamma \int_{\mathbb{R}^2} \boldsymbol{v}(\boldsymbol{x},t)\rho(\boldsymbol{x},t)\cdot\nabla\big([\phi_\varepsilon * \rho(.,t)](\boldsymbol{x}) - \rho(\boldsymbol{x},t)\big)\,d\boldsymbol{x}$$

$$= -\gamma \int_{\mathbb{R}^2} \operatorname{div}(\boldsymbol{v}(\boldsymbol{x},t)\rho(\boldsymbol{x},t))\big([\phi_\varepsilon * \rho(.,t)](\boldsymbol{x}) - \rho(\boldsymbol{x},t)\big)\,d\boldsymbol{x}$$

$$= \gamma \int_{\mathbb{R}^2} \rho_t(\boldsymbol{x},t)\big([\phi_\varepsilon * \rho(.,t)](\boldsymbol{x}) - \rho(\boldsymbol{x},t)\big)\,d\boldsymbol{x}$$

$$= -\frac{d}{dt}\int_{\mathbb{R}^2}\frac{\gamma}{4}\int_{\mathbb{R}^2}\phi_\varepsilon(\boldsymbol{x}-\boldsymbol{y})(\rho(\boldsymbol{x},t)) - \rho(\boldsymbol{y},t))^2\,d\boldsymbol{y}d\boldsymbol{x}.$$

To derive the last equation observe

$$\frac{d}{dt}\int_{\mathbb{R}^2}\frac{\gamma}{4}\int_{\mathbb{R}^2}\phi_\varepsilon(\boldsymbol{x}-\boldsymbol{y})(\rho(\boldsymbol{x},t)) - \rho(\boldsymbol{y},t))^2\,d\boldsymbol{y}d\boldsymbol{x}$$

$$= \frac{\gamma}{2}\int_{\mathbb{R}^2}\int_{\mathbb{R}^2}\phi_\varepsilon(\boldsymbol{x}-\boldsymbol{y})(\rho(\boldsymbol{y},t) - \rho(\boldsymbol{x},t))\rho_t(\boldsymbol{y},t)\,d\boldsymbol{y}d\boldsymbol{x}$$

$$+ \frac{\gamma}{2}\int_{\mathbb{R}^2}\int_{\mathbb{R}^2}\phi_\varepsilon(\boldsymbol{y}-\boldsymbol{x})(\rho(\boldsymbol{x},t) - \rho(\boldsymbol{y},t))\rho_t(\boldsymbol{x},t)\,d\boldsymbol{y}d\boldsymbol{x}$$

$$= \gamma\int_{\mathbb{R}^2}\int_{\mathbb{R}^2}\phi_\varepsilon(\boldsymbol{x}-\boldsymbol{y})(\rho(\boldsymbol{y},t) - \rho(\boldsymbol{x},t))\rho_t(\boldsymbol{y},t)\,d\boldsymbol{y}d\boldsymbol{x}$$

$$= -\gamma\int_{\mathbb{R}^2}\big([\phi_\varepsilon * \rho(.,t)](\boldsymbol{x}) - \rho(\boldsymbol{x},t)\big)\rho_t(\boldsymbol{x},t)\,d\boldsymbol{x}.$$

Note that we used the symmetry of ϕ and $\int_{\mathbb{R}^2}\phi(\boldsymbol{x})\,d\boldsymbol{x} = 1$. □

2.2.3 Notes on local-in-time existence theorems. The most reliable analytical background for numerics would be to have a global-in-time existence theory. This cannot be delivered by state-of-the-art methods. We discuss here the question of local-in-time-existence for classical solutions of the Cauchy problem with inital data

$$\rho(.,t) = \rho_0, \quad \boldsymbol{v} = \boldsymbol{v}_0 \text{ in } \mathbb{R}^2 \tag{2.9}$$

satisfying

$$\text{there exists } \hat{\rho} > 0 \text{ such that } \rho_0 - \hat{\rho},\ \boldsymbol{v}_0 \in H^5(\mathbb{R}^2),\ \rho_0 > 0. \tag{2.10}$$

Sobolev embedding ensures that the initial datum is smooth enough such that all functions and their derivatives appearing are smooth.

Let us start with the Euler case $\alpha = \gamma = 0$ and view the system as a first-order conservation law in the unknown $\boldsymbol{u} = (\rho, \rho\boldsymbol{v})^T$. If we denote the two flux functions with f_1, f_2 the three eigenvalues of the Jacobian $n_1 Df_1(\boldsymbol{u}) + n_2 Df_2(\boldsymbol{u})$ ($\boldsymbol{n} = (n_1, n_2) \in \mathcal{S}_1^2$, $u \in (0,b) \times \mathbb{R}^2$) are given by

$$\lambda_1(\boldsymbol{u},\boldsymbol{n}) = \boldsymbol{n}\cdot\boldsymbol{v} - \sqrt{p'(\rho)}, \quad \lambda_2(\boldsymbol{u},\boldsymbol{n}) = \boldsymbol{n}\cdot\boldsymbol{v}, \quad \lambda_3(\boldsymbol{u},\boldsymbol{n}) = \boldsymbol{n}\cdot\boldsymbol{v} + \sqrt{p'(\rho)}.$$

We observe (cf. Figure 1) that the system is *not* hyperbolic since the above eigenvalues may fail to be real-valued. Therefore the standard theory does not apply, there cannot be classical solutions if the smooth initial data takes values in the elliptic phase. Phase boundaries are ruled out. For the numerics we need another well-posedness theory. This will be the concept of entropy solutions supplemented with the kinetic relation (Section 4).

For the diffuse interface models (2.8) with $\alpha, \gamma > 0$ the higher-order differential operators dominate (at least in the critical momentum equations). The following results have been proven in [HL94], [Roh04].

Theorem 2.3. *Let* (2.10) *be satisfied. Then there exists a constant $t_* > 0$ such that the Cauchy problem for* (2.8) *has a classical solution $\boldsymbol{u}$ in Ω_{t_*}. The classical solution $\boldsymbol{u}$ is unique in the class of classical solutions.*

We assume that also in the large there exists a unique classical solution and will base the numerics on this assumption.

3 Diffuse interface model

We consider the initial boundary value problem for the *local* Navier–Stokes–Korteweg system (2.8) in this section. Here we deal with the numerical approximation of the solution. Standard methods fail to give accurate results in the situation where the solution is close to the static equilibrium, i.e. the numerical solution given by a standard scheme does not satisfy (2.3) on the discrete level as time tends to infinity. The second problem is that most standard schemes cannot be applied because of the lack of hyperbolicity of the first-order part of the Navier–Stokes–Korteweg system.

We tackle the first problem when addressing the momentum equation in nonconservative form (see also [Jac99] and [JT02] for a related approach) and we circumvent the lack of hyperbolicity by using a relaxation formulation of the original model.

3.1 Problems with standard schemes

In this section we apply a very simple scheme to solve the initial boundary value problem for the Navier–Stokes–Korteweg system (2.8) numerically. We discretize the first order terms in equation (2.8) by using the Lax–Friedrichs flux, the higher order terms are discretized by central differences. We will see that this simple scheme gives very bad results in the situation where the solution is close to the static equilibrium. Further we know from Proposition 2.1 that the total energy E given by the relation

$$E(t) = \int_{\Omega} W(\rho) + \frac{\varepsilon^2\gamma}{2}|\nabla\rho|^2 + \frac{1}{2}\rho|\boldsymbol{v}|^2 d\boldsymbol{x}$$

decreases with time. Thus, a reasonable scheme should give a numerical solution that shows the same behavior, but from Figure 3 we can see that the simple scheme does not have this behavior. For the numerical experiment we have chosen the computational

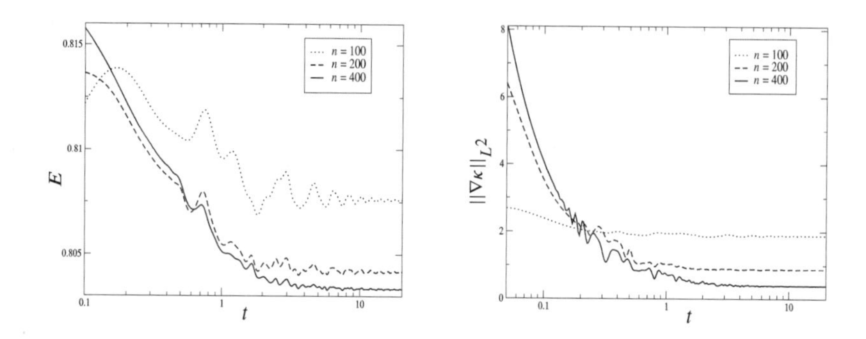

Figure 3. Graphs of the total energy (left) and graphs of $||\nabla\kappa||_{L^2}$ (right).

domain $\Omega = [-1, 1]^2$ with an underlying Cartesian mesh with $n \times n$ cells, $n = 100, 200, 400$. Boundary conditions are chosen periodically and the initial values are given by

$$\rho_0(x, y) = \begin{cases} 0.3 \text{ if } \max\{|x|, |y|\} \le 0.4 \\ 1.8 \text{ else,} \end{cases} \qquad \boldsymbol{v}_0 = \boldsymbol{0}.$$

Figure 3 shows the time evolution of the total energy (left), and the function $t \mapsto ||\nabla\kappa\big(\rho(\cdot, t), \Delta\rho(\cdot, t)\big)||_{L^2(\Omega)}$ (right) where κ is defined by

$$\kappa(\rho, \Delta\rho) = W'(\rho) - \varepsilon^2\gamma\Delta\rho. \tag{3.1}$$

The solution should converge to the static equilibrium as time tends to infinity. This means κ should converge to a constant as $t \to \infty$, but the numerical solution behaves differently. Let us also note that the discrete total energy is not strictly decreasing in time.

The numerical solution at time $t = 20$ (from this time the numerical solution does not change essentially) is presented Figure 4 for three different mesh sizes. At this time the exact solution should be very close to the static equilibrium and the velocity field should have almost vanished, but we observe that the scheme produces a velocity field of order of the mesh size inside the interface. Other schemes have similar problems with spurious velocities inside the interface at the static equilibrium state. In [JLCD01] these velocities are called parasitic currents.

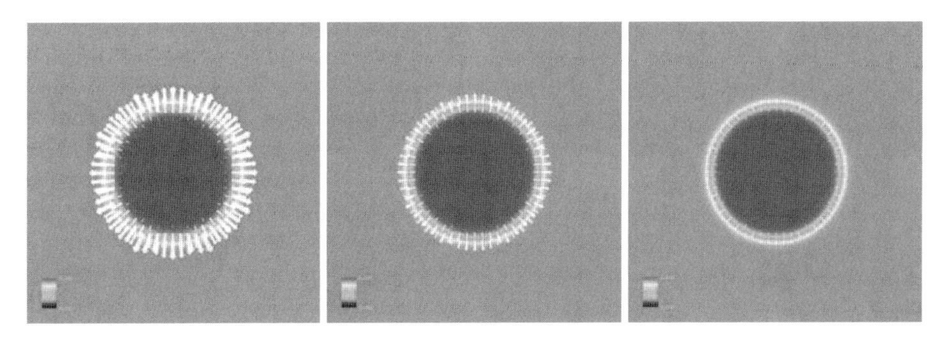

Figure 4. Density component of the approximate solution at time $t = 20$ for $n = 100, 200, 400$. The white arrows represent the velocity. This display method is used throughout this section.

The rest of this section is dedicated to the development of a scheme that gives more accurate results at the static equilibrium state and which gives solutions with decreasing energy (at least in all of our numerical test cases).

3.2 Formulation of the relaxation system

With the definition of κ in (3.1) and the use of the thermodynamic relation $p'(\rho) = \rho W''(\rho)$ we can rewrite system (2.8) as a nonconservative system (see also [Jac99] and [JT02]):

$$\left.\begin{aligned} \rho_t + \operatorname{div}(\rho \boldsymbol{v}) &= 0 \\ (\rho \boldsymbol{v})_t + \operatorname{div}(\rho \boldsymbol{v} \otimes \boldsymbol{v}) + \rho \nabla \kappa(\rho, \Delta\rho) &= \varepsilon \alpha \Delta \boldsymbol{v} \end{aligned}\right\} \quad \text{in } \Omega \times (0, T). \tag{3.2}$$

In the next step we understand $\kappa = \kappa(\boldsymbol{x}, t) \in \mathbb{R}$ as a *new independent unknown* and consider the following relaxation approximation for (3.2). We search for $(\rho, \boldsymbol{v}, \kappa)^T : \mathbb{R}^2 \times (0, T) \to (0, \infty) \times \mathbb{R}^3$ such that

$$\begin{aligned} \rho_t + \operatorname{div}(\rho \boldsymbol{v}) &= 0, \\ (\rho \boldsymbol{v})_t + \operatorname{div}(\rho \boldsymbol{v} \otimes \boldsymbol{v}) + \rho \nabla \kappa &= \varepsilon \alpha \Delta \boldsymbol{v}, \\ \kappa_t + \boldsymbol{v} \cdot \nabla \kappa + \frac{a^2}{\rho^2} \nabla \boldsymbol{v} &= \frac{\tilde{\mu}(\rho, \Delta\rho) - \kappa}{d} \end{aligned} \tag{3.3}$$

holds in $\Omega \times (0, T)$. The parameter $d > 0$ is the (small) relaxation parameter and

$$\tilde{\mu}(\rho, \Delta\rho) = W'(\rho) - \varepsilon^2 \gamma \Delta\rho.$$

The constant a is chosen according to a generalized Whitham condition:

$$a^2 > \rho^2 c^2, \quad c^2 = p'(\rho) + \frac{\varepsilon^2 \gamma \rho}{\Delta x^2}. \tag{3.4}$$

Note that the choice of the parameter a depends on the up-to-now free parameter $\Delta x > 0$ which later stands for the mesh width of the numerical mesh. At first Δx has to be chosen so small such that the last term in (3.4) is nonnegative. In fact condition (3.4) is derived from a Fourier analysis for a linearized version of (3.3): If (3.4) holds for given $\Delta x > 0$ all *discrete waves* that can occur on a uniform mesh with mesh size Δx are slower than the characteristic speeds of the (then automatically hyperbolic) system (3.3). In this sense it is a generalized Whitham condition. Finally we note that our relaxation approach can be considered as Sulicius relaxation method in Eulerian coordinates [Sul90]. See also [CP98] and, for details on the choice of a the companion paper [CDR04].

Before we discuss the discretization let us note basic facts on the left hand side of system (3.3). Due to rotational invariance it suffices for all our analytical issues to consider the one-dimensional version only. The one-dimensional system is (of course) also a nonconservative system. Omitting the righthand side in (3.3) we get in primitive variables the first-order system

$$\begin{aligned} \rho_t + (\rho v)_x &= 0, \\ v_t + v v_x + \kappa_x &= 0, \\ \kappa_t + v \kappa_x + \frac{a^2}{\rho^2} v_x &= 0. \end{aligned} \tag{3.5}$$

Let us summarize the primitive unknowns ρ, v, κ of (3.5) into the vector

$$u = (\rho, v, \kappa)^T.$$

System (3.5) is a strictly hyperbolic system in $\mathcal{U} = (0, \infty) \times \mathbb{R}^2$ with the property that all characteristic fields are linearly degenerate. The corresponding characteristic speeds are given by

$$\lambda_1(u) = v - \frac{a}{\rho}, \quad \lambda_2(u) = v, \quad \lambda_3(u) = v + \frac{a}{\rho}, \quad (u \in \mathcal{U}). \tag{3.6}$$

For numerical purpose it is important to know about the structure of the Riemann problem for system (3.5), i.e. for states $u_L, u_R \in \mathcal{U}$ we consider the initial datum

$$u_0(x) = \begin{cases} u_L, & x < 0, \\ u_R, & x > 0. \end{cases}$$

This Riemann problem cannot be treated by routine methods since the system (3.5) is in nonconservative form. However due to the linear degeneracy of (3.5) it is possible to give meaning to the nonconservative products and of course due to the linear degeneracy the solution of the Riemann problem has a very simple structure which is important for numerical efficiency. We summarize the results in the following theorem. For details see [CDR04].

Theorem 3.1 (Solution of the Riemann problem). *Let the states $u_L, u_R \in \mathcal{U}$ be given. Then there exists a generalized solution $u : \mathbb{R} \times [0, T] \to \mathcal{U}$ of the corresponding Riemann problem (in the sense of* [DMLM95]*).*

The solution u consists of the four states $u_L, u_L^*, u_R^*, u_R \in \mathcal{U}$ which are separated by three contact discontinuities which travel with speeds $s_1, s_2, s_3 \in \mathbb{R}$ given by

$$
\begin{aligned}
s_1 &= v_L - \frac{a}{\rho_L},\\
s_2 &= v^*,\\
s_3 &= v_R + \frac{a}{\rho_R}.
\end{aligned}
\tag{3.7}
$$

The states $u_L^* = (\rho_L^*, v^*, \kappa^*)^T$ and $u_R^* = (\rho_R^*, v^*, \kappa^*)^T \in \mathcal{U}$ are defined by

$$
\begin{aligned}
v^* &= \frac{\tau_l v_l + \tau_r v_r - \frac{1}{a}(\kappa_r - \kappa_l)}{\tau_l + \tau_r},\\
\kappa^* &= \kappa_l + a\tau_l(v_l - v^*),\\
\rho_L^* &= \frac{a}{v^* - s_1},\\
\rho_R^* &= -\frac{a}{v^* - s_3},
\end{aligned}
\tag{3.8}
$$

with $\tau = \frac{1}{\rho}$.

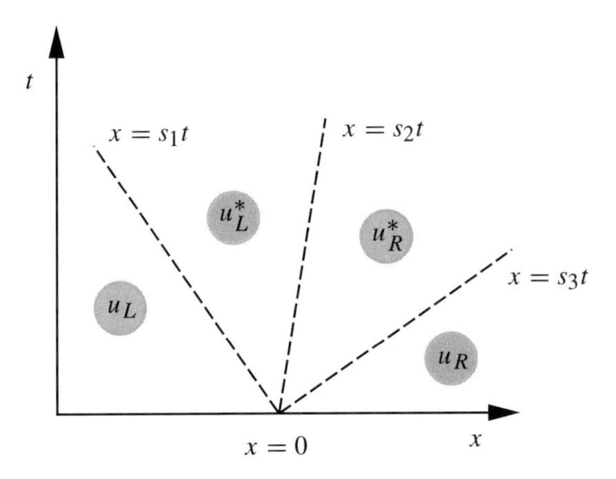

Figure 5. The structure of the self-similar solution of the Riemann problem in the (x, t)-halfspace.

3.3 The relaxation scheme in 1-D

For simplicity we assume that the underlying mesh is of uniform size $\Delta x > 0$. First we provide the discretized initial data

$$
\begin{aligned}
\rho_j^0 &= \frac{1}{\Delta x}\int_{x_{j-\frac{1}{2}}}^{x_{j+\frac{1}{2}}} \rho_0(x)dx, \\
(\rho u)_j^0 &= \frac{1}{\Delta x}\int_{x_{j-\frac{1}{2}}}^{x_{j+\frac{1}{2}}} (\rho_0 v_0)(x)dx, \\
\kappa_j^0 &= W'(\rho_j^0) - \frac{\varepsilon^2\gamma}{\Delta x^2}(\rho_{j+1}^0 - 2\rho_j^0 + \rho_{j-1}^0).
\end{aligned}
$$

The update procedure from one timestep to another consists of four parts. First we choose the artificial parameter a locally at the cell interfaces according to the generalized Whitham condition (3.4).

Second, we neglect the right hand side in (3.3) and consider the first-order system

$$
\begin{aligned}
\rho_t + (\rho v)_x &= 0, \\
(\rho v)_t + (\rho v^2)_x + \rho\kappa_x &= 0, \\
\kappa_t + v\kappa_x + \frac{a^2}{\rho^2}v_x &= 0.
\end{aligned}
$$

In the third step we take the viscous part of the equation into account

$$
\begin{aligned}
\rho_t &= 0, \\
(\rho v)_t &= \varepsilon\alpha v_{xx}, \\
\kappa_t &= 0.
\end{aligned}
$$

The fourth step is the relaxation step, we solve the ordinary differential equations

$$
\begin{aligned}
\rho_t &= 0, \\
(\rho v)_t &= 0, \\
\kappa_t &= \frac{\tilde{\mu}(\rho, \Delta\rho) - \kappa}{d}
\end{aligned}
$$

as the relaxation parameter d tends to zero. As initial data we take the data from the third step. This means this is a projection of κ back to the *equilibrium manifold.* Because d tends to zero we get

$$
\kappa = \tilde{\mu}(\tilde{\rho}, \Delta\tilde{\rho}),
$$

where $\tilde{\rho}$ is the data that comes from the third step. We summarize the update procedure from timestep n to $n + 1$ as follows:

1) Choose the parameter a locally at the cell interfaces according to the generalized Whitham condition (3.4)

$$a^2_{j+\frac{1}{2}} = \max_{i=j,j+1} \left\{ (\rho_i^n)^2 \left(p'(\rho_i^n) + \frac{\varepsilon^2 \gamma \rho_i^n}{\Delta x^2} \right) \right\}.$$

2, 3) We combine the second and third step. Solve the Riemann problem at each cell interface $x_{j+\frac{1}{2}}$ with initial data $(\rho_j^n, v_j^n, \kappa_j^n)$, $(\rho_{j+1}^n, v_{j+1}^n, \kappa_{j+1}^n)$. Let $(\tilde{\rho}_{j+\frac{1}{2}}, \tilde{v}_{j+\frac{1}{2}}, \tilde{\kappa}_{j+\frac{1}{2}})$ denote the solution of the corresponding Riemann problem. Then approximate the viscous term.

$$\begin{aligned}
\rho_j^{n+1} &= \rho_j^n - \frac{\Delta t}{\Delta x} \left(\tilde{\rho}_{j+\frac{1}{2}}(0) \tilde{v}_{j+\frac{1}{2}}(0) - \tilde{\rho}_{j-\frac{1}{2}}(0) \tilde{v}_{j-\frac{1}{2}}(0) \right), \\
(\rho v)_j^{n+1} &= (\rho v)_j^n - \frac{\Delta t}{\Delta x} \left(\tilde{\rho}_{j+\frac{1}{2}}(0) \tilde{v}_{j+\frac{1}{2}}(0)^2 - \tilde{\rho}_{j-\frac{1}{2}}(0) \tilde{v}_{j-\frac{1}{2}}(0)^2 \right) \\
&\quad - \frac{\Delta t}{\Delta x} (\nu^R_{j-\frac{1}{2}} + \nu^L_{j+\frac{1}{2}}) \\
&\quad + \varepsilon \alpha \frac{\Delta t}{\Delta x^2} (v_{j+1}^n - 2 v_j^n + v_{j-1}^n),
\end{aligned}$$

with

$$\begin{aligned}
\nu^R_{j-\frac{1}{2}} &= \int_{x_{j-\frac{1}{2}}}^{x_j} \tilde{\rho}_{j-\frac{1}{2}}(x) \partial_x \tilde{\kappa}_{j-\frac{1}{2}}(x) dx, \\
\nu^L_{j+\frac{1}{2}} &= \int_{x_j}^{x_{j+\frac{1}{2}}} \tilde{\rho}_{j+\frac{1}{2}}(x) \partial_x \tilde{\kappa}_{j+\frac{1}{2}}(x) dx.
\end{aligned}$$

4) Finally perform the relaxation step

$$\kappa_j^{n+1} = W'(\rho_j^{n+1}) - \frac{\varepsilon^2 \gamma}{\Delta x^2} (\rho_{j+1}^{n+1} - 2\rho_j^{n+1} + \rho_{j-1}^{n+1}).$$

Note: Let $a_1 < a_2 < a_3$ and

$$\rho(x) = \begin{cases} \rho_l, & x \in (a_1, a_2) \\ \rho_r, & x \in (a_2, a_3), \end{cases} \qquad \kappa(x) = \begin{cases} \kappa_l, & x \in (a_1, a_2) \\ \kappa_r, & x \in (a_2, a_3). \end{cases}$$

Then we set $\int_{a_1}^{a_2} \rho(x) \kappa_x(x) dx = \frac{1}{2}(\rho_l + \rho_r)(\kappa_r - \kappa_l)$. Note that the solution of the Riemann problem can have zero, one or two jumps in the intervals $(x_{j-\frac{1}{2}}, x_j)$ and $(x_j, x_{j+\frac{1}{2}})$.

3.4 Extension to 2-D

For the sake of simplicity we neglect the viscous terms in this section. In order to describe the scheme in two space dimensions on a Cartesian mesh we have to consider planar waves solving the system (3.3). Due to rotational invariance it is sufficient to consider planar waves that propagate in x-direction only. These waves satisfy the equation

$$\begin{aligned} \rho_t + (\rho v_1)_x &= 0, \\ (\rho v_1)_t + (\rho v_1^2)_x + \rho \kappa_x &= 0, \\ (\rho v_2)_t + (\rho v_1 v_2)_x &= 0, \\ \kappa_t + v_1 \kappa_x + \frac{a^2}{\rho^2} v_{1,x} &= \frac{\tilde{\mu}(\rho, \Delta\rho) - \kappa}{d}. \end{aligned} \tag{3.9}$$

We can easily verify that the left hand side of system (3.9) is a hyperbolic (but not strictly hyperbolic) system in $\mathcal{U} = (0, \infty) \times \mathbb{R}^3$ and all characteristic fields are linearly degenerate. The characteristic speeds are given by

$$\lambda_1(u) = v_1 - \frac{a}{\rho}, \quad \lambda_2(u) = \lambda_3(u) = v_1, \quad \lambda_4(u) = v_1 + \frac{a}{\rho} \qquad (u \in \mathcal{U}).$$

Now the solution of the Riemann problem for the left hand side of equation (3.9) has almost the same structure as the Riemann problem for the 1-D equation. We generalize Theorem 3.1.

Theorem 3.2 (Solution of the Riemann problem). *Let the states $u_L, u_R \in \mathcal{U}$ be given. Then there exists a generalized solution $u : \mathbb{R}^2 \times [0, T] \to \mathcal{U}$ of the corresponding planar Riemann problem (in the sense of* [DMLM95]).

The solution u consists of the four states $u_L, u_L^*, u_R^*, u_R \in \mathcal{U}$ which are separated by four contact discontinuities which travel with speeds $s_1, s_2 = s_2', s_3 \in \mathbb{R}$ given by (3.7). *The states $u_L^*, u_R^* \in \mathcal{U}$ are defined by* (3.8) *and*

$$\begin{aligned} v_{2,L}^* &= v_{2,L}, \\ v_{2,R}^* &= v_{2,R}. \end{aligned} \tag{3.10}$$

Thus, the formulation of the scheme on Cartesian meshes is straightforward. For the approximation of the Laplacians we use the five point stencil. We omit the details.

3.5 Numerical experiments with the relaxation scheme

This section is dedicated to numerical experiments with the relaxation scheme described in the previous section. The computational domain and the parameters are chosen as in Section 3.1.

3.5.1 Test case 1: return to equilibrium. This is the same numerical experiment as described in Section 3.1, but now with the relaxation scheme. As in Section 3.1 we have used three different mesh sizes in the computation ($n = 100, 200, 400$). Figure 6 shows the numerical solution at time $t = 0, 1, 20$ for each computation. We can see that the relaxation scheme does not produce these velocities inside the interface we have seen in the experiment with the simple scheme which leads to a more accurate result at the static equilibrium. The right sub-figure of Figure 7 indicates that κ tends to zero if $t \to \infty$ which means we end up with the correct equilibrium state on the numerical level. The left part of this figure shows that the total energy decreases with time (the behavior we have required for a good scheme).

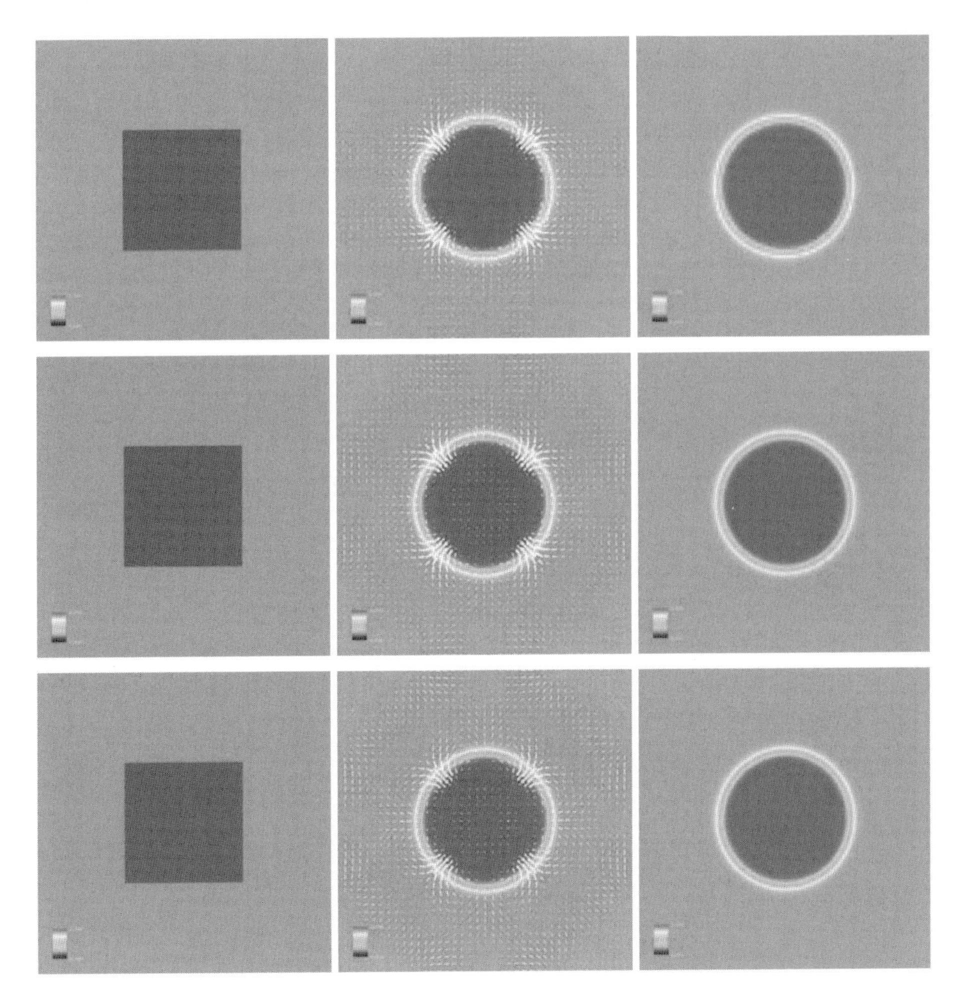

Figure 6. Return to equilibrium: solution at $t = 0, 1, 20$ for $n = 100, 200, 400$ respectively.

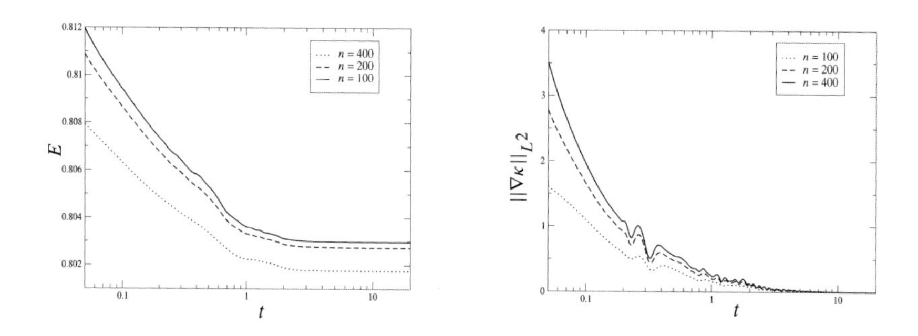

Figure 7. Return to equilibrium: graphs of the total energy (left) and graphs of $||\nabla\kappa||_{L^2}$ (right).

3.5.2 Test case 2: return to equilibrium for two bubbles. We start with two *square* bubbles, one slightly smaller than the other. First these bubbles form spherical bubbles, then the smaller bubble shrinks and the larger bubble increases without having contact. At the end it remains only one bubble that reaches a static equilibrium state (at $t = \infty$). This time behaviour fully agrees with the physics (see [CDR04]). The velocity field is initially chosen equal to zero. For this computation we used only one mesh size ($n = 200$).

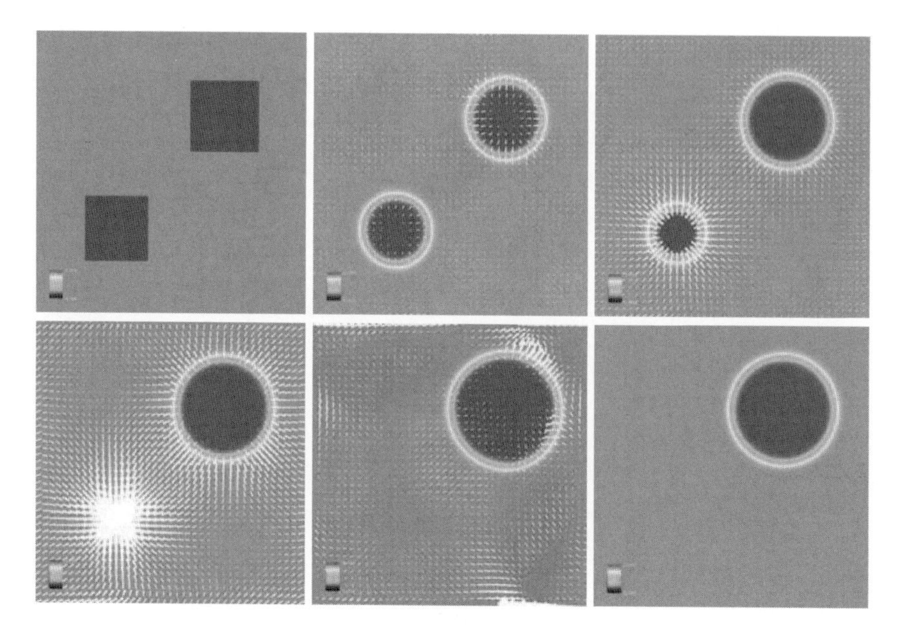

Figure 8. Merging Bubbles: Solution at time $t = 0, 1, 4, 5, 6, 20$.

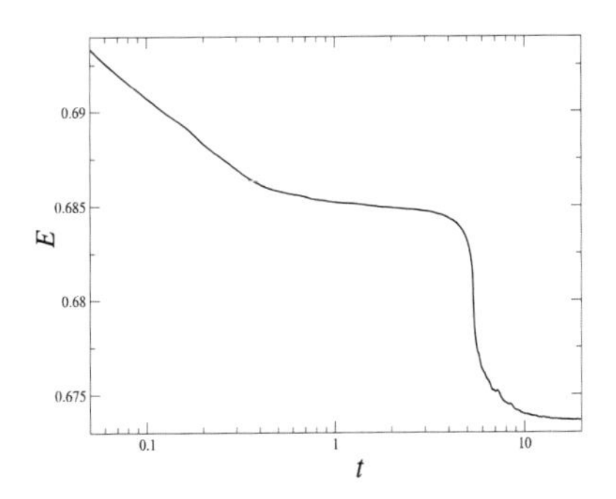

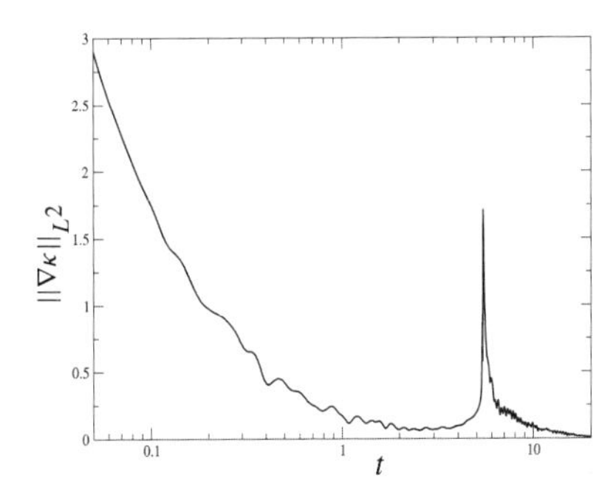

Figure 9. Merging Bubbles: Graph of the total energy (left) and graph of $||\nabla\kappa||_{L^2}$ (right).

4 Sharp-interface model

In this section we are dealing with the sharp interface model as discussed earlier in (2.8) with $\alpha = \gamma = 0$. All our considerations are done in 1D in Lagrangian coordinates. First we discuss the differences to the classical hydrodynamical case. In Section 4.2 we regularize the system to overcome problems of the nonhyperbolic case. Finally, we present a numerical example.

4.1 Entropy solution and kinetic relation

An isothermal model of a liquid-vapour phase transition is described in Lagrangian coordinates in 1D by the following system of partial differential equations (cf. the sharp interface model (2.8) with $\alpha = \gamma = 0$ in Section 2.2.1):

$$\begin{aligned} \tau_t - v_x &= 0 \quad \text{in } \mathbb{R} \times (0, T), \\ v_t + p(\tau)_x &= 0 \quad \text{in } \mathbb{R} \times (0, T), \end{aligned} \tag{4.1}$$

with initial conditions

$$(\tau, v)(x, 0) = (\tau_0, v_0)(x) \quad \text{in } \mathbb{R}.$$

Here, v is the velocity and $\tau = \frac{1}{\rho}$ the specific volume. The function $p = p(\tau)$ is given by the negative derivative of the internal energy function: $p(\tau) = -w'(\tau)$. Note that p is *not* the same function as in the previous sections. It should be denoted by e. g. $\tilde{p}$ and we have then the relation

$$\tilde{p}(\tau) = p\left(\frac{1}{\rho}\right),$$

which relates $\tilde{p}$ to the original pressure p given in (1.1) as a function of density ρ. But for the simplicity of the notation we use the symbol p for $\tilde{p}$ in (4.1) and in the sequel.

In order to let the model (4.1) describe a phase transition, the free energy function $W = w(\tau)/\tau$ has to have the shape as in Figure 1.

The first order conservation law (4.1) is hyperbolic, if the Jacobian of the flux

$$D\begin{pmatrix} -v \\ p(\tau) \end{pmatrix} = \begin{pmatrix} 0 & -1 \\ p'(\tau) & 0 \end{pmatrix}, \qquad (\tau, v) \in U := (1/b, \infty) \times \mathbb{R}$$

only has real eigenvalues and the corresponding eigenvectors are linearly independent. This is the case, if and only if $p'(\tau) < 0$. Remember (cf. Figure 1) that the associated phase of values of $\tau \in (1/\alpha_1, \infty)$ is called *vapour phase* and of values of $\tau \in (1/b, 1/\alpha_2)$ *liquid phase*.

For the hyperbolic case ($p'(\tau) < 0$) local existence of a unique classical solution of the Cauchy problem can be shown (see i.e. [Maj84] and the discussion in Section 2.2.3). Weak solutions no longer have to be unique, but by introducing the entropy solution existence and uniqueness of the entropy solution of the Riemann problem for small data has been shown by Lax (see e.g. [GR91]) in this case.

Definition 4.1 (Weak solution). A function $(\tau, v) \in L^\infty_{\text{loc}}(\mathbb{R} \times \mathbb{R}_{\geq 0}, U)$ is called a *weak solution* of (4.1), if and only if

$$\int_0^\infty \int_{\mathbb{R}} (\tau, v)^T \phi_t + (-v, p(\tau))^T \phi_x \, dx\, dt + \int_{\mathbb{R}} (\tau_0, v_0)^T \phi(\cdot, 0)\, dx = 0$$

for all $\phi \in C_0^\infty(\mathbb{R} \times \mathbb{R}_{\geq 0})$.

The standard entropy pair (E, F) for (4.1) is given by:

$$E(\tau, v) = \frac{v^2}{2} + \int_0^\tau -p(s)\, ds = \frac{v^2}{2} + w(\tau),$$
$$F(\tau, v) = vp(\tau),$$

for $(\tau, v) \in U$.

Definition 4.2 (Entropy solution). A weak solution $(\tau, v) \in L^\infty_{\text{loc}}(\mathbb{R} \times \mathbb{R}_{\geq 0}, U)$ of the Cauchy-problem (4.1) is called *entropy solution*, if and only if

$$\int_0^\infty \int_{\mathbb{R}} E(\tau, v)\phi_t + F(\tau, v)\phi_x \geq -\int_{\mathbb{R}} E(\tau_0, v_0)\phi(\cdot, 0)$$

for all $\phi \in C_0^\infty(\mathbb{R} \times \mathbb{R}_{\geq 0})$, $\phi \geq 0$ holds, i.e. if in the weak (distributional) sense holds:

$$\partial_t E(\tau, v) + \partial_x F(\tau, v) \leq 0.$$

This inequality is called *entropy inequality*.

Definition 4.3 (Shock wave, phase boundary). Let $s \in \mathbb{R}$ and $(\tau^+, v^+), (\tau^-, v^-) \in U$. A function

$$(\tau, v)(x, t) = \begin{cases} (\tau^+, v^+), & x < st, \\ (\tau^-, v^-), & x > st, \end{cases}$$

is called *shock wave* or travelling wave with speed s if and only if it is a weak solution of (4.1).

A shock wave is called *phase boundary* if and only if τ^+ and τ^- are located in different phases.

In contrast, for the nonhyperbolic case, local existence of classical solutions as well as existence and uniqueness of entropy solutions cannot be shown. Even more, solving the system (4.1) with a convex–concave function $p(\tau)$ with different "classical" schemes for hyperbolic conservation laws one gets different weak solutions.

Example 4.4 (Nonuniqueness of entropy solutions). In Figure 10 we print the numerical solutions of the system (4.1) with Riemann initial conditions chosen such that the left and right states are in different phases:

$$(\tau_0, v_0) := \begin{cases} (2.1, 0.5), & x < 0, \\ (1.0908, 0.5963), & x > 0. \end{cases} \tag{4.2}$$

w is chosen as described later in Section 4.2. We use the Lax–Friedrichs scheme and two schemes which we describe later in this paper (see Remark 4.9). We see that all the schemes produce different weak solutions. One can check that all of them are entropy solutions. They differ in different speeds of the phase boundary and corresponding left and right states satisfying the Rankine–Hugoniot jump conditions. Therefore, the entropy inequality is obviously not strong enough to single out a unique weak solution of (4.1).

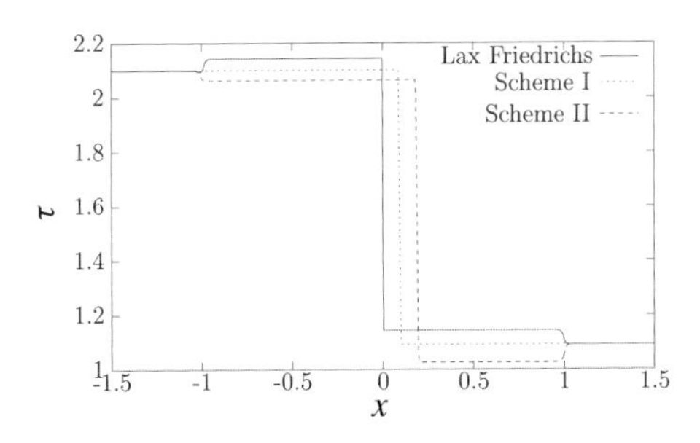

Figure 10. Nonuniqueness of weak solutions. Different numerical solutions of (4.1).

The difference of a classical hydrodynamic shock wave and a phase boundary can be seen when we look at the characteristic curves of the problem. A hydrodynamic shock wave is a Lax-shock (Figure 11(a)): three characteristic curves are entering into the shock line $s = \frac{x}{t}$. For a phase boundary (Figure 11(b)) only two characteristic curves enter the shock wave. This kind of shock is called undercompressive.

To overcome this nonuniqueness an additional constraint has to be added at the location of phase transitions. Following the presentation in [LeF02], we do this by

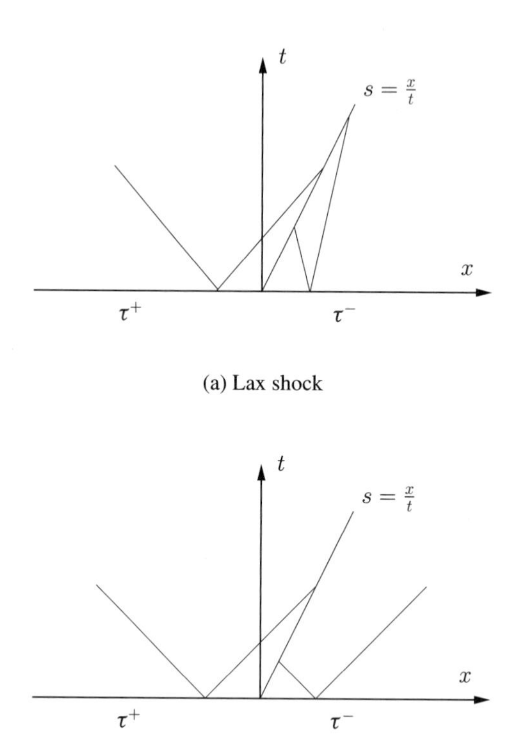

Figure 11. Characteristic curves of the system (4.1) in the classical and undercompressive case.

introducing the so-called kinetic relation. For earlier discussions on the kinetic relation see the pioneering work of Abeyaratne and Knowles [AK90] and [AK91].

Definition 4.5 (Φ-admissible phase boundary). Let $\Phi \in C^{0,1}(U^2, \mathbb{R}_{\leq 0})$. An entropy solution (τ, v) of (4.1) is called Φ-admissible solution, if the following strengthened version of the entropy inequality is satisfied for phase boundaries:

$$-s[\![E(\tau, v)]\!] + [\![F(\tau, v)]\!] = \Phi\left(\left(\tau^+, v^+\right), \left(\tau^-, v^-\right)\right) \leq 0,$$

Here, the brackets $[\![\cdot]\!]$ replace the jump of a function across a discontinuity. The function Φ usually is called *kinetic relation*.

Theorem 4.6 (Existence and uniqueness for the Riemann problem). *Under suitable assumptions on the pressure function* p *and on* Φ *the Riemann problem of* (4.1) *admits a unique piecewise smooth, self-similar solution made of rarefaction fans and* *shock waves, satisfying the entropy inequality, Φ-admissible phase boundaries and the condition: The Riemann solution uses nonclassical shocks whenever available.*

The proof of this theorem is done in [LT00].

In the following we use a special version of the kinetic relation Φ, which is motivated in [HRL99] and only depends on τ^- and τ^+:

$$\Phi(\tau^+, \tau^-) = -g(f(\tau^+, \tau^-))f(\tau^+, \tau^-)$$

Here, f is defined as

$$f(\tau^+, \tau^-) := -[\![w(\tau)]\!] - \frac{p(\tau^+) + p(\tau^-)}{2}[\![\tau]\!]. \tag{4.3}$$

We postulate that $g(f)$ corresponds to the velocity s of the phase boundary:

$$s = g(f)$$

Then of course g has to fulfill $g(f)f \geq 0$ to satisfy the entropy inequality.

Hence, a solution of the problem (4.1) including a phase boundary, not only has to fulfill the entropy inequality, but also the kinetic relation.

4.2 Regularization

As we have seen in Example 4.4 it is not possible to solve the system (4.1) with a usual numerical scheme for hyperbolic systems, because not only the entropy inequality has to be fulfilled but also the kinetic relation must hold at phase boundaries. To ensure the kinetic relation, it is convenient to have the function f (4.3) defined globally and not only on the phase interface. To reach this, we are regularizing the system (4.1) following the idea of [HRL99].

We want to solve the Riemann problem for (4.1)

$$\begin{aligned} \tau_t - v_x &= 0 \quad \text{in } \mathbb{R} \times (0, T), \\ v_t + p(\tau)_x &= 0 \quad \text{in } \mathbb{R} \times (0, T), \\ (\tau, w)(\cdot, 0) &= \begin{cases} (\tau^+, v^+), & x \leq 0, \\ (\tau^-, v^-), & x > 0. \end{cases} \end{aligned} \tag{P}$$

To construct the numerical scheme and to calculate an exact solution to test this scheme we look at the following simplified case: Let the internal energy function $w(\tau)$ be given by (cf. Figure 12(a)):

$$w(\tau) = \begin{cases} w_0(\tau) = \frac{1}{2}|\tau - 1|^2 - \tau, & \text{for } \tau \leq \frac{3}{2}, \\ w_1(\tau) = \frac{1}{2}|\tau - 2|^2 - \tau, & \text{for } \tau > \frac{3}{2}. \end{cases}$$

Hence, we get linear pressure relations in both phases (cf. Figure 12(b)):

$$p(\tau) = \begin{cases} p_0(\tau) = 2 - \tau, & \text{for } \tau \leq \frac{3}{2}, \\ p_1(\tau) = 3 - \tau, & \text{for } \tau > \frac{3}{2}. \end{cases} \tag{4.4}$$

For simplicity, we choose the kinetic response function g to be

$$g(f) := Mf, \quad M > 0. \tag{4.5}$$

First we extend the system (4.1) to a bigger system by introducing a level-set function

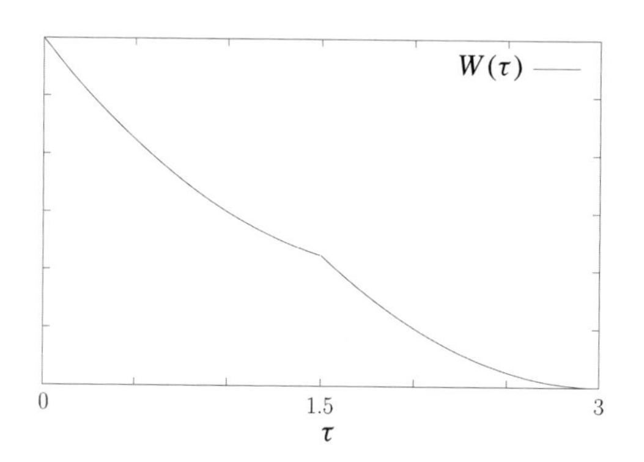

(a) Internal energy w

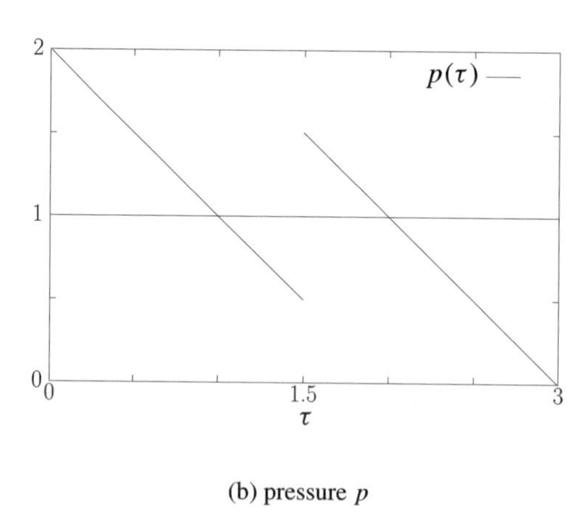

(b) pressure p

Figure 12. Simplified internal energy functions and resulting pressure p.

$\varphi = \varphi(x,t)$. We want to use φ to distinguish between the liquid and the vapour phase. φ is supposed to be positive in the vapour and negative in the liquid phase. φ is moving according to a given velocity field $V = V(x,t)$, where the velocity at the phase interface location ($\varphi = 0$) is supposed to match the velocity of the interface ($V|_{\text{interface}} = g(f) = s$). Hence, φ has to satisfy the Hamilton–Jacobi equation:

$$\varphi_t - V(x,t)|\varphi_x| = 0.$$

Then, φ stays always positive in the vapour region and stays always negative in the liquid region. The location of the phase boundary is given by $\varphi = 0$. We end up with the extended system:

$$\begin{aligned} \tau_t - v_x &= 0 \quad \text{in } \mathbb{R} \times (0,T), \\ v_t + p(\tau)_x &= 0 \quad \text{in } \mathbb{R} \times (0,T), \end{aligned} \tag{$\tilde{P}_1$}$$

$$\varphi_t - V(x,t)|\varphi_x| = 0 \quad \text{in } \mathbb{R} \times (0,T), \tag{$\tilde{P}_2$}$$

From now on we are using the following initial conditions:

$$\begin{aligned} (\tau, w)(\cdot, 0) &= \begin{cases} (\tau^+, v^+), & x \leq 0, \\ (\tau^-, v^-), & x > 0, \end{cases} \\ \varphi(x,0) &= -x \quad \text{in } \mathbb{R}. \end{aligned} \tag{4.6}$$

Up to now there is no coupling between ($\tilde{P}_1$) and ($\tilde{P}_2$). We will couple the equations in the following way.

For $h \in [0,1]$ let $\hat{w}(w,h)$ be a smooth function interpolating between the branch w_0 and the branch w_1 of w, i.e.

$$\hat{w}(\tau,h) := w_0(\tau) + h\,(w_1(\tau) - w_0(\tau))$$

and

$$\hat{p}(\tau,h) := -\frac{\partial \hat{w}}{\partial \tau}(\tau,h).$$

In our case the regularized version of the (bi-)quadratic internal energy function w is

$$\begin{aligned} \hat{w}(w,h) &= w_0(w) + h(w_1(w) - w_0(w)) \\ &= \frac{1}{2}\big(|\tau - 1|^2 - 2\tau + h(3 - 2\tau)\big) \end{aligned}$$

and hence, the regularized pressure relation is given by

$$\hat{p}(\tau,h) = 2 + h - \tau.$$

We now replace in ($\tilde{P}$) w by $\hat{w}$, p by $\hat{p}$ and put the Heaviside function $H(\varphi)$ in the second argument of $\hat{w}$ and $\hat{p}$. We deduce:

$$\begin{aligned} \tau_t - v_x &= 0, \\ v_t + \hat{p}(\tau, H(\varphi))_x &= 0, \\ \varphi_t - V(x,t)|\varphi_x| &= 0. \end{aligned} \tag{$\hat{P}$}$$

Remark 4.7. So far we did not change the solution in the following sense: Let (τ, v) be a weak travelling wave solution of the first two equations of $(\tilde{P})$ and let $\varphi = \varphi(x,t) := -(x - st)$ where s is the velocity of the travelling wave. Then (τ, v) is a weak solution of the first two equations of $(\hat{P})$ and $\varphi(x,t)$ is a classical solution of the third equation of $(\hat{P})$ with $V(x,t) = s$.

We would like to replace $V(x,t)$ in $(\hat{P})$ by the velocity of the phase transition, i.e. $V(x,t) = g(f)$. Since the function f only is defined at the location of the phase boundary, this is not possible. By defining a regularized version of f as in [HRL99]

$$f_\varepsilon := -\frac{\partial \hat{w}}{\partial h}(\tau, H_\varepsilon(\varphi))$$

we get in the simplified (bi-)quadratic model

$$f_\varepsilon(w, \varphi) = \tau - \frac{3}{2}.$$

We end up with a regularized system of partial differential equations:

$$\begin{aligned} \tau_t - v_x &= 0, \\ v_t + \hat{p}(\tau, H_\varepsilon(\varphi))_x &= 0, \\ \varphi_t - g(f_\varepsilon(\tau, \varphi))|\varphi_x| &= 0, \end{aligned} \tag{P_ε}$$

where $H_\varepsilon(\varphi)$ is a regularized version of the Heaviside function H. We choose the regularized Heaviside function to be (cf. Figure 13(a))

$$H_\varepsilon(z) := \begin{cases} 0, & z < -\varepsilon, \\ \dfrac{z+\varepsilon}{2\varepsilon} + \dfrac{\sin(\frac{\pi z}{\varepsilon})}{2\pi}, & |z| \le \varepsilon, \\ 1, & z > \varepsilon, \end{cases}$$

which leads to a regularized Dirac delta function (cf. Figure 13(b))

$$\delta_\varepsilon(z) := \begin{cases} \dfrac{1 + \cos(\frac{\pi z}{\varepsilon})}{2\varepsilon}, & |z| \le \varepsilon, \\ 0, & |z| > \varepsilon. \end{cases} \tag{4.7}$$

Finally, we arrive at the system

$$\begin{aligned} \tau_t - v_x &= 0, \\ v_t - (\tau - (2 + H_\varepsilon(\varphi)))_x &= 0, \end{aligned} \tag{4.8}$$

$$\varphi_t - M\left(\tau - \frac{3}{2}\right)|\varphi_x| = 0. \tag{4.9}$$

We are now looking for an exact solution of the above system (4.8), (4.9).

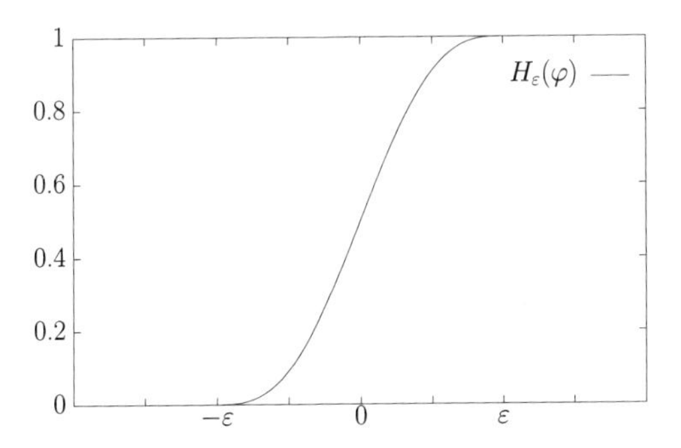

(a) Regularized Heaviside function H_ε.

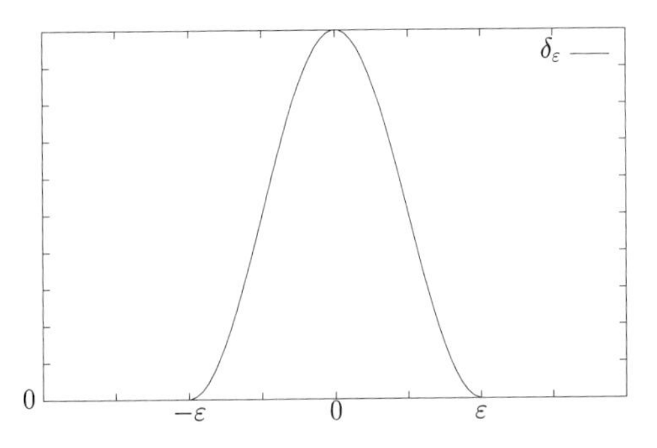

(b) Regularized Dirac delta function δ_ε.

Figure 13. Regularized Heaviside and Dirac delta function.

Proposition 4.8 (Traveling wave solution). *Let $(\tau^+, v^+) \in U$ be a state in the vapour phase and $(\tau^-, v^-) \in U$ be a state in the liquid phase such that*

$$(\tau, v) = \begin{cases} (\tau^+, v^+), & x \leq st, \\ (\tau^-, v^-), & x > st, \end{cases}$$

with some $s \in \mathbb{R}$ is a Φ-admissible solution of (P) for p as in (4.4)*. Let $\varphi_\varepsilon(x, t) = -(x - st)$. Then, for $\varepsilon > 0$, the following statements hold:*

(i) *There exist functions $\tilde{\tau}_\varepsilon, \tilde{v}_\varepsilon \in C^1(\mathbb{R})$ with $(\tilde{\tau}_\varepsilon, \tilde{v}_\varepsilon)(\mp\infty) = (\tau^\pm, v^\pm)$, such that for*

$$(\tau_\varepsilon, v_\varepsilon)(x, t) = (\tilde{\tau}_\varepsilon, \tilde{v}_\varepsilon)\left(\frac{x - st}{\varepsilon}\right)$$

$(\tau_\varepsilon, v_\varepsilon)$ is a classical solution of (4.8)*, and $\varphi_t - s|\varphi_x| = 0$.*

(ii) *The sequence* $\{(\tau_\varepsilon, v_\varepsilon)\}$ *converges to* (τ, v) *almost everywhere in* $\mathbb{R} \times \mathbb{R}_{\geq 0}$.

Proof. Let $\varphi_\varepsilon(x, t) = -(x - st)$. (4.8) can be transformed into an equation of τ_ε alone and by using $\tau_\varepsilon(x, t) = \tilde{\tau}_\varepsilon(\xi)$ and $v_\varepsilon(x, t) = \tilde{v}_\varepsilon(\xi)$ with $\xi = \frac{x-st}{\varepsilon}$ to

$$\tilde{\tau}_\varepsilon'(\xi) = \frac{\varepsilon \delta_\varepsilon(\varphi)}{s^2 - 1}.$$

Substituting $\delta_\varepsilon(\varphi)$ by (4.7) we see that $\tilde{\tau}_\varepsilon$ has to be constant for $|\xi| > 1$. We choose $\tilde{\tau}_\varepsilon(\xi) = \tau^+$ for $\xi < -1$.

For $|\xi| \leq 1$ integration leads to:

$$\begin{aligned}\tilde{\tau}_\varepsilon(\xi) &= \frac{\varepsilon}{s^2 - 1} \int_{-1}^{\xi} \frac{1 + \cos(\pi z)}{2\varepsilon} \, dz + C \\ &= \frac{\pi \xi + \sin(\pi \xi) + \pi}{2\pi(s^2 - 1)} + C.\end{aligned}$$

From above we get $\tilde{\tau}_\varepsilon(-1) = \tau^+$, which leads to $C = \tau^+$.

Therefore, $\tilde{\tau}_\varepsilon(1) = \tau^+ + \frac{1}{s^2-1} = \tau^-$ using the Rankine–Hugoniot jump conditions. Resubstituting we end up with

$$\tau_\varepsilon(x, t) = \begin{cases} \tau^+, & x - st < -\varepsilon, \\ \frac{\sin(\frac{\pi}{\varepsilon}(x-st)) + \frac{\pi}{\varepsilon}(x-st) + 2\pi(s^2-1)\tau^+ + \pi}{2\pi(s^2-1)}, & |x - st| < \varepsilon, \\ \tau^- : & x - st > \varepsilon. \end{cases}$$

Using the same method we get the solution for v

$$v_\varepsilon(x, t) = \begin{cases} v^+, & x - st < -\varepsilon, \\ \frac{-\sin(\frac{\pi}{\varepsilon}(x-st))s - s\frac{\pi}{\varepsilon}(x-st) + 2\pi(s^2-1)v^+ + s\pi}{2\pi(s^2-1)}, & |x - st| < \varepsilon, \\ v^-, & x - st > \varepsilon. \end{cases}$$

Statement (ii) follows by construction. □

Remark. The assumption $\varphi_\varepsilon(x, t) = -(x - st)$ makes sense since the level set function is reinitialised in this way each time step.

4.3 Numerical example

In this section we are presenting numerical results of the solution to (4.8), (4.9). As initial conditions we choose $(\tau_\varepsilon, v_\varepsilon, \varphi_\varepsilon)(x, 0)$ from Proposition 4.8 with

$$(\tau^+, v^+) = (2.1, 0.5)$$

and

$$(\tau^-, v^-) = (1.0908, 0.5963)$$

such that the exact solution is a travelling wave. The function g of the kinetic relation is chosen to be $g(f) = f$. For solving (4.8) we use the local Lax–Friedrichs scheme with linear reconstruction and the minmod limiter. For (4.9) we are using a first order scheme described in [KT00]. Additionally, we are reinitialising the level-set function every timestep so that it remains the signed distance function.

In Figure 14 we display the exact solution of the sharp interface model (P) as well as the numerical solution of (P_ε) with $\varepsilon = 0.1$ and $\varepsilon = 0.01$. The convergence of the numerical solution of (P_ε) to the exact solution of (P) can also be seen in Table 2. In Figure 15 we present the numerical solution of (P_ε) for different times.

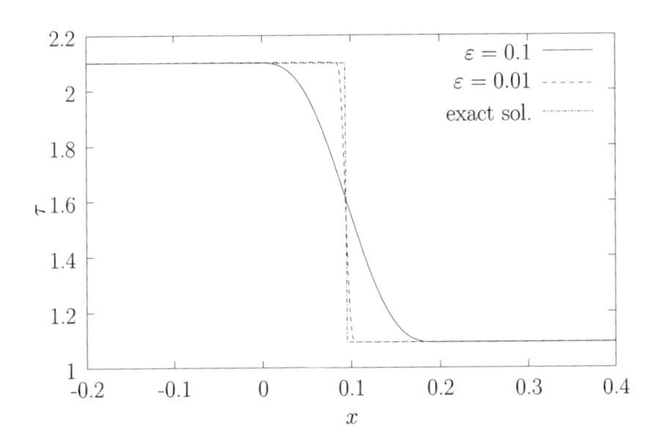

Figure 14. Time $t = 1.0$, exact solution and numerical solution for different values of ε.

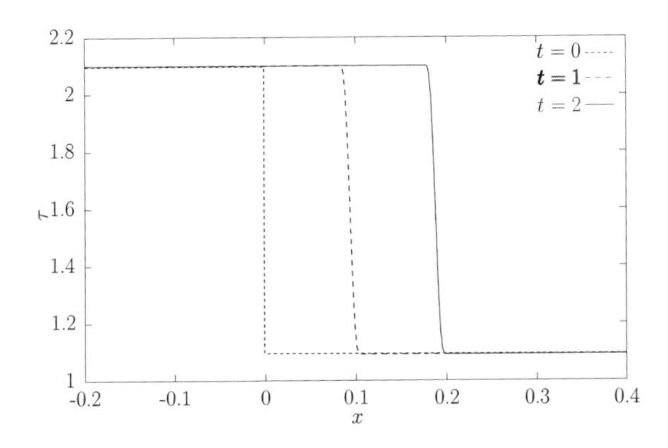

Figure 15. Initial data and numerical solution for different times.

In Table 1 the error of the numerical scheme to the exact solution of (P_ε) is shown. Notice that for fixed ε the experimental order of convergence is approximately 2. In Table 2 we can see the error of the numerical scheme to the exact solution of the sharp interface model, which we assume can be estimated in our case in the following way (where u is either τ or v):

$$\|u - u_h^\varepsilon\|_{L_1} < ||u - u^\varepsilon||_{L_1} + ||u^\varepsilon - u_h^\varepsilon||_{L_1} < C\varepsilon + C(\varepsilon)h^2.$$

The second constant is depending on ε in the following way. For fixed h and increasing ε we get a experimental order of convergence of -1. Therefore, we assume $C(\varepsilon) = \frac{1}{\varepsilon}$. If we choose $\varepsilon = Ch$ we are expecting

$$||u - u_h^\varepsilon||_{L_1} < Ch.$$

This is exactly the rate of convergence we get out of Table 2 going down the diagonal, which corresponds to $\varepsilon = Ch$.

Table 1. $||u_h^\varepsilon - u^\varepsilon||_{L_1}$, where the upper value in each cell corresponds to $u = \tau$ and the lower to $u = v$.

NOE	$\varepsilon = 0.08$	0.04	0.02	0.01	0.005
1000	0.000155255 0.000203839	0.000371509 0.000462668	0.00123758 0.00117728	0.00461472 0.00340628	0.00812824 0.00579192
2000	0.0000386991 0.0000499565	0.0000844802 0.00010874	0.000260971 0.00027862	0.00107627 0.000840271	0.0043847 0.00288085
4000	0.00000939659 0.0000119813	0.0000199859 0.0000256544	0.0000511942 0.0000608323	0.00021626265 0.000185998	0.0010009 0.000672474
8000	0.00000231987 0.00000285554	0.00000471506 0.00000598947	0.0000105783 0.0000134101	0.0000359364 0.0000367194	0.000195095 0.000140528
16000	0.000000581827 0.00000066381	0.0000012166 0.00000147023	0.00000236219 0.00000300074	0.00000607337 0.00000732471	0.0000290553 0.0000244986

Table 2. $||u_h^\varepsilon - u||_{L_1}$, where the upper value in each cell corresponds to $u = \tau$ and the lower to $u = v$. In the last line the exact model error of τ is shown.

NOE	$\varepsilon = 0.08$	0.04	0.02	0.01	0.005
1000	0.0241503 0.00240583	0.0123352 0.00143798	0.00688164 0.00146395	0.00591979 0.00334653	0.00760631 0.00579192
2000	0.024046 0.00231845	0.0120885 0.00120952	0.00621976 0.000764756	0.00367274 0.000960309	0.00427156 0.00282156
4000	0.0240171 0.00229709	0.0120248 0.00115988	0.00605274 0.000610962	0.00316669 0.000426167	0.00220068 0.000737241
8000	0.0240093 0.00229202	0.012008 0.0011485	0.00601188 0.0005804	0.00302966 0.000311205	0.00161663 0.000255159
16000	0.0240078 0.00229083	0.012005 0.00114608	0.00600459 0.000574314	0.00300761 0.000290874	0.00152236 0.0001613
$\|\|\tau^\varepsilon - \tau\|\|_{L_1}$	0.02400713516	0.01200356764	0.006001783802	0.003000891886	0.001500445956

Remark 4.9. In Example 4.4 we presented three numerical schemes to solve the Riemann problem with initial datum (4.2) and obtained three different solutions (see Figure 10). One of the schemes was the Lax–Friedrichs scheme applied directly to (4.1). The other schemes – called Scheme I and Scheme II in Example 4.4 – are exactly the scheme for (4.8), (4.9) as described above but for two different choices for M in (4.5). Precisely we have chosen

$$M_{\text{Scheme I}} = 1, \quad M_{\text{Scheme II}} = 5.$$

Acknowledgement. D. D. acknowledges the support from the EU financed network no. HPRN-CT-2002-00282.

References

[AB98] G. Alberti and G. Bellettini, A non-local anisotropic model for phase transitions: Asymptotic behaviour of rescaled energies. *European J. Appl. Math.* 9 (3) (1998), 261–284.

[AK90] R. Abeyaratne and K. Knowles, On the Driving Traction Acting on a Surface of Strain Discontinuity in a Continuum. *J. Mech. Phys. Solids* 38 (1990), 345–360.

[AK91] R. Abeyaratne and K. Knowles, Kinetic relations and the propagation of phase boundaries in solids. *Arch. Rational Mech. Anal.* 114 (2) (1991), 119–154.

[AMW98] D. M. Anderson, G. B. McFadden, and A. A. Wheeler, Diffuse interface methods in fluid mechanics. *Ann. Rev. Fluid Mech.* 30 (1998) ,139–165.

[CDR04] F. Coquel, D. Diehl, and C. Rohde, Static Equilibrium Solutions of the Navier-Stokes-Korteweg System and Relaxation Schemes, Technical report, Math. Institut, Albert-Ludwigs-Universität Freiburg, 2004, preprint.

[CP98] F. Coquel and B. Perthame. Relaxation of energy and approximate riemann solvers for general pressure laws in fluid dynamics. *SIAM J. Numer. Anal.* 35 (6) (1998), 2223–2249.

[DMLM95] G. Dal Maso, P. G. LeFloch, and F. Murat, Definition and weak stability of nonconservative products, *J. Math. Pures Appl.*, 74 (6) (1995), 483–548.

[FS02] H. Fan and M. Slemrod, Dynamic flows with liquid/vapor phase transitions, in *Handbook of Mathematical Fluid Mechanics* (S. Friedlander and D. Serre, eds.), Elsevier Science, 2002, 373–420.

[GR91] E. Godlewski and P.-A. Raviart, *Hyperbolic Systems of Conservation Laws*, Ellipses, 1991.

[GR96] E. Godlewski and P.-A. Raviart, *Numerical approximation of hyperbolic systems of conservation laws*, Springer-Verlag, 1996.

[Her55] J. W. Herivel, The derivation of the equations of motion of an ideal fluid by Hamilton's principle, *Proc. Cambridge Philos. Soc.* 51 (1955), 344–349.

[HL94] H. Hattori and D. Li, Solutions for two-dimensional system for materials of Korteweg type, *SIAM J. Math. Anal.* 25 (1) (1994), 85–98.

[HRL99] T. Y. Hou, P. Rosakis, and P. LeFloch, A Level–Set Approach to the Computation of Twinning and Phase–Transition Dynamics, *J. Comput. Phys.* 150 (1999), 302–331.

[Jac99] D. Jacqmin, Calculation of two phase Navier-Stokes flows using phase field modeling, *J. Comput. Phys.*, 155 (1999), 96–127.

[JLCD01] D. Jamet, O. Lebaigue, N. Coutris, and J. M. Delhaye, The second gradient method for the direct numerical simulation of liquid-vapor flows with phase change, *J. Comput. Phys.*, 169 (2) (2001), 624–651.

[JT02] D. Jamet and J. U. Torres, D. Brackbill, On the theory and computation of surface tension: The elimination of parasitic currents through energy conservation in the second-gradient method, *J. Comput. Phys.*, 182 (2002), 262–276.

[Krö97] D. Kröner, *Numerical Schemes for Conservation Laws*, Verlag Wiley & Teubner, Stuttgart 1997.

[KT00] A. Kurganov and E. Tadmor, New High–Resolution Semi-discrete Central Schemes for Hamilton–Jacobi Equations, *J. Comput. Phys.* 160 (2000), 720–742.

[LeF02] P. G. LeFloch, *Hyperbolic Systems of Conservation Laws: The Theory of Classical and Nonclassical Shock Waves*, Birkäuser, 2002.

[Lev02] R. J. Leveque, *Finite volume methods for hyperbolic problems*, Cambridge University Press, 2002.

[LT00] P. G. LeFloch and M. D. Thanh, Nonclassical Riemann solvers and kinetic relations III: A nonconvex hyperbolic model for Van der Waals fluids, *Electron. J. Differential Equations* 72 (2000), 1–19.

[Maj84] A Majda, *Compressible fluid flow and systems of conservation laws in several space variables,* Appl. Math. Sci. 53, Springer-Verlag, 1984.

[Mod87] L. Modica, The gradient theory of phase transitions and the minimal interface criterion, *Arch. Rational Mech. Anal.* 98 (1987), 123–142.

[Roh04] C. Rohde, On local and non-local Navier-Stokes-Korteweg systems for liquid-vapour phase transitions, Technical report, Math. Institut, Albert-Ludwigs-Universität Freiburg, 2004.

[Ser59] J. Serrin, Mathematical principles of classical fluid mechanics, in *Handbuch der Physik, Band VIII/1 Strömungsmechanik* (D. Flügge and C. Truesdell, eds.), Springer-Verlag, 1959, 1–263.

[Sul90] I. Suliciu, On modelling phase transitions by means of rate type constitutive equations, *Int. J.Engng. Sci.* 28 (1990), 827–841.

[vdW94] J. D. van der Waals, Thermodynamische Theorie der Kapillarität unter Voraussetzung stetiger Dichteänderung, *Z. Phys. Chem.* 13 (1894), 657–725.

Geometric Eddington factor for radiative transfer problems

J. Cartier and A. Munnier

Commissariat à l'Energie Atomique - Direction des Applications Militaires BP 12
91680 Bruyères le Châtel, France
email: `julien.cartier@cea.fr`

Université Dauphine Paris IX, Place du Maréchal De Lattre De Tassigny
75775 Paris Cedex 16, France
email: `munnier@ceremade.dauphine.fr`

Abstract. In this paper a geometric closure method for radiative transfer equations has been developed and investigated using particular geometric configurations. Then we propose a new formulation of the Eddington factor (and the related flux limiter) adapted to radiative transfer calculations, whereas classical Eddington's approximation cannot be applied. Moreover, a numerical scheme and numerical results for the new flux limiter are presented in two dimensions configurations.

Key words. Radiative transfer, diffusion approximation, Eddington factor, flux limiter.

1 Introduction

Inertial Confinement Fusion (ICF) is a way to achieve thermonuclear fusion in a laboratory: it consists in imploding a small target of hydrogen so that high temperature and high density lead to the burn of a large fraction of fuel. A more efficient way to obtain a quasi-isotropic implosion is to use the radiation flux created by conversion of Laser energy into X-ray in the wall of a Hohlraum (see Figure 1).

Radiative transfer equations describe the transport of X-ray energy in the Hohlraum, neglecting hydrodynamic motion it writes as

$$\frac{1}{c}\partial_t I + \vec{\Omega}\cdot\vec{\nabla}_x I + \sigma_\nu I = \frac{c}{4\pi}\sigma_\nu S(\nu, T), \tag{1.1}$$

where $I(x, \vec{\Omega}, \nu, t)$ is the radiative intensity ($\vec{\Omega}$ takes its values in the unit sphere $\mathcal{S}^2$ and the frequency ν is strictly positive), c is the speed of light and σ_ν is the opacity. The

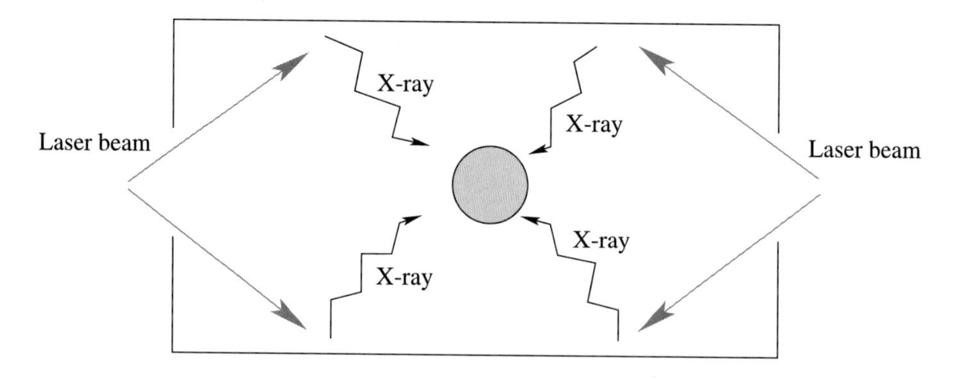

Figure 1. Hohlraum cavity.

intensity $I(x, \vec{\Omega}, \nu, t)$ is related to the distribution function of a photon gas that is of a population of particles travelling in straight lines at the speed of light c. The source function $S(\nu, T)$ is the so-called Planck function which depends on the temperature T so that equation (1.1) is coupled with an energy balance equation:

$$\partial_t E_i(T) + \int_{\mathcal{S}^2} d\vec{\Omega} \int_{\nu>0} d\nu\, \sigma_\nu (I - \frac{c}{4\pi} S(\nu, T)) = 0, \tag{1.2}$$

where E_i is the internal energy.

The usual method for solving system (1.1), (1.2) consists in using an implicit Monte-Carlo method (see [3] for example). However, Monte-Carlo oscillations of the solution alter the required symmetry of the radiation field and can cause an incorrect calculation of the implosion. On the other hand, deterministic methods (see [2]) are very costly and cannot be used for parametric studies. This is why it is useful to obtain approximate models for radiative transfer problems. The goal of this paper is to present a new closure for (1.1), (1.2) which can be used in the ICF context.

In order to simplify the analysis, we will restrict ourselves to a simplified version of (1.1), (1.2): first we consider a stationary problem (i.e. we solve the system on one single time step). Secondly, we assume that the source function is given so that frequency variable can be removed and (1.1) becomes

$$\vec{\Omega} \cdot \vec{\nabla}_x I + \sigma I = \frac{c}{4\pi} \sigma S. \tag{1.3}$$

Finally, we impose boundary conditions that account for albedo of the gold wall. For each x on the boundary and each incoming direction $\vec{\Omega}$ we impose

$$I(x, \vec{\Omega}) = \frac{1-\omega}{\pi} \int_{\vec{\Omega}.\vec{n}_b>0} I(x, \vec{\Omega}')\vec{\Omega}'.\vec{n}_b\, d\vec{\Omega}', \tag{1.4}$$

where $\vec{n}_b$ is the unit normal outward vector at the boundary point x. For the sake of simplicity, we only consider a model geometry (see Figure 2). When opacity is large,

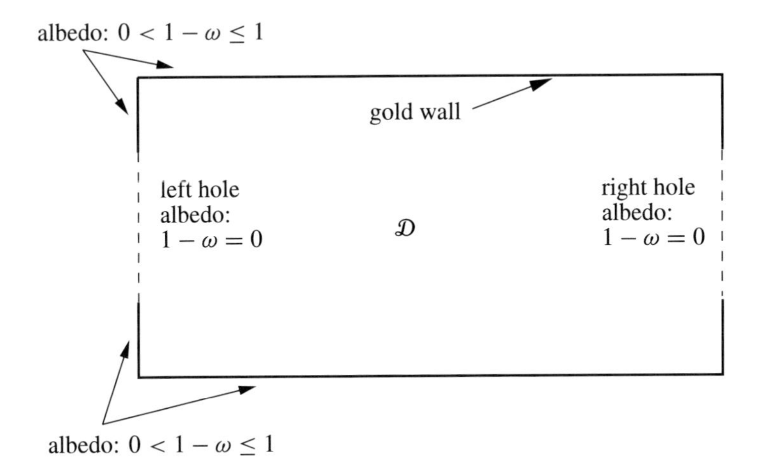

Figure 2. Model geometry.

it is well known (see [4]) that diffusion approximation can be applied to system (1.3), (1.4). In this limit, radiative energy E_r is solution of a diffusion equation:

$$-\mathrm{div}\Big(\frac{1}{3\sigma}\nabla E_r\Big) + \sigma E_r = \sigma S, \quad E_r := \frac{1}{c}\int_{\mathcal{S}^2} I(x,\vec{\Omega})\,d\vec{\Omega}. \tag{1.5}$$

In a Hohlraum this assumption does not hold: the cavity is usually filled with a low density gas and the photon mean free path in the gas (σ^{-1}) is large compared to typical dimension of the device. So limit (1.5) which corresponds to the small mean free path limit is no longer valid.

However, it is always possible (at least formally) to define a corrected diffusion limit

$$-\mathrm{div}\Big(\frac{\lambda}{\sigma}\nabla E_r\Big) + \sigma E_r = \sigma S, \quad E_r := \frac{1}{c}\int_{\mathcal{S}^2} I(x,\vec{\Omega})\,d\vec{\Omega}, \tag{1.6}$$

where λ is related to the Eddington factor γ so that equation (1.6) provides a correct approximation of the solution. It is our purpose to discuss this method and to propose a calculation of the Eddington factor adapted to our problem.

The outline of the paper is as follows. In the next section we describe the calculation of γ in the Hohlraum. Then we analyse its properties with respect to the general theory of Eddington factors of [5]. In Section 4 we give some details on the numerical scheme for the non-linear diffusion equation. Finally, we present numerical results showing the significance of our approach and we compare them with usual flux-limited diffusion theory.

2 Calculation of the Eddington factor

Let us consider the radiative transfer equation:

$$\left.\begin{aligned} &\vec{\Omega}\cdot\vec{\nabla} I(x,\vec{\Omega}) + \sigma\, I(x,\vec{\Omega}) = \frac{c}{4\pi}\,\sigma\, S(x), && (x,\vec{\Omega}) \in \mathcal{D}\times\mathcal{S},\\ &I(x,\vec{\Omega}) = \frac{1-\omega}{\pi}\int_{\vec{\Omega}'\cdot\vec{n}>0} I(x,\vec{\Omega}')\,\vec{\Omega}'\cdot\vec{n}\; d\vec{\Omega}', && x\in\partial\mathcal{D},\ \vec{\Omega}\cdot\vec{n}<0. \end{aligned}\right\} \tag{2.1}$$

Here, $1-\omega$ denotes the albedo of the wall and is defined as the ratio between incoming and outcoming radiative fluxes at the boundary. The geometrical configuration that we consider in the sequel is described in Figure 3.

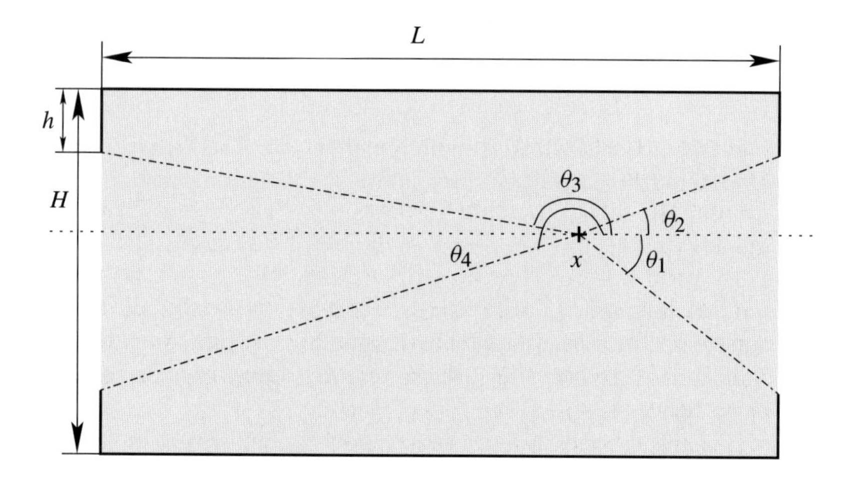

Figure 3. Geometric configuration.

We also introduce

$$E(x) := \frac{1}{c}\int_{4\pi} I(x,\vec{\Omega})\, d\vec{\Omega}, \quad \text{(radiative energy density)}, \tag{2.2}$$

$$\vec{F}(x) := \int_{4\pi} I(x,\vec{\Omega})\,\vec{\Omega}\, d\vec{\Omega}, \quad \text{(radiative flux)}, \tag{2.3}$$

$$[P(x)] := \frac{1}{c}\int_{4\pi} I(x,\vec{\Omega})\,\vec{\Omega}\otimes\vec{\Omega}\, d\vec{\Omega}, \quad \text{(radiative pressure tensor)},$$

and the Eddington tensor $[\gamma]$ which is defined by the relation

$$[\gamma] := \frac{[P]}{E}.$$

We apply the so-called "moment method" which consists in integrating the first equation of (2.1) over the solid angles $\vec{\Omega}$. We get

$$\frac{1}{c}\vec{\nabla}\cdot\vec{F}+\sigma\,E=\sigma\,S\quad\text{in }\mathcal{D}. \tag{2.4}$$

Multiplying the first line of (2.1) by $\vec{\Omega}$ and integrating over directions $\vec{\Omega}$, we obtain

$$\operatorname{div}[P]+\frac{\sigma}{c}\vec{F}=0\quad\text{in }\mathcal{D}, \tag{2.5}$$

that is,

$$-\operatorname{div}\left(\frac{1}{\sigma}\operatorname{div}([\gamma]\,E)\right)+\sigma\,E=\sigma\,S\quad\text{in }\mathcal{D}. \tag{2.6}$$

When σ is large, diffusion approximation applies (see [4]) and the Eddington factor becomes constant, where [Id] denotes the identity tensor:

$$[\gamma]=\frac{1}{3}\,[\mathrm{Id}]. \tag{2.7}$$

In this case, radiative transfer equation can be replaced by a diffusion equation (the so-called Rosseland approximation).

In our situation the diffusion approximation (1.5) is not valid but it is always possible to replace formally the transport equation by a diffusion equation provided that the Eddington tensor is correctly defined: we refer here to the review paper [5] which explains how a flux-limited diffusion equation can in some sense approximate a transport equation. We want to apply this method in our case, but using specific information about the geometry we can improve the calculation of the Eddington factor (and hence of the flux-limiter) and obtain better results.

We have to consider now the following model with a flux-limiter λ:

$$\left.\begin{aligned} -\operatorname{div}\left(\frac{\lambda}{\sigma}\vec{\nabla}\,E\right)+\sigma\,E=\sigma\,S &\quad \text{in }\mathcal{D},\\ \omega\,E+2\,(2-\omega)\frac{\lambda}{\sigma}\vec{\nabla}E\cdot\vec{n}_b=0 &\quad \text{on }\partial\mathcal{D}, \end{aligned}\right\} \tag{2.8}$$

where λ is related to the Eddington factor $[\gamma]$ (see Section 3.2). Our aim is to determine an algebraic expression of λ in such a way that the solution of the approximated problem (2.8) is close (in a formal sense) to the solution of the transport problem. We impose that λ (and the related Eddington factor) only depends on

- a gradient length $\dfrac{|\vec{\nabla}E|}{E}$,
- geometrical data.

We briefly describe how the boundary condition for the diffusion equation has been derived.

We first introduce the quantities $F := \vec{F}.\vec{n}_b$ on $\partial\mathcal{D}$ and $F^+ := \int_{\vec{\Omega}'\cdot\vec{n}_b>0} I(x,\vec{\Omega}')\vec{\Omega}'\cdot\vec{n}_b\, d\vec{\Omega}'$ and $F^- := \int_{\vec{\Omega}'.\vec{n}_b<0} I(x,\vec{\Omega}')\vec{\Omega}'.\vec{n}_b\, d\vec{\Omega}'$ such that $F = F^+ - F^-$. According to the definition of the albedo, $F^- = (1-\omega)F^+$ and hence, $F^+ - F^- = \omega F^+ = F$.

On the other hand,

$$F = \vec{F}\cdot\vec{n}_b = -\frac{\lambda c}{\sigma}\vec{\nabla}E\cdot\vec{n}_b. \tag{2.9}$$

Furthermore, we assume the following quasilinear P1 approximation:

$$I(x,\vec{\Omega}) = \frac{cE}{4\pi} + \frac{3}{4\pi}\vec{\Omega}\cdot\vec{F}. \tag{2.10}$$

Hence we get

$$\omega F^+ = \omega\int_{\vec{\Omega}'\cdot\vec{n}_b>0}\left(\frac{cE}{4\pi} + \frac{3}{4\pi}\vec{\Omega}'\cdot\vec{F}\right)\vec{\Omega}'.\vec{n}_b\, d\vec{\Omega}', \tag{2.11}$$

and

$$\omega F^+ = 2\pi\omega\int_0^1\left(\frac{cE}{4\pi} - \frac{3\mu\lambda c}{4\pi\sigma}\vec{\nabla}E\cdot\vec{n}_b\right)\mu d\mu = \omega\frac{cE}{4} - \frac{\omega}{2}\frac{c\lambda}{\sigma}\vec{\nabla}E\cdot\vec{n}_b. \tag{2.12}$$

Summarizing (2.9) and (2.12), we obtain the boundary condition for the energy E:

$$\omega E + 2(2-\omega)\frac{\lambda}{\sigma}\vec{\nabla}E\cdot\vec{n}_b = 0. \tag{2.13}$$

Remark 2.1. The quantity $\dfrac{2(2-\omega)}{\omega}\dfrac{\lambda}{\sigma}$ is called the extrapolation length.

3 Computation of the Eddington tensor $[\gamma]$

In order to compute $[\gamma]$, we apply a method inspired by the one developped in [6] for spherical problems: we assume that the radiative intensity at a given point x in $\mathcal{D}$ is piecewise constant with respect to the solid angle variable $\vec{\Omega}$.

3.1 The two intensities model

We first introduce some notations: let us define $S_T(x)$ as the part of $\mathcal{S}^2$ containing all the solid angles that link point x to the holes and $S_T{}^c(x) = \mathcal{S}^2 \setminus S_T(x)$. We make the following modelling assumption:

$$I(x,\vec{\Omega}) = \begin{cases} I_1(x) & \text{when } \vec{\Omega}\in S_T(x), \\ \\ I_2(x) & \text{when } \vec{\Omega}\in S_T{}^c(x). \end{cases} \tag{3.1}$$

This modelling can be justified by considering the numerical solution of the transport equation obtained with the classical DSN method: Figure 4 presents the intensity as a function of $\vec{\Omega}$ at a given point x in $\mathcal{D}$ close to the right hole. We clearly see that it is piecewise constant over S_T and S_T^c.

The intensity $I_1(x)$ is the mean radiative intensity resulting from the emission of the holes (and the medium) and $I_2(x)$ is the intensity resulting from the emission of the wall (and the medium). We also define

$$\alpha_1 := \int_{S_T} d\vec{\Omega} \quad \text{and} \quad \alpha_2 := \int_{S_T{}^c} d\vec{\Omega}, \tag{3.2a}$$

the vectors

$$\vec{v}_1 := \int_{S_T} \vec{\Omega}\, d\vec{\Omega} \quad \text{and} \quad \vec{v}_2 := \int_{S_T{}^c} \vec{\Omega}\, d\vec{\Omega}, \tag{3.2b}$$

and finally the tensors

$$[m_1] := \int_{S_T} \vec{\Omega} \otimes \vec{\Omega}\, d\vec{\Omega} \quad \text{and} \quad [m_2] := \int_{S_T{}^c} \vec{\Omega} \otimes \vec{\Omega}\, d\vec{\Omega}. \tag{3.2c}$$

The analytical expressions of all these data will be given in the Appendix A. Using our assumption we obtain the following system:

$$\left.\begin{aligned} E &= \frac{1}{c} I_1 \int_{S_T} d\vec{\Omega} + \frac{1}{c} I_2 \int_{S_T{}^c} d\vec{\Omega}, \\ \vec{F} &= I_1 \int_{S_T} \vec{\Omega}\, d\vec{\Omega} + I_2 \int_{S_T{}^c} \vec{\Omega}\, d\vec{\Omega}, \\ [\gamma]\, E &= \frac{1}{c} I_1 \int_{S_T} \vec{\Omega} \otimes \vec{\Omega}\, d\vec{\Omega} + \frac{1}{c} I_2 \int_{S_T{}^c} \vec{\Omega} \otimes \vec{\Omega}\, d\vec{\Omega}, \end{aligned}\right\} \tag{3.3}$$

where the unknowns are I_1, I_2 and $[\gamma]$. Nevertheless, it is not possible to solve this system directly because $[\gamma]$ is a tensor. In order to solve the system, we will project the second and third equations along an appropriate direction.

3.1.1 Solution of the system. It is easily seen that the following relations hold:

$$\alpha_1 + \alpha_2 = 4\pi, \tag{3.4a}$$

$$\vec{v}_1 + \vec{v}_2 = 0, \tag{3.4b}$$

and

$$[m_1] + [m_2] = \frac{4}{3}\pi\, [\text{Id}]. \tag{3.4c}$$

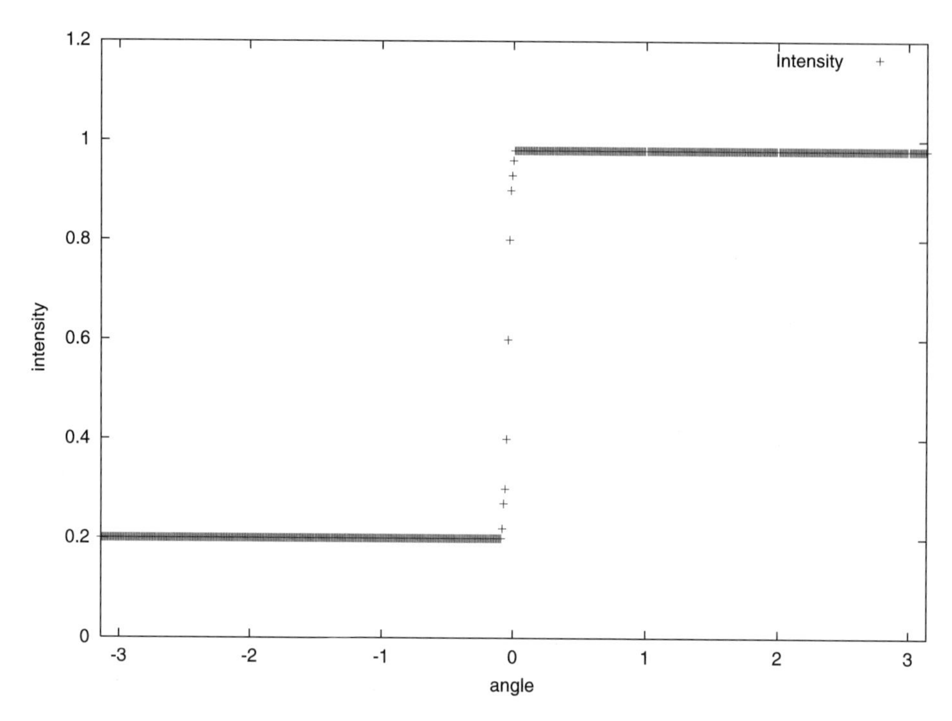

Figure 4. Intensity with respect to angle, close to the right hole of the domain $\mathcal{D}$ (results obtained on a uniform 100×100 grid in space and S_{64} quadrature for velocities).

Applying (3.4) to (3.3), the system under considerations reads:

$$\left.\begin{aligned} 1 &= \alpha_1 \bar{u}_1 + 4\pi u_2, \\ \vec{f} &= \vec{v}_1 \bar{u}_1, \\ [\gamma] &= [m_1]\bar{u}_1 + \frac{4\pi}{3}[\mathrm{Id}]u_2, \end{aligned}\right\} \tag{3.5}$$

where

$$\vec{f} := \frac{\vec{F}}{c\,E}. \tag{3.6}$$

The new unknowns are

$$\bar{u}_1 := \frac{1}{c\,E}(I_1 - I_2), \quad u_2 := \frac{1}{c\,E} I_2 \quad \text{and} \quad [\gamma].$$

Observe that f has to satisfy, as required in the flux-limited diffusion theory,

$$|\vec{f}| \leq 1. \tag{3.7}$$

Moreover, $\vec{f}$ and $[\gamma]$ (the first and second moment of a nonnegative unit intensity $\frac{I}{cE}$) must satisfy the following constraints:

$$\mathrm{tr}([\gamma]) = 1, \tag{3.8}$$

$$[\gamma] - \vec{f} \otimes \vec{f} \geq 0. \tag{3.9}$$

Multiplying by $\frac{1}{3}[\mathrm{Id}]$ the first equation of the system (3.5) and subtracting to the third equation we get

$$\left.\begin{aligned} 1 &= \alpha_1 \bar{u}_1 + 4\pi\, u_2, \\ \vec{f} &= \vec{v}_1\, \bar{u}_1, \\ [\gamma] &= \frac{1}{3}[\mathrm{Id}] + \left([m_1] - \frac{\alpha_1}{3}\,[\mathrm{Id}]\right) \bar{u}_1. \end{aligned}\right\} \tag{3.10}$$

3.1.2 Computation of a scalar Eddington factor γ. According to [5], we are interested in finding a scalar Eddington factor γ defined by

$$\gamma\, \vec{n} = [\gamma]\vec{n}, \tag{3.11}$$

where $\vec{n}$ is the unit vector $\vec{f}/|\vec{f}|$. This leads to the following relation (see [5]):

$$[\gamma] = \frac{1-\gamma}{2}\,[\mathrm{Id}] + \frac{3\gamma - 1}{2}\vec{n} \otimes \vec{n}. \tag{3.12}$$

This model can be seen as a generalization of the classical Eddington approximation (2.7). In order to determine γ, we shall use the tensor $[\gamma]$ introduced in the previous subsection.

First of all, observe that $\vec{v}_1$ is an eigenvector of the matrix $[m_1]$ and denote by m_1 the associated eigenvalue. Hence, we have

$$[m_1]\frac{\vec{v}_1}{|\vec{v}_1|} = m_1 \frac{\vec{v}_1}{|\vec{v}_1|}. \tag{3.13}$$

The third equation of (3.10) shows that $\vec{v}_1$ is also an eigenvector of $[\gamma]$. Using (3.11) we obtain

$$\gamma = \frac{1}{3} + m_1\, \bar{u}_1 - \frac{\alpha_1}{3}\,\bar{u}_1. \tag{3.14}$$

Then we express $\bar{u}_1$ in terms of $\vec{f}$. The second equation of (3.10) shows that $\vec{v}_1$ has the same direction as $\vec{n}$ ($\vec{n} = -\frac{\vec{v}_1}{|\vec{v}_1|}$) and we get

$$\bar{u}_1 = \frac{|\vec{f}|}{|\vec{v}_1|}.$$

Finally, we obtain

$$\gamma = \frac{1}{3} + \frac{m_1 - \frac{\alpha_1}{3}}{|\vec{v}_1|} \frac{|\vec{F}|}{cE}. \tag{3.15}$$

We shall rewrite this expression as

$$\gamma(f) = \frac{1}{3} - h_1 f, \tag{3.16}$$

where h_1 is defined by

$$h_1 = \frac{\frac{\alpha_1}{3} - m_1}{|\vec{v}_1|} = \frac{\frac{1}{3}\int_{S_T} d\vec{\Omega} - \int_{S_T} (\vec{\Omega}\cdot\vec{n})^2 \, d\vec{\Omega}}{\left|\int_{S_T} \vec{\Omega}\cdot\vec{n} \, d\vec{\Omega}\right|}, \tag{3.17}$$

and

$$f = \frac{|\vec{F}|}{cE}.$$

3.2 Computation of a flux limiter λ

In this section we relate the Eddington factor γ to a flux-limiter λ. Let us consider again the moment system:

$$\vec{\nabla}\cdot(\vec{f}E) + \sigma(E - S) = 0, \tag{3.18}$$

$$\vec{\nabla}\cdot([\gamma]E) + \sigma\vec{f}E = 0. \tag{3.19}$$

Multiplying the equation (3.18) by $\vec{f}$ and subtracting to equation (3.19), one obtains

$$\vec{\nabla}\cdot\left[\left([\gamma] - \vec{f}\otimes\vec{f}\right)E\right] + \sigma\vec{f}S = 0. \tag{3.20}$$

From now on we assume that the spatial variations of $[\gamma]$ and $\vec{f}$ are small, and we deduce, according to [5], an algebraic relation between $[\gamma]$, $\vec{f}$, E and $\vec{\nabla}E$:

$$\left([\gamma] - \vec{f}\otimes\vec{f}\right)\vec{R} = \vec{f}, \tag{3.21}$$

where $\vec{R}$ is the dimensionless gradient defined by

$$\vec{R} = -\frac{\vec{\nabla}E}{\sigma S}. \tag{3.22}$$

So $R := |\vec{R}|$ and f are related by the relation:

$$R = \frac{f}{\gamma(f) - f^2}. \tag{3.23}$$

This allows us to define the flux limiter λ as

$$f = \lambda(R)R \quad \text{or} \quad \lambda(R) = \gamma(f) - f^2, \tag{3.24}$$

which yields the flux

$$\vec{F} = -\frac{c\,\lambda(R)\vec{\nabla}E}{\sigma}. \tag{3.25}$$

We compute the flux limiter $\lambda(R)$ related to the Eddington factor $\gamma(f)$. We have:

$$\lambda(R) = \frac{1}{3} - h_1\lambda(R)R - \lambda^2(R)R^2, \tag{3.26}$$

which yields

$$\lambda(R) = \frac{2}{3\,(h_1 R - 1)^2 + \sqrt{9(1 - h_1 R)^2 + 12R^2}}. \tag{3.27}$$

4 Properties of γ and λ

This section is devoted to the behaviour of the scalar γ (and λ) with respect to σ and f.
In [5], Levermore gives four conditions that γ and λ must satisfy:

(C1) $\lambda(R) + \lambda(R)^2 R^2 \leq 1$.

(C2) $\gamma(0) = \frac{1}{3}$ and $\lambda(0) = \frac{1}{3}$.

(C3) $\lambda(R) + \lambda(R)^2 R^2$ is an increasing function of R.

(C4) $\gamma(f) - f^2$ is a decreasing function of f.

The conditions (C1), (C2) and (C4) are satisfied, but our flux limiter violates the condition (C3). In order to satisfy (C3), the Eddington factor must be increasing as a function of f and it is not the case here because h_1 is always positive, hence $\lambda(R) + \lambda(R)^2 R^2$ is decreasing as a function of R.

Influence of geometrical parameter h_1. We can see that, when h_1 is equal to zero, our geometric limiter is equal to the first part of the Minerbo limiter related to the Eddington factor $\gamma(f) = \frac{1}{3}$.

Behaviour of λ for diffusive and void cases. For diffusive cases, i.e. $\sigma >> 1$, the Eddington approximation is valid and we can take $\lambda = \frac{1}{3}$. For our flux limiter, we get: $\lim_{\sigma\to+\infty}\lambda = \frac{1}{3}$, and for the void case: $\lim_{\sigma\to 0}\lambda = 0$.

Influence of gradient. A main property of flux limiter is to decrease when the norm of the radiative energy gradient increase, thus we have: $\lim_{R\to 0}\lambda = \frac{1}{3}$, $\lim_{R\to+\infty}\lambda = 0$. When $f = 0$ there is no preferred direction, the intensity is isotropic and the Eddington approximation is valid: $\gamma(0) = \frac{1}{3}$ and $\lambda(0) = \frac{1}{3}$. Let us now compare in Figure 5 our geometric flux limiter with other classical flux limiters for a fixed value of h_1 ($h_1 = 1$).

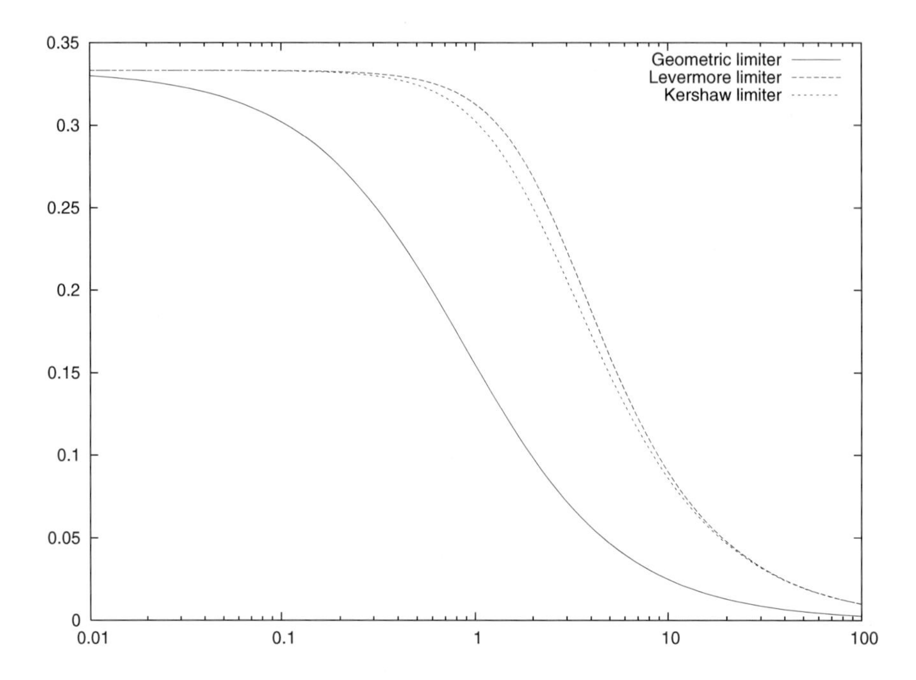

Figure 5. Behavior of flux limiters when R increases.

5 Numerical scheme

In order to solve the photon transport equation, we use the standard S_N method with the spatial diamond-differencing scheme. This is the reference transport method for a two-dimensional Cartesian fine mesh.

To solve the diffusion equation with a fixed flux limiter, we use a two-dimensional finite differences scheme coupled with a Jacobi algorithm. We also use a mixed-hybrid finite elements method (see [8]) in order to obtain more accuracy and to reduce the CPU time which becomes prohibitive when σ is small.

Let us remark that no general result claiming the well-posedness of the nonlinear diffusion problem is available because of nonlinear dependence of the diffusion coefficient as a function of the gradient of the solution. Nevertheless, the flux limiter is bounded and positive and we observe convergence of the following fixed-point iterative method.

Step 1. Initialization: solve the diffusion equation with $\lambda(R) = \frac{1}{3}$ (Eddington's approximation):

$$\left.\begin{aligned} -\mathrm{div}\,(\frac{1}{3\sigma}\,\nabla E^0) + \sigma\, E^0 &= \sigma\, S \quad \text{in } \mathcal{D}, \\ \omega\, E^0 + 2\,(2-\omega)\frac{1}{3\sigma}\vec{\nabla} E^0 \cdot \vec{n}_b &= 0 \qquad \text{on } \partial\mathcal{D}. \end{aligned}\right\} \tag{5.1}$$

Step 2. We compute $R^0 = -\frac{|\vec{\nabla} E^0|}{S}$.

Step 3. For all $k \geq 0$, we compute a solution E^k of the diffusion problem with λ given by the expression of flux limiter, function of R^{k-1}:

$$\left.\begin{aligned} -\mathrm{div}\,(\frac{\lambda}{\sigma}\,(R^{k-1})\nabla\, E^k) + \sigma\, E^k &= \sigma\, S \quad \text{in } \mathcal{D}, \\ \omega\, E^k + 2\,(2-\omega)\frac{\lambda(R^{k-1})}{\sigma}\vec{\nabla} E^k \cdot \vec{n}_b &= 0 \qquad \text{on } \partial\mathcal{D}. \end{aligned}\right\} \tag{5.2}$$

Step 4. Then we compute $R^k = -\frac{|\vec{\nabla} E^k|}{S}$ and we loop to step 3 while $\|E^k - E^{k-1}\|_1$ becomes larger than a small arbitrary parameter.

6 Numerical results

We first consider the case of two plane plates with reflecting conditions on the plane plates, entering flux equal to zero on the holes and a source equal to 1 ($S = 1$). We compare in the Figures 6, 7 and 8 a reference transport solution, a solution obtained by Eddington's approximation, a solution obtained with our model and a solution obtained with other flux limiters: Wilson's limiter [5], Chapman–Enskog's limiter [5] and Kershaw's limiter [5] for different values of σ. Let us give here the expressions of these flux limiters:

Wilson's limiter: $\lambda(R) = \frac{1}{3+R}$ and the related $\gamma(f) = \frac{1-f+3f^2}{3}$.

Chapman–Enskog's limiter: $\lambda(R) = \frac{1}{R}\left(\coth R - \frac{1}{R}\right)$ and the related $\gamma = \coth R \left(\coth R - \frac{1}{R}\right)$ with $f = \coth R - \frac{1}{R}$.

Kershaw's limiter: $\lambda(R) = \frac{2}{3+\sqrt{9+4R^2}}$ and the related $\gamma(f) = \frac{1+2f^2}{3}$.

The influence of these flux limiters will be compared with our geometrical model.

These numerical results are computed in two dimensions (on a 50×50 grid) but we just give a one-dimensional profile section of the energy in the figure and only in a quarter of the domain $\mathcal{D}$ because of symmetry reasons. According to the notations of Figure 3, these numerical results were computed with the following parameters: $L = 2$, $H = 1$, $h = 0$ and the albedo of the wall $\omega = 0$.

One observes that the energy density is, in all cases considered, better approximated by the geometric method than by the diffusion's Eddington approximation (P1) or by the diffusion equation with classical flux limiters. This is particularly true for the region close to the hole where our geometric model clearly seems to be the most competitive method.

Figure 6 presents the profile of the energy in the case where σ is large. That means an isotropic case and this is why every model gives the same results. In Figure 7 we

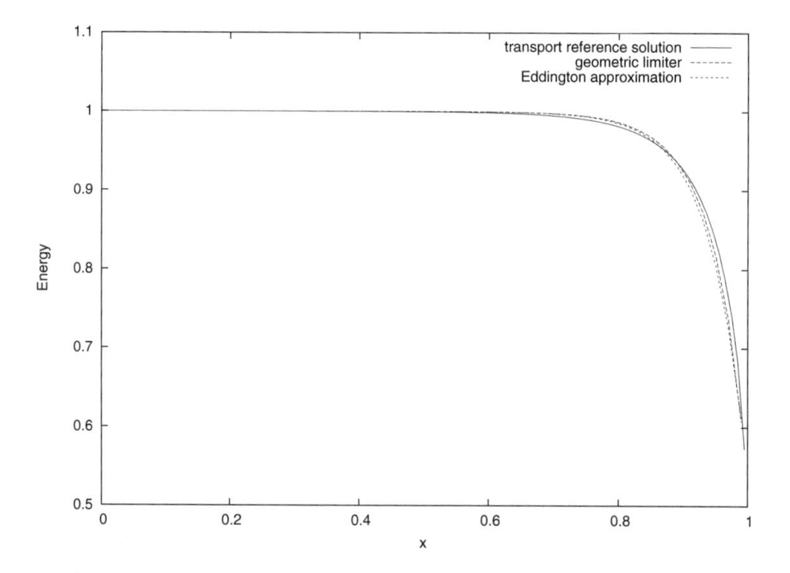

Figure 6. Section of energy for plane plates configuration in the case $\sigma = 10$.

show the case where opacity is smaller than in the previous one. It seems that the geometrical model is better than the others. When the opacity σ is close to zero, the geometrical model always seems to be more efficient than the others, but the flux-limiter results are too far away from the transport solution (as we can see in Figure 8) because of the limits of validity of the model.

Indeed, we can compute, using the transport program, the effective ideal flux limiter given by the following relation:

$$\vec{F} = -\frac{c\lambda(R)\vec{\nabla}E}{\sigma}. \tag{6.1}$$

This flux limiter is optimal when working with our model; it does not, however, provide better results than the geometric flux limiter.

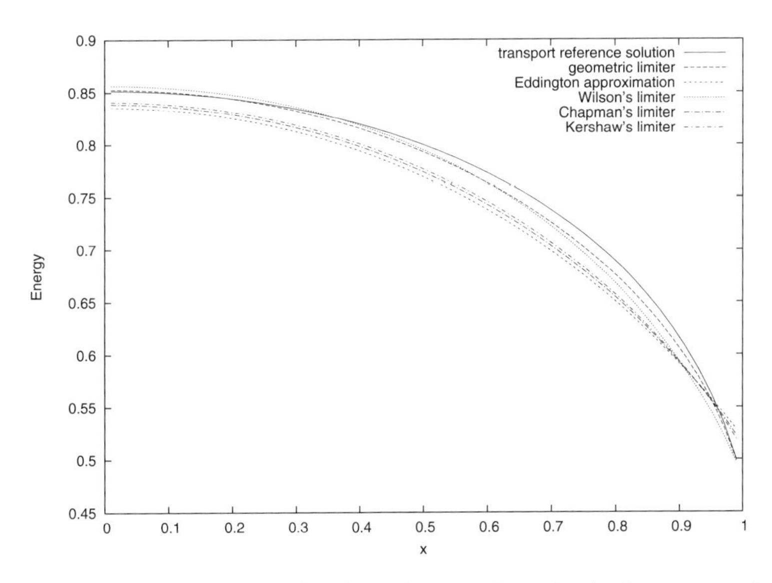

Figure 7. Section of energy for plane plates configuration in the case $\sigma = 1$.

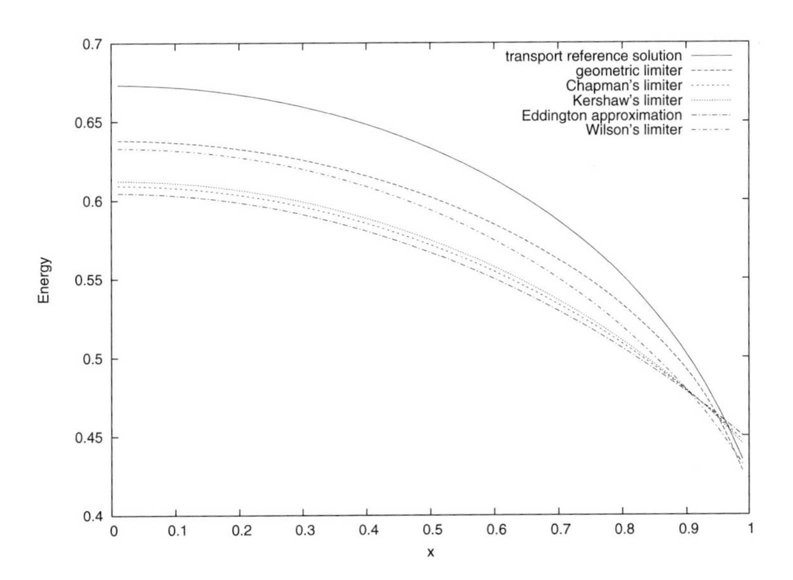

Figure 8. Section of energy for plane plates configuration in the case $\sigma = 0.5$.

Let us now compare the results obtained with the "transport flux limiter" and the other ones in Figure 9.

We can see that even with the "transport flux limiter" the results are not close to the reference transport solution and our geometric limiter seems as good as the "transport

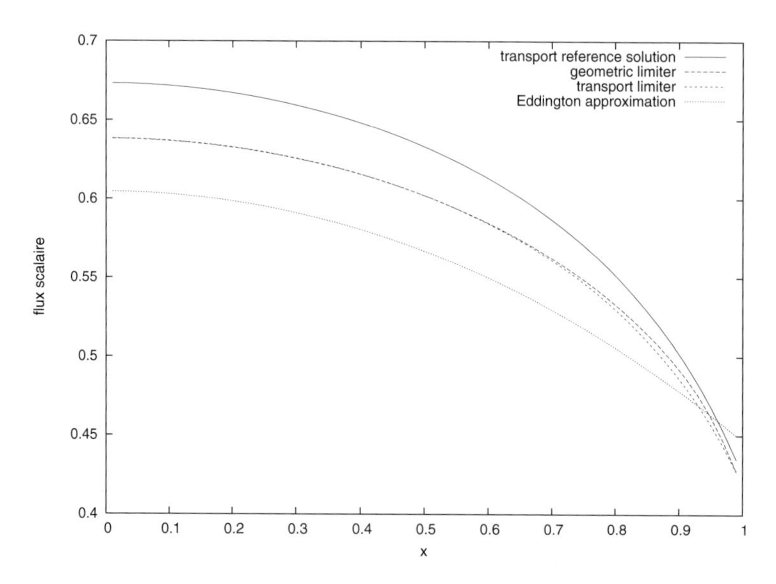

Figure 9. Section of energy for plane plates configuration in the case $\sigma = 0.5$.

flux limiter". When σ is close to zero, the diffusion model does not seem to be valid. Indeed, the "transport flux limiter" being obtained directly from the transport equation is optimal for our model.

We can also compare the profiles of the flux limiters as shown in Figure 10:

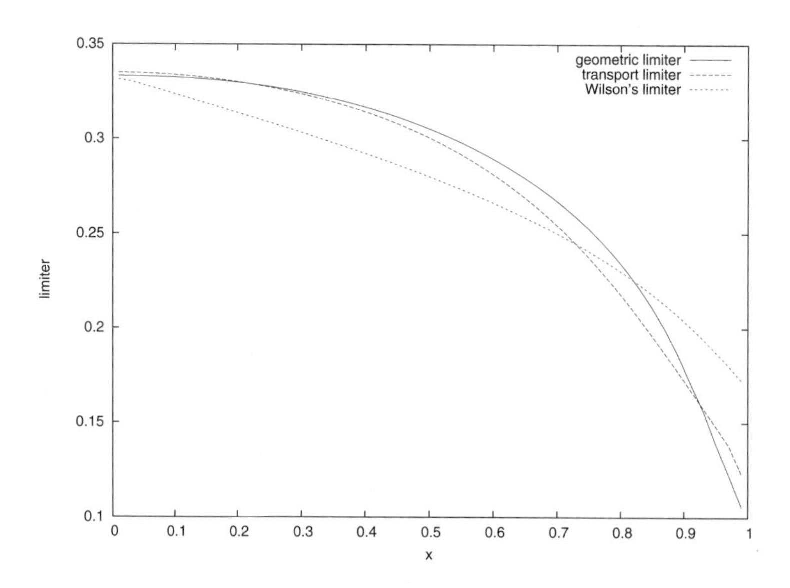

Figure 10. Profiles of flux limiters in the case $\sigma = 0.5$.

Relative errors

Here we present a table containing the relative errors $\frac{\|E_{\text{dif}}-E_{\text{tra}}\|_1}{\|E_{\text{tra}}\|_1}$ between the transport reference solution and the solutions of our diffusion model for different flux limiters and some values of σ in the case of two plane plates.

	Geometric limiter	Wilson's limiter	Kershaw's limiter	Eddington's approx. (P1)
$\sigma = 10$	$2,63.10^{-3}$	$2,75.10^{-3}$	$3,02.10^{-3}$	$3,27.10^{-3}$
$\sigma = 1$	$7,94.10^{-3}$	$1,19.10^{-2}$	$2,66.10^{-2}$	$3,45.10^{-2}$
$\sigma = 0,5$	$4,42.10^{-2}$	$5,77.10^{-2}$	$8,22.10^{-2}$	$9,30.10^{-2}$

7 Conclusion

In this paper we have proposed a new formulation of the Eddington factor (and the underlaying flux limiter) adapted to radiative transfer calculations in a cavity. The main point was that classical flux-limited diffusion theory did not apply because photon mean free path and gradient length of the solution are large in front of the size of the cavity so that geometric effects are dominant. Hence, the Eddington factor includes geometric features of the domain.

Although our modelling of the Eddington factor is correct, the corresponding results on the diffusion equation are not totally satisfying: our understanding is that the treatment of the boundary condition should be questioned. Future work will consist in applying this factor to realistic configurations (i.e. radiation hydrodynamic flows in an ICF cavity). In this case it may be necessary to improve the model in order to account for a more detailed description of the cavity (presence of the target, distinction between left and right laser entry holes, etc.).

Acknowledgements

This proceeding could not have been written without J. F. Clouët and G. Samba from CEA-DAM. The authors gratefully thank S. Cordier and B. Dubroca for their contributions to the present work.

A Analytical expression of the data

The angles $\theta_i, i = 1, 2, 3, 4$ are defined as shown in Figure 3. We may then compute, according to the definitions (3.2), the following expressions:

$$\alpha_1 = 2(\theta_2 - \theta_1) + 2(\theta_4 - \theta_3) \tag{1.1}$$

$$\alpha_2 = 2\pi - \alpha_1. \tag{1.2}$$

We compute as well:

$$\vec{v}_1 = \int_{\theta_1}^{\theta_2}\int_{-1}^{1}\begin{pmatrix} -\sqrt{1-z^2}\cos\theta \\ -\sqrt{1-z^2}\sin\theta \end{pmatrix} dz\, d\theta + \int_{\theta_3}^{\theta_4}\int_{-1}^{1}\begin{pmatrix} -\sqrt{1-z^2}\cos\theta \\ -\sqrt{1-z^2}\sin\theta \end{pmatrix} dz\, d\theta.$$

Applying the change of variables $z = \sin t$, $dz = \cos t\, dt$, we get

$$\int_{-1}^{1}\sqrt{1-z^2}\, dz = \int_{-\pi/2}^{\pi/2}(\cos t)^2\, dt = \left[\frac{t}{2} + \frac{\cos 2t}{2}\right]_{-\pi/2}^{\pi/2} = \frac{\pi}{2},$$

and therefore we obtain

$$\vec{v}_1 = -\frac{\pi}{2}\int_{\theta_1}^{\theta_2}\begin{pmatrix} \cos\theta \\ \sin\theta \end{pmatrix} d\theta - \frac{\pi}{2}\int_{\theta_3}^{\theta_4}\begin{pmatrix} \cos\theta \\ \sin\theta \end{pmatrix} d\theta, \tag{1.3}$$

that is,

$$\vec{v}_1 = \pi\sin\left(\frac{\theta_2-\theta_1}{2}\right)\begin{pmatrix} \cos\left(\frac{\theta_2+\theta_1}{2}\right) \\ \sin\left(\frac{\theta_2+\theta_1}{2}\right) \end{pmatrix} + \pi\sin\left(\frac{\theta_4-\theta_3}{2}\right)\begin{pmatrix} \cos\left(\frac{\theta_4+\theta_3}{2}\right) \\ \sin\left(\frac{\theta_4+\theta_3}{2}\right) \end{pmatrix}. \tag{1.4}$$

Then we compute the expressions of $[m_1]$. The definition of $[m_1]$ is

$$[m_1] = \int_{\theta_1}^{\theta_2}\int_{-1}^{1}[M(\theta, z)]\, dz\, d\theta + \int_{\theta_3}^{\theta_4}\int_{-1}^{1}[M(\theta, z)]\, dz\, d\theta,$$

where

$$[M(\theta, z)] = (1-z^2)\begin{pmatrix} (\cos\theta)^2 & \cos\theta\sin\theta \\ \cos\theta\sin\theta & (\sin\theta)^2 \end{pmatrix}.$$

It is a simple matter to obtain $\displaystyle\int_{-1}^{1}(1-z^2)\, dz = \frac{4}{3}$ and

$$\int_{\theta_1}^{\theta_2}(\cos\theta)^2\, d\theta = \int_{\theta_1}^{\theta_2}\left[\frac{1}{2} + \frac{\cos 2\theta}{2}\right]_{\theta_1}^{\theta_2} = \frac{\theta_2-\theta_1}{2} + \left[\frac{\sin 2\theta}{4}\right]_{\theta_1}^{\theta_2}$$

and

$$\int_{\theta_1}^{\theta_2}(\sin\theta)^2\, d\theta = \frac{\theta_2-\theta_1}{2} - \left[\frac{\sin 2\theta}{4}\right]_{\theta_1}^{\theta_2}.$$

We also have

$$\int_{\theta_1}^{\theta_2} \sin\theta \cos\theta \, d\theta = \left[\frac{1}{2}(\sin\theta)^2\right]_{\theta_1}^{\theta_2} = \frac{(\sin\theta_2)^2 - (\sin\theta_1)^2}{2}.$$

Finally, we get

$$[m_1] = \frac{4}{3}\begin{bmatrix} \frac{\theta_2-\theta_1}{2} + \frac{\sin 2\theta_2 - \sin 2\theta_1}{4} & \frac{(\sin\theta_2)^2 - (\sin\theta_1)^2}{2} \\ \frac{(\sin\theta_2)^2 - (\sin\theta_1)^2}{2} & \frac{\theta_2-\theta_1}{2} - \frac{\sin 2\theta_2 - \sin 2\theta_1}{4} \end{bmatrix} + \frac{4}{3}\begin{bmatrix} \frac{\theta_4-\theta_3}{2} + \frac{\sin 2\theta_4 - \sin 2\theta_3}{4} & \frac{(\sin\theta_4)^2 - (\sin\theta_3)^2}{2} \\ \frac{(\sin\theta_4)^2 - (\sin\theta_3)^2}{2} & \frac{\theta_4-\theta_3}{2} - \frac{\sin 2\theta_4 - \sin 2\theta_3}{4} \end{bmatrix}. \tag{1.5}$$

The expressions above simplify, using the trigonometric relations:

$$\begin{aligned} &\sin 2\theta_2 - \sin 2\theta_1 \\ &\quad = 2\sin(\theta_2-\theta_1)\cos(\theta_1+\theta_1) \\ &\quad = 4\sin\left(\frac{\theta_2-\theta_1}{2}\right)\cos\left(\frac{\theta_2-\theta_1}{2}\right)\cos(\theta_1+\theta_2), \end{aligned}$$

and

$$\begin{aligned} &(\sin\theta_2)^2 - (\sin\theta_1)^2 \\ &\quad = (\sin\theta_2 - \sin\theta_1)(\sin\theta_2 + \sin\theta_1) \\ &\quad = 4\sin\left(\frac{\theta_2-\theta_1}{2}\right)\cos\left(\frac{\theta_1+\theta_2}{2}\right)\sin\left(\frac{\theta_2+\theta_1}{2}\right)\cos\left(\frac{\theta_2-\theta_1}{2}\right) \\ &\quad = 2\sin\left(\frac{\theta_2-\theta_1}{2}\right)\cos\left(\frac{\theta_2-\theta_1}{2}\right)\sin(\theta_1+\theta_2). \end{aligned}$$

Hence, we obtain

$$[\bar{m}_1] = \frac{4}{3}\sin\left(\frac{\theta_2-\theta_1}{2}\right)\cos\left(\frac{\theta_2-\theta_1}{2}\right)\begin{bmatrix} \cos(\theta_1+\theta_2) & \sin(\theta_1+\theta_2) \\ \sin(\theta_1+\theta_2) & -\cos(\theta_1+\theta_2) \end{bmatrix} + \frac{4}{3}\sin\left(\frac{\theta_4-\theta_3}{2}\right)\cos\left(\frac{\theta_4-\theta_3}{2}\right)\begin{bmatrix} \cos(\theta_4+\theta_3) & \sin(\theta_4+\theta_3) \\ \sin(\theta_4+\theta_3) & -\cos(\theta_4+\theta_3) \end{bmatrix}. \tag{1.6}$$

Combining expressions (1.3) and (1.6) of $\vec{v}_1$ and $[\bar{m}_1]$ with relation (3.3) from the previous section, we obtain explicitly:

$$[\gamma] = [\gamma](\vec{R}, \theta_1, \theta_2, \theta_3, \theta_4).$$

We can also compute the angles $\theta_i, i = 1, 2, 3, 4$ with respect to the position (x, y) of the point, setting the origin $(0, 0)$ at the center of the domain:

$$\theta_1(x, y) = \arctan\left(\frac{-y - H/2 + h}{L/2 - x}\right)$$

$$\theta_2(x, y) = \arctan\left(\frac{H/2 - y - h}{L/2 - x}\right)$$

$$\theta_3(x, y) = \pi - \arctan\left(\frac{-y - H/2 + h}{L/2 + x}\right)$$

$$\theta_4(x, y) = \pi - \arctan\left(\frac{H/2 - y - h}{L/2 + x}\right).$$

B Figures of numerical results in two dimensions

Cases between two plane plates ($h = 0$ and $\omega = 0$)

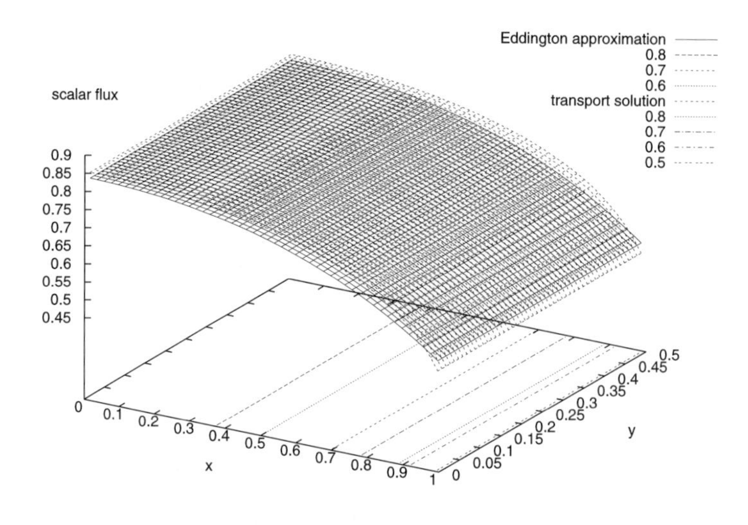

Figure 11. Transport versus Eddington's approximation.

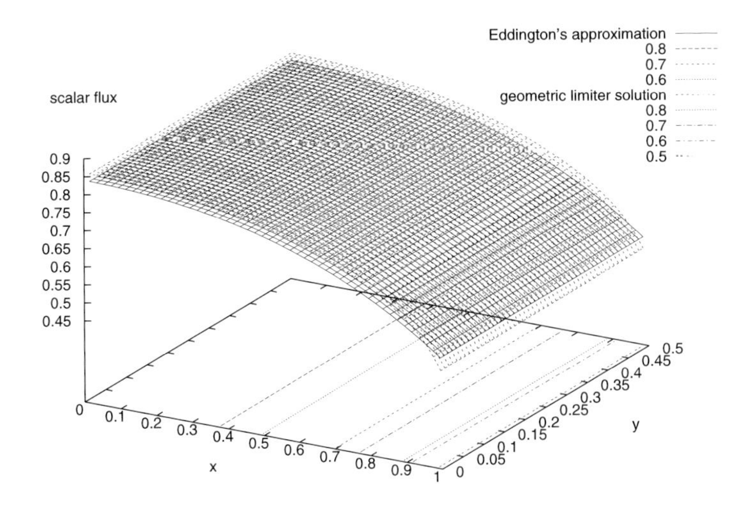

Figure 12. Geometric limiter versus Eddington's approximation.

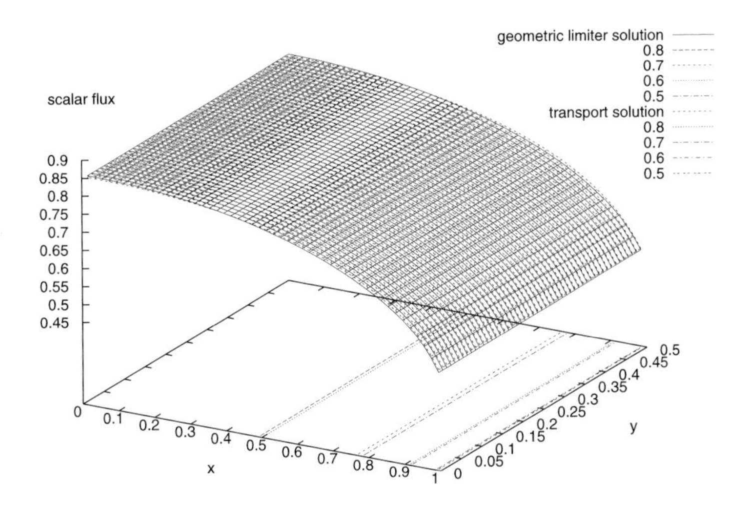

Figure 13. Transport versus geometric limiter.

Cases with edges ($h = 0.2$ and $\omega = 0.5$)

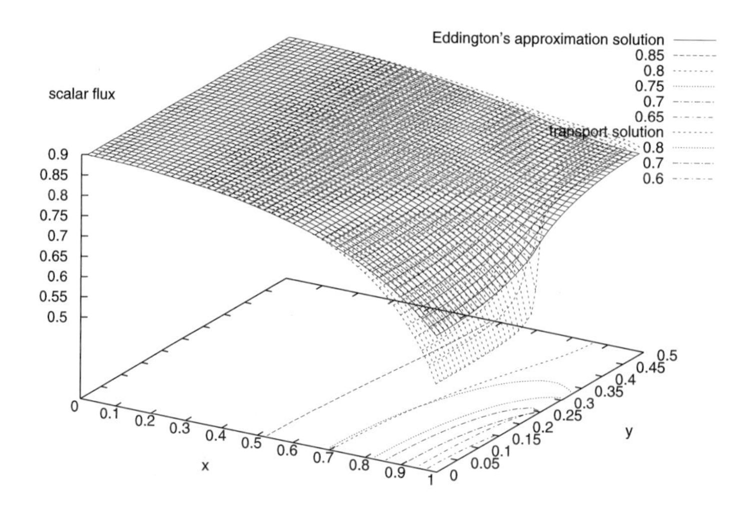

Figure 14. Transport versus Eddington's approximation.

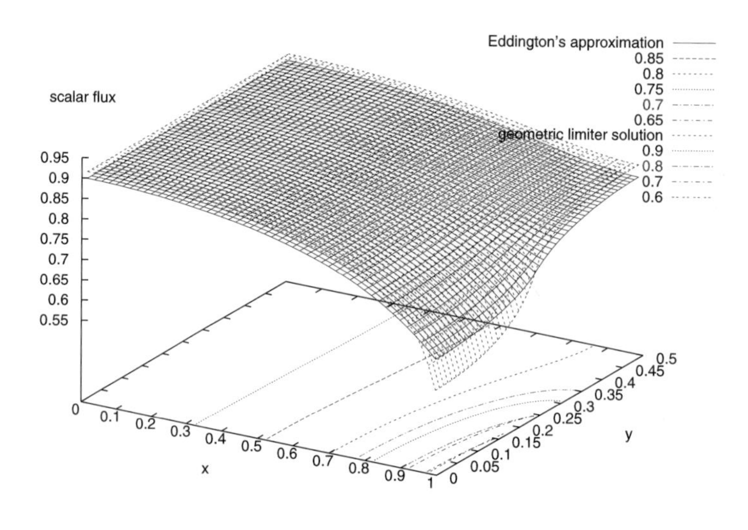

Figure 15. Geometric limiter versus Eddington's approximation.

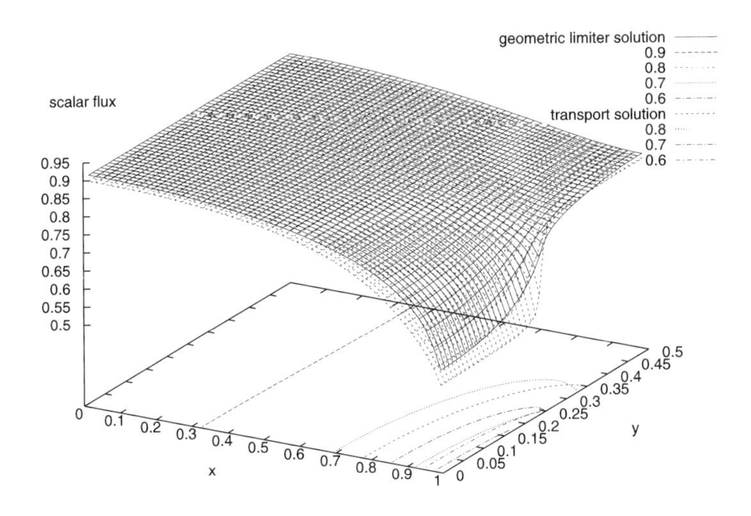

Figure 16. Transport versus geometric limiter.

References

[1] C. G. Pomraning, *The Equation of Radiation Hydrodynamics*, Pergamon Press, 1973.

[2] R. E. Alcouffe, B. A. Clark and E. W. Larsen, The diffusion synthetic acceleration of transport iterations, in *Multiple Time Scales* (J. Brackbill and B. Cohen ed.), Comput. Tech. 3, Academic Press, New-York 1985, 73–111.

[3] J. A. Fleck and J. D. Cummings, An implict Monte Carlo scheme for calculating time and frequency dependent non linear radiation transport, *J. Comput. Phys.* 8 (1971), 313–342.

[4] E. W. Larsen, V. C. Badham and G. C. Pomraning, Asymptotic analysis of radiative transfer problems, *J. Quant. Spec. Radia. Transfer* 29 (1983), 285–310.

[5] D. Levermore, Relating Eddington factors to flux limiters, *J. Quant. Spec. Radia. Transfer* 31 (1984), 149–160.

[6] M. Rampp and H.-Th. Janka, Radiation hydrodynamics with neutrinos: Variable Eddington factor method for core-collapse supernova simulations, *A&A* 396 (2002), 361–392.

[7] G. N. Minerbo, Maximum entropy Eddington factors, *J. Quant. Spec. Radia. Transfer* 20 (1978), 541–545.

[8] F. Brezzi and M. Fortin, *Mixed and hybrid finite element methods*, Springer Ser. in Comput. Math. 15, Springer-Verlag, New York 1991.

Arbitrary high order discontinuous Galerkin schemes

M. Dumbser and C.-D. Munz

Institut für Aerodynamik und Gasdynamik, Pfaffenwaldring 21
70550 Stuttgart, Germany
email: `michael.dumbser@iag.uni-stuttgart.de`
`munz@iag.uni-stuttgart.de`

Abstract. In this paper we apply the ADER one step time discretization to the discontinuous Galerkin framework for hyperbolic conservation laws. In the case of linear hyperbolic systems we obtain a quadrature-free explicit single-step scheme of arbitrary order of accuracy in *space and time* on Cartesian and triangular meshes. The ADER-DG scheme does not need more memory than a first order explicit Euler time-stepping scheme. This becomes possible because of an extensive use of the governing equations inside the numerical scheme. In the nonlinear case, quadrature of the ADER-DG scheme in space and time is performed with Gaussian quadrature formulae of suitable order of accuracy. We show numerical convergence results for the linearized Euler equations up to 10th order of accuracy in space and time on Cartesian and triangular meshes. Numerical results for the nonlinear Euler equations up to 6th order of accuracy in space and time are provided as well. In this paper we also show the possibility of applying a linear reconstruction operator of the order $3N + 2$ to the degrees of freedom of the DG method resulting in a numerical scheme of the order $3N + 3$ on Cartesian grids where N is the order of the original basis functions before reconstruction.

1 Introduction

In [16], [17] Schwartzkopff et al. successfully constructed a finite volume scheme of arbitrary high order of accuracy in space and time for linear two-dimensional hyperbolic systems using Toro's ADER approach for time discretization [21], [22], [19]. The ADER approach is based on solving generalized Riemann problems (GRPs) at the cell boundaries and the application of the Lax–Wendroff procedure for highly accurate time integration of the numerical flux. In the present paper we will show the application of the ADER time discretization method to the discontinuous Galerkin (DG) framework as developed by Cockburn and Shu [3], [4], [5], [6] using TVD Runge–Kutta schemes for time integration. However, Runge–Kutta time integration schemes have to face the so-called Butcher barriers [2] for schemes of higher order than four, where the number of substages becomes greater than the order of the scheme.

The aim of our ongoing research [9], [10] is to obtain an arbitrarily accurate scheme in space and time even on unstructured grids, which should be useful e.g. in aeroacoustics for accurate noise propagation in the time domain or other wave propagation in complex geometries. In order to avoid a cumbersome reconstruction on unstructured meshes, the combination of the ADER and DG approaches seems to be highly attractive to reach this goal because all space-derivatives are readily available at each point inside and on the borders of the finite elements. This makes the application of the Lax–Wendroff procedure quite easy, see Section 2.

Since for linear hyperbolic equations the solution remains smooth, if the initial conditions are smooth, we try to further increase the accuracy of the ADER-DG method on Cartesian grids for linear hyperbolic systems without increasing the number of degrees of freedom by using a generalization of conservative reconstruction formulae known from finite volume schemes. We can show numerically, that the resulting scheme is indeed of the order $3N+3$ of accuracy, if we use N-th order basis functions and a reconstruction operator of the order $3N+2$, see Section 3.

In the nonlinear case, we follow the same procedure as described in [19], i.e. solving GRPs on the element boundaries and then carrying out time integration of the numerical flux by using the Lax–Wendroff procedure. We will show the results of our numerical experiments for one-dimensional Riemann problems computed with ADER-DG schemes up to 6th order of accuracy in space and time. Results of a simple smooth two-dimensional test problem ([11]) up to sixth order of accuracy, showing that the designed order of the method has been reached in numerical experiments also in the nonlinear case, are presented in Section 4.

2 ADER-DG schemes for linear hyperbolic equations

2.1 Semi-discrete form of the scheme

In the following section we apply the ADER approach of time-integration originally developed by Toro et al. [21], [22], [19], [16], [17] in the finite volume framework to the discontinuous Galerkin method for solving a two-dimensional system of linear hyperbolic equations:

$$\frac{\partial u_p}{\partial t} + A_{pq}\frac{\partial u_q}{\partial x} + B_{pq}\frac{\partial u_q}{\partial y} = 0. \tag{2.1}$$

Here, we use classical tensor notation which implies summation over each index appearing twice. The resulting scheme of this section will be a generalization of the approach presented in [9] for the simple one-dimensional scalar advection equation.

The computational domain is divided in conforming triangles $T^{(m)}$ being addressed by a unique index (m). The numerical solution u_h of (2.1) is approximated inside each triangle $T^{(m)}$ by a linear combination of some time independent orthogonal basis functions $\Phi_l(\xi, \eta) \in V \subset L^2$ with support $T^{(m)}$ and with time dependent degrees of freedom $\hat{u}_{pl}^{(m)}(t)$:

$$\left(u_h^{(m)}\right)_p (\xi, \eta, t) = \hat{u}_{pl}^{(m)}(t)\Phi_l(\xi, \eta). \tag{2.2}$$

ξ and η are coordinates in a reference element and are defined by (2.9) and (2.10). Eqn. (2.1) is multiplied by a test function $\Phi_k \in V$ and is integrated over a triangle $T^{(m)}$:

$$\int_{T^{(m)}} \Phi_k \frac{\partial u_p}{\partial t} dV + \int_{T^{(m)}} \Phi_k \left(A_{pq} \frac{\partial u_q}{\partial x} + B_{pq} \frac{\partial u_q}{\partial y} \right) dV = 0. \tag{2.3}$$

Integration by parts yields

$$\int_{T^{(m)}} \Phi_k \frac{\partial u_p}{\partial t} dV + \int_{\partial T^{(m)}} \Phi_k F_p^h dS - \int_{T^{(m)}} \left(\frac{\partial \Phi_k}{\partial x} A_{pq} u_q + \frac{\partial \Phi_k}{\partial y} B_{pq} u_q \right) dV = 0 \tag{2.4}$$

where a numerical flux F_p^h has been introduced in the surface integral since u_h may be discontinuous at an element boundary. We suppose rotational invariance of system (2.1), so the flux can be written very easily in a coordinate system which is aligned with the outward pointing unit normal vector $\vec{n} = (n_x, n_y)^T$ on the boundary. The transformation of the vector u_p from the global system to the vector u_q^T in an edge-aligned coordinate system is given by

$$u_p = T_{pq} u_q^T. \tag{2.5}$$

For the (linearized) two-dimensional Euler equations the transformation matrix is

$$T_{pq} = \begin{pmatrix} 1 & 0 & 0 & 0 \\ 0 & n_x & -n_y & 0 \\ 0 & n_y & n_x & 0 \\ 0 & 0 & 0 & 1 \end{pmatrix}. \tag{2.6}$$

We use Godunov's exact flux as a numerical flux between two triangles $T^{(m)}$ and $T^{(k_j)}$. For the global $x - y$ system, we finally obtain

$$F_p^h = \frac{1}{2} T_{pq} (A_{qr}^T + |A_{qr}^T|)(T_{rs})^{-1} \hat{u}_{sl}^{(m)} \Phi_l^{(m)} + \frac{1}{2} T_{pq} (A_{qr}^T - |A_{qr}^T|)(T_{rs})^{-1} \hat{u}_{sl}^{(k_j)} \Phi_l^{(k_j)} \tag{2.7}$$

in the linear case, where $\hat{u}_{sl}^{(m)} \Phi_l^{(m)}$ and $\hat{u}_{sl}^{(k_j)} \Phi_l^{(k_j)}$ are the boundary extrapolated values of the numerical solution from element (m) and the j-th side neighbor (k_j), respectively, since both elements adjacent to a boundary contribute to the numerical flux. A_{qr}^T means the evaluation of matrix A_{qr} in the local edge-aligned system. Inserting (2.2) and (2.7) into (2.4) and splitting the boundary integral into the contributions of each edge j of the triangle $T^{(m)}$, we obtain

$$
\begin{aligned}
&\frac{\partial}{\partial t}\hat{u}_{pl}^{(m)} \int_{T^{(m)}} \Phi_k \Phi_l dV \\
&+ \sum_{j=1}^{3} T_{pq}^{j} \frac{1}{2}(A_{qr}^{T} + |A_{qr}^{T}|)(T_{rs}^{j})^{-1} \hat{u}_{sl}^{(m)} \int_{(\partial T^{(m)})_j} \Phi_k^{(m)} \Phi_l^{(m)} dS \\
&+ \sum_{j=1}^{3} T_{pq}^{j} \frac{1}{2}(A_{qr}^{T} - |A_{qr}^{T}|)(T_{rs}^{j})^{-1} \hat{u}_{sl}^{(k_j)} \int_{(\partial T^{(m)})_j} \Phi_k^{(m)} \Phi_l^{(k_j)} dS \\
&- A_{pq}\hat{u}_{ql}^{(m)} \int_{T^{(m)}} \frac{\partial \Phi_k}{\partial x} \Phi_l dV - B_{pq}\hat{u}_{ql}^{(m)} \int_{T^{(m)}} \frac{\partial \Phi_k}{\partial y} \Phi_l dV = 0.
\end{aligned}
\tag{2.8}
$$

Equation (2.8) is written in physical $x - y$ space, but, if we transform each physical triangle $T^{(m)}$ to a canonical reference triangle T_E in a $\xi - \eta$ reference space (see Figure 1), the method can be implemented much more efficiently since all integrals can be precomputed beforehand in the reference space. The transformation is defined by

$$
\begin{aligned}
x &= x_1 + (x_2 - x_1)\,\xi + (x_3 - x_1)\,\eta, \\
y &= y_1 + (y_2 - y_1)\,\xi + (y_3 - y_1)\,\eta
\end{aligned}
\tag{2.9}
$$

with the inverse transformation

$$
\begin{aligned}
\xi &= \frac{1}{|J|}\left((x_3 y_1 - x_1 y_3) + x\,(y_3 - y_1) + y\,(x_1 - x_3)\right), \\
\eta &= \frac{1}{|J|}\left((x_1 y_2 - x_2 y_1) + x\,(y_1 - y_2) + y\,(x_2 - x_1)\right)
\end{aligned}
\tag{2.10}
$$

where $|J| = (x_2 - x_1)(y_3 - y_1) - (x_3 - x_1)(y_2 - y_1)$ is the determinant of the Jacobian matrix of the transformation being equal to the double of the triangle's

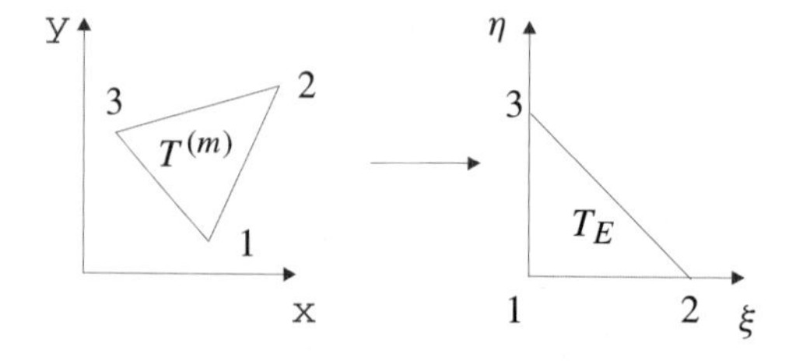

Figure 1. Transformation from the physical triangle $T^{(m)}$ to the canonical reference triangle T_E with nodes (0, 0), (1, 0) and (0, 1).

surface,

$$J = \begin{pmatrix} x_\xi & y_\xi \\ x_\eta & y_\eta \end{pmatrix}. \tag{2.11}$$

With respect to the transformation, we have furthermore

$$dxdy = |J|\, d\xi d\eta, \tag{2.12}$$

$$\begin{pmatrix} u_x \\ u_y \end{pmatrix} = J^{-1} \begin{pmatrix} u_\xi \\ u_\eta \end{pmatrix}. \tag{2.13}$$

With integration in the reference space, the semi-discrete DG formulation reads as

$$\begin{aligned}
&\frac{\partial}{\partial t}\hat{u}_{pl}^{(m)} |J| \int_{T_E} \Phi_k \Phi_l d\xi d\eta \\
&+ \sum_{j=1}^{3} T_{pq}^j \frac{1}{2}(A_{qr}^T + |A_{qr}^T|)(T_{rs}^j)^{-1} \hat{u}_{sl}^{(m)} |S_j| \int_0^1 \Phi_k^{(m)}(\chi_j) \Phi_l^{(m)}(\chi_j) d\chi_j \\
&+ \sum_{j=1}^{3} T_{pq}^j \frac{1}{2}(A_{qr}^T - |A_{qr}^T|)(T_{rs}^j)^{-1} \hat{u}_{sl}^{(k_j)} |S_j| \int_0^1 \Phi_k^{(m)}(\chi_j) \Phi_l^{(k_j)}(\chi_j) d\chi_j \\
&- A_{pq}^* \hat{u}_{ql} |J| \int_{T_E} \frac{\partial \Phi_k}{\partial \xi} \Phi_l d\xi d\eta - B_{pq}^* \hat{u}_{ql} |J| \int_{T_E} \frac{\partial \Phi_k}{\partial \eta} \Phi_l d\xi d\eta = 0
\end{aligned} \tag{2.14}$$

with

$$A_{pq}^* = A_{pq}\xi_x + B_{pq}\xi_y, \tag{2.15}$$

$$B_{pq}^* = A_{pq}\eta_x + B_{pq}\eta_y. \tag{2.16}$$

In (2.14) $0 \le \chi_j \le 1$ is a parameter to parametrize the j-th edge of the reference triangle and $|S_j|$ is the length of this edge in physical space. The following integrals can all be calculated beforehand, e.g. by using a computer algebra system:

$$M_{kl} = \int_{T_E} \Phi_k \Phi_l d\xi d\eta, \tag{2.17}$$

$$F_{kl}^{j,0} = \int_0^1 \Phi_k^{(m)}(\chi_j) \Phi_l^{(m)}(\chi_j) d\chi_j, \tag{2.18}$$

$$F_{kl}^{j,i} = \int_0^1 \Phi_k^{(m)}(\chi_j) \Phi_l^{(k_j)}(\chi_j) d\chi_j, \tag{2.19}$$

$$K_{kl}^{\xi} = \int_{T_E} \frac{\partial \Phi_k}{\partial \xi} \Phi_l d\xi d\eta, \tag{2.20}$$

$$K_{kl}^{\eta} = \int_{T_E} \frac{\partial \Phi_k}{\partial \eta} \Phi_l d\xi d\eta. \quad (2.21)$$

On the unstructured grid, $F_{kl}^{j,0}$ accounts for the contribution of the element (m) itself to the fluxes over all three edges j and $F_{kl}^{j,i}$ accounts for the contribution of the element's direct side neighbors (k_j) to the fluxes over the edges j. Index $1 \leq i \leq 3$ indicates the local number of the common edge in neighbor (k_j) and depends on the mesh generator. On a given mesh, where index i is known, only three of the nine possible matrices $F_{kl}^{j,i}$ are used per element. Be aware not to confound the tensor indices l and k with the indices (m) and (k_j) of the element and its neighbors on the unstructured grid.

If the semi-discrete equation (2.14) is integrated in time with a Runge–Kutta method, we obtain the quadrature-free Runge–Kutta discontinuous Galerkin approach developed by Atkins et al., see [1].

2.2 The ADER time discretization

Toro et al. [21], [22], [19] developed the ADER approach of highly accurate time discretization in the finite volume framework. The main ingredients of the method are a Taylor expansion in time of the solution, the Lax–Wendroff procedure for replacing time derivatives by space derivatives and the solution of generalized Riemann problems (GRP) to approximate the space derivatives. We shall explain this in more detail in Section 4 for the nonlinear case. In this section we show how the ADER approach can be used for high order time integration of the discontinuous Galerkin method, called ADER-DG method in the following, for linear hyperbolic systems.

For the development of linear ADER-DG schemes, we first need a general formula for the Lax–Wendroff procedure in order to replace the k-th time derivative by pure space derivatives. Since all our basis functions are given in the $\xi - \eta$ space, we need a Lax–Wendroff procedure which makes use of the spatial derivatives in the reference space. Therefore we rewrite our original PDE (2.1) with the use of (2.13) in the following way

$$\frac{\partial u_p}{\partial t} + A_{pq} \left(\frac{\partial u_q}{\partial \xi} \xi_x + \frac{\partial u_q}{\partial \eta} \eta_x \right) + B_{pq} \left(\frac{\partial u_q}{\partial \xi} \xi_y + \frac{\partial u_q}{\partial \eta} \eta_y \right) = 0 \quad (2.22)$$

and finally obtain

$$\frac{\partial u_p}{\partial t} + \left(A_{pq} \xi_x + B_{pq} \xi_y \right) \frac{\partial u_q}{\partial \xi} + \left(A_{pq} \eta_x + B_{pq} \eta_y \right) \frac{\partial u_q}{\partial \eta} = 0 \quad (2.23)$$

which is, using the definitions (2.15) and (2.16),

$$\frac{\partial u_p}{\partial t} + A_{pq}^* \frac{\partial u_q}{\partial \xi} + B_{pq}^* \frac{\partial u_q}{\partial \eta} = 0. \quad (2.24)$$

The k-th time derivative as a function of pure space derivatives in the $\xi - \eta$ reference space is the result of the Lax–Wendroff procedure applied to (2.24) and is given by

$$\frac{\partial^k u_p}{\partial t^k} = (-1)^k \left(A^*_{pq} \partial_\xi + B^*_{pq} \partial_\eta \right)^k u_q. \tag{2.25}$$

It is a key point to use the Lax–Wendroff procedure in the form (2.25) since it allows us to pre-calculate many matrices beforehand, as we will see in the following.

We develop the solution of (2.1) in a Taylor series in time up to order N

$$u_p(x,t) = \sum_{k=0}^{N} \frac{t^k}{k!} \frac{\partial^k}{\partial t^k} u_p(x,0) \tag{2.26}$$

and replace time derivatives by space derivatives, using equation (2.25).

$$u_p(x,t) = \sum_{k=0}^{N} \frac{t^k}{k!} (-1)^k (A^*_{pq} \partial_\xi + B^*_{pq} \partial_\eta)^k u_q(x,0). \tag{2.27}$$

We now introduce the discontinuous Galerkin approximation (2.2) and obtain

$$u_p(x,t) = \sum_{k=0}^{N} \frac{t^k}{k!} (-1)^k (A^*_{pq} \partial_\xi + B^*_{pq} \partial_\eta)^k \Phi_l(\xi) \hat{u}_{ql}(0). \tag{2.28}$$

This approximation can now be projected onto each basis function in order to get an approximation of the evolution of the degrees of freedom during one time step from time level n to time level $n+1$. We obtain

$$\hat{u}_{pl}(t) = \frac{\left\langle \Phi_n, \sum_{k=0}^{N} \frac{t^k}{k!} (-1)^k (A^*_{pq} \partial_\xi + B^*_{pq} \partial_\eta)^k \Phi_m(\xi) \right\rangle}{\langle \Phi_n, \Phi_l \rangle} \hat{u}_{qm}(0) \tag{2.29}$$

where $\langle .\,,. \rangle$ denotes the inner product over the reference triangle T_E and the division by $\langle \Phi_n, \Phi_l \rangle$ denotes the multiplication with the inverse of the mass matrix. This reduces indeed to division by its diagonal entries since the mass matrix is diagonal due to the supposed orthogonality of the basis functions. Equation (2.29) can be integrated analytically in time and we obtain

$$\int_0^{\Delta t} \hat{u}_{pl}(t) dt = \frac{\left\langle \Phi_n, \sum_{k=0}^{N} \frac{\Delta t^{(k+1)}}{(k+1)!} (-1)^k (A^*_{pq} \partial_\xi + B^*_{pq} \partial_\eta)^k \Phi_m(\xi) \right\rangle}{\langle \Phi_n, \Phi_l \rangle} \hat{u}_{qm}(0). \tag{2.30}$$

With the definition

$$I_{plqm}(\Delta t) = \frac{\left\langle \Phi_n, \sum_{k=0}^{N} \frac{\Delta t^{(k+1)}}{(k+1)!} (-1)^k (A^*_{pq} \partial_\xi + B^*_{pq} \partial_\eta)^k \Phi_m(\xi) \right\rangle}{\langle \Phi_n, \Phi_l \rangle} \tag{2.31}$$

equation (2.30) simply becomes

$$\int_0^{\Delta t} \hat{u}_{pl}(t) dt = I_{plqm}(\Delta t) \hat{u}_{qm}(0) \tag{2.32}$$

and we finally obtain the fully discrete ADER-DG scheme by integration of (2.14) in time:

$$
\begin{aligned}
&[(\hat{u}_{pl}^{(m)})^{n+1} - (\hat{u}_{pl}^{(m)})^{n}]|J|M_{kl} \\
&\quad + \frac{1}{2}\sum_{j=1}^{3} T_{pq}^{j}(A_{qr}^{T} + |A_{qr}^{T}|)(T_{rs}^{j})^{-1}|S_j|F_{kl}^{j,0} \cdot I_{slmn}(\Delta t)(\hat{u}_{mn}^{(m)})^{n} \\
&\quad + \frac{1}{2}\sum_{j=1}^{3} T_{pq}^{j}(A_{qr}^{T} - |A_{qr}^{T}|)(T_{rs}^{j})^{-1}|S_j|F_{kl}^{j,i} \cdot I_{slmn}(\Delta t)(\hat{u}_{mn}^{(k_j)})^{n} \\
&\quad - A_{pq}^{*}|J|K_{kl}^{\xi} \cdot I_{qlmn}(\Delta t)(\hat{u}_{mn}^{(m)})^{n} - B_{pq}^{*}|J|K_{kl}^{\eta} \cdot I_{qlmn}(\Delta t)(\hat{u}_{mn}^{(m)})^{n} = 0.
\end{aligned}
\tag{2.33}
$$

2.3 Numerical results

For the following numerical convergence tests we solve the two-dimensional system of linearized Euler equations (LEE). The vector $u_p = (\ \rho' \ \ u' \ \ v' \ \ p'\ )^T$ is the vector of the fluctuations of the primitive variables density, velocity and pressure, the matrices A_{pq} and B_{pq} are the Jacobian matrices and are functions of the background flow ρ_0, u_0, v_0 and p_0. They read as

$$
A_{pq} = \begin{pmatrix} u_0 & \rho_0 & 0 & 0 \\ 0 & u_0 & 0 & \frac{1}{\rho_0} \\ 0 & 0 & u_0 & 0 \\ 0 & \gamma p_0 & 0 & u_0 \end{pmatrix}, \quad B_{pq} = \begin{pmatrix} v_0 & 0 & \rho_0 & 0 \\ 0 & v_0 & 0 & 0 \\ 0 & 0 & v_0 & \frac{1}{\rho_0} \\ 0 & 0 & \gamma p_0 & v_0 \end{pmatrix}. \tag{2.34}
$$

Example 1. We now solve the following problem given by (2.1) with the Jacobians (2.34) evaluated at the constant values $\rho_0 = 1$, $u_0 = v_0 = 1$, $p_0 = \frac{1}{\gamma}$, $\gamma = 1.4$ and the initial condition at time $t = 0$:

$$
\begin{aligned}
\rho'(x) &= e^{-\frac{1}{2}\frac{x^2+y^2}{\sigma^2}}, \\
u'(x) &= v'(x) = p'(x) = 0.
\end{aligned}
\tag{2.35}
$$

The computational domain has the extents $[-50, 50] \times [-50, 50]$ and has 4-periodic boundary conditions. The problem is solved in the time interval $0 \le t \le 100$, i.e. we calculate the advection of the Gaussian density distribution during one period. The exact solution at $t = 100$ is then given by the initial condition (2.35). The numerical convergence calculations are performed on a purely Cartesian grid, see Figure 2a, as well as on a regular triangular mesh, see Figure 2b. We set the half-width parameter $\sigma = 3$ on the Cartesian grid and $\sigma = 5$ on the triangular mesh. The results are given in Tables 3 to 16. We can clearly see that the designed order has been reached very well up to tenth order in *space and time*. We remind that the ADER-DG scheme is a

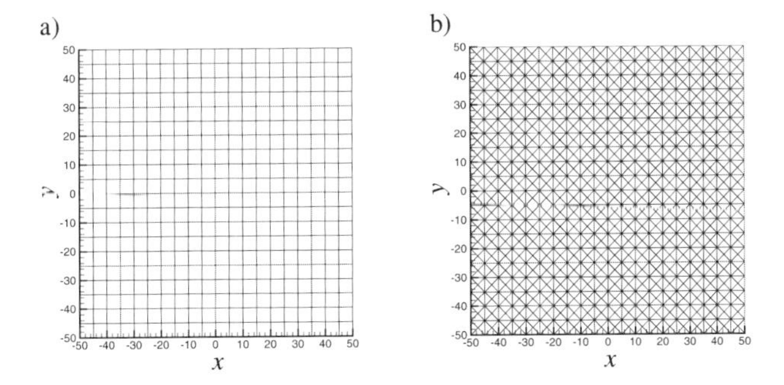

Figure 2. View of the Cartesian grid a) and the regular triangular mesh b) for the numerical convergence studies.

single-step scheme and thus uses the same computer memory as the explicit first order Euler time-stepping.

Example 2. In the following we solve the problem of a slotted cylinder performing a rigid-body rotation, see e.g. [16]. Therefore we have to solve the locally linearized Euler equations where the non-uniform background flow is given by

$$\begin{pmatrix} \rho_0 \\ u_0 \\ v_0 \\ p_0 \end{pmatrix} = \begin{pmatrix} 1 \\ -\omega y \\ \omega x \\ 1 \end{pmatrix}. \tag{2.36}$$

In our example, we set $\omega = \pi$, so one period of rotation is completed after the time $t = 2$. The computational domain has the extents $[-50, 50] \times [-50, 50]$ and on the boundary we impose the value zero for all the fluctuations. The center of the slotted cylinder at $t = 0$ lies at $\vec{x} = (25, 0)^T$ and its radius is $R = 15$. The density fluctuation is $\rho' = 4$ in the cylinder and $\rho' = 0$ elsewhere, the slot has the width $W = 6$. For the sake of simplicity, we suppose that the background flow is piecewise constant in each element which makes application of the Lax–Wendroff procedure quite easy, since no spatial derivatives of the Jacobians have to be considered. For the locally linearized Euler equations with non-uniform mean flow, we use the CIR flux [8] as numerical flux at the element boundaries which reduces to Godunov's exact flux if the Jacobians A_{pq} and B_{pq} are globally constant. All computations have been performed on a machine with one single Athlon XP 2500+ processor and 1 GB of RAM. The numerical results of the ADER-DG method for P1 up to P4 elements after one period of rotation ($t = 2$) on the regular unstructured mesh (Figure 2b) with $50 \times 50 \times 4 = 10000$ triangles are presented in Figure 3. As a reference, we show as well the exact solution which is identical to the initial condition and the results obtained with a RK-DG method using P4 elements and fourth order Runge–Kutta time-stepping. We can clearly see that the spurious oscillations in the numerical solution decrease, if we increase the order of the basis polynomials. Another computation is performed on an irregular mesh consisting of approximately 13000 triangles. The mesh and the results are shown in Figure 4.

A comparison of CPU time of the ADER-DG scheme with a quadrature-free RK-DG method on both meshes is given in Tables 1 and 2. Time is normalized in each table with respect to the second order quadrature-free RK-DG scheme. Up to fourth order, time integration is performed with the same order of accuracy as spatial integration, but we note that the P4 elements of the RK-DG method (fifth order in space) have

Table 1. CPU time comparison between ADER-DG and RK-DG on the regular unstructured mesh.

Elements	RK–DG	ADER–DG	Elements	RK–DG	ADER–DG
$P1$	1.00	0.56	$P3$	9.82	3.52
$P2$	3.50	1.61	$P4$	19.9	8.04

Table 2. CPU time comparison between ADER-DG and RK-DG on the irregular unstructured mesh.

Elements	RK–DG	ADER–DG	Elements	RK–DG	ADER–DG
$P1$	1.00	0.58	$P3$	10.5	3.75
$P2$	3.75	1.64	$P4$	20.8	8.42

been integrated only with a four-stage fourth order Runge–Kutta scheme, whereas the ADER-DG scheme automatically uses fifth order time integration of the P4 elements. With the ADER approach, spatial and temporal discretization order are intrinsically coupled and have always the same order of accuracy. We remind that a fifth order Runge–Kutta method would at least have six stages due to the Butcher barriers and would thus lead to a further increase of CPU time of the P4 RK-DG elements by a factor of $\frac{6}{4}$. Therefore for very high order of accuracy, the gain in CPU time of the ADER-DG method is quite considerable as well as the savings of computer storage. This is due to the fact that for the ADER-DG scheme it is not necessary to compute and store any intermediate Runge–Kutta stage. We note that the maximum CFL numbers for the Runge–Kutta DG schemes as well as for ADER-DG schemes are functions of the order N of the basis functions. For RK-DG schemes in one space dimension, we have the following maximum CFL numbers: 0.33 (P1), 0.20 (P2), 0.14 (P3) and 0.11 (P4). For one-dimensional ADER-DG schemes, a von Neumann stability analysis yields the following lower CFL stability limits: 0.33 (P1), 0.17 (P2), 0.10 (P3) and 0.068 (P4). In Tables 1 and 2, the computations for RK-DG and ADER-DG, respectively,

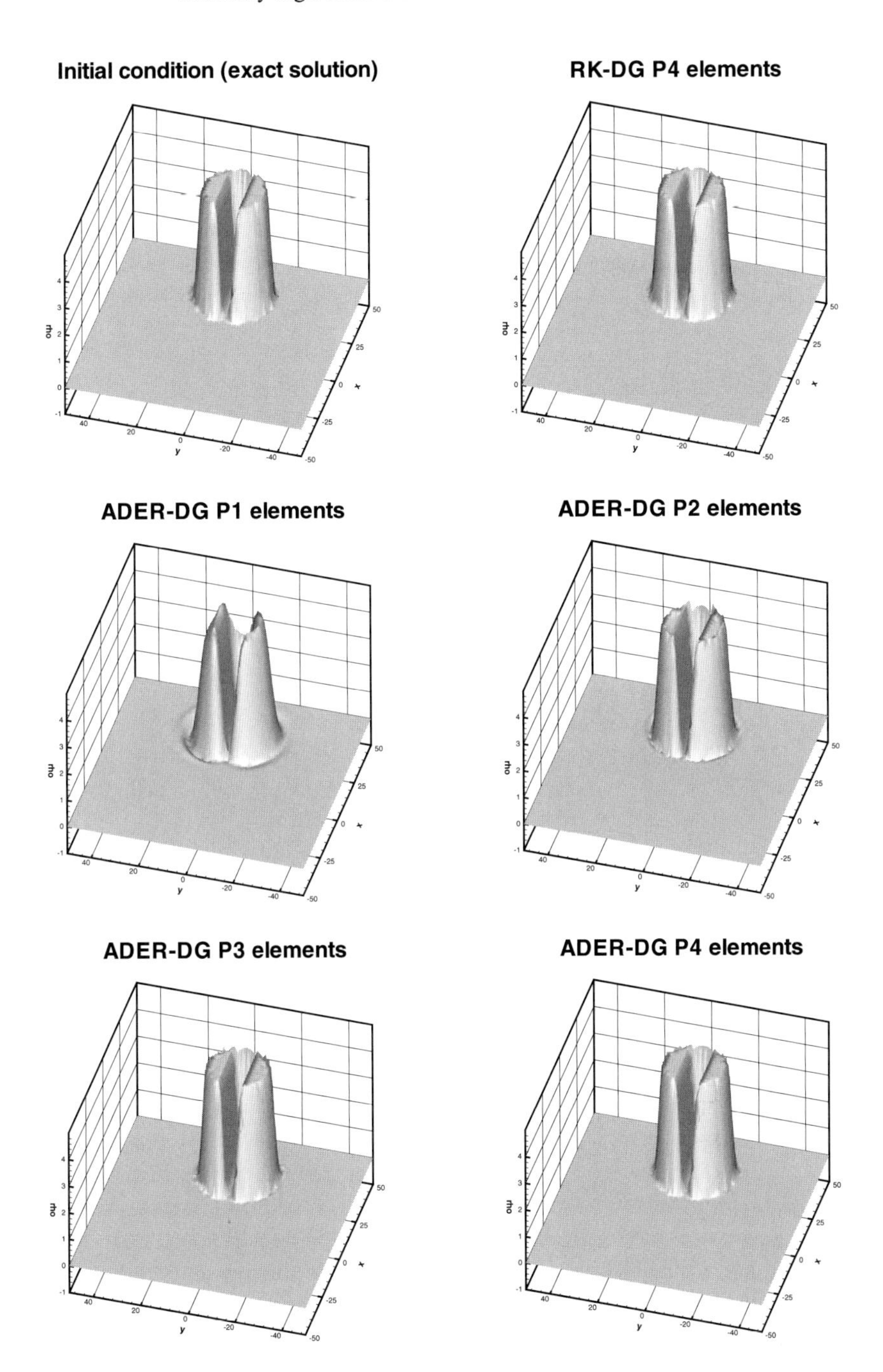

Figure 3. Numerical results of the ADER-DG method for the slotted cylinder example after one rotation ($t = 2$) on the regular triangular mesh.

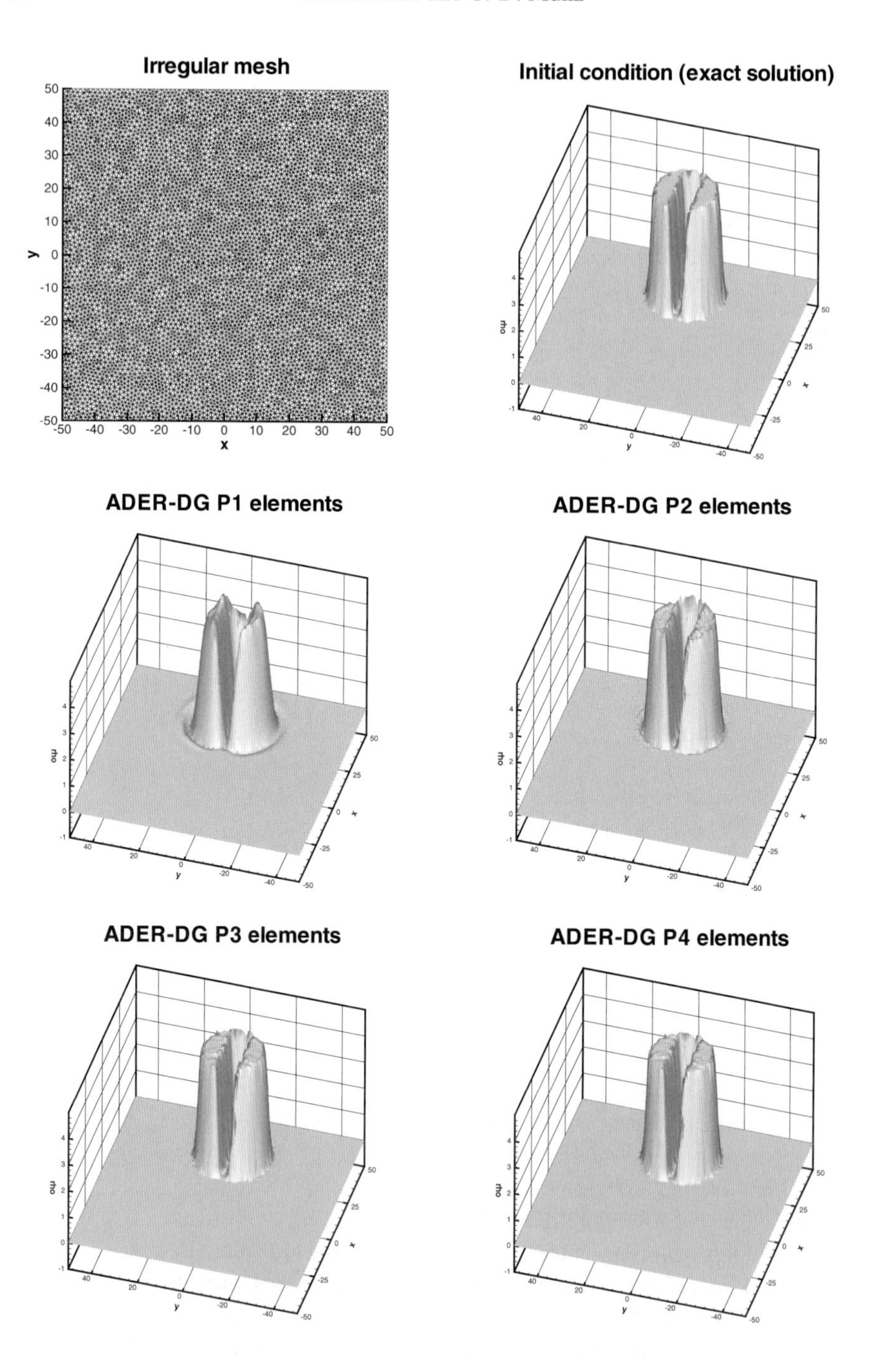

Figure 4. Numerical results of the ADER-DG method for the slotted cylinder example after one rotation ($t = 2$) on an irregular mesh.

have been performed with the *same* CFL numbers. We have chosen 50 % of the maximum RK-DG CFL number. We remark that even if we take also into account the lower stability limit of the ADER-DG schemes, the fifth order ADER-DG scheme (P4 elements) uses still 35 % less CPU time than a RK-DG scheme with P4 elements using fourth order time integration and uses 50 % less CPU time than a RK-DG P4 scheme using fifth order time integration.

Considering the CPU savings we can conclude that the Lax–Wendroff procedure for the (locally) linearized Euler equations is relatively cheap compared to the rest of the DG method. For the efficient evaluation of equation (2.31), the necessary projections of all space derivatives of the basis functions onto all basis functions are calculated once on the reference element, e.g. with the aid of a computer algebra system, and are then stored.

We underline the fact that the Lax–Wendroff procedure (2.25) and furthermore equation (2.31) can be coded very efficiently in a completely automatic manner using an unrolled recursive algorithm, so that the order of accuracy becomes merely a numerical parameter which may be chosen by the user.

3 Reconstructed ADER-DG schemes for linear hyperbolic systems

In the previous section we derived arbitrary high order explicit single-step discontinuous Galerkin schemes for linear hyperbolic systems in two space dimensions on triangular meshes. Since for linear equations like the aeroacoustical system of the linearized Euler equations (LEE) we expect smooth solutions for smooth initial data, we will further enhance the accuracy of the ADER-DG method by adding a linear reconstruction operator to the scheme. Cockburn et al. [7] developed a one-dimensional reconstruction operator based on a B-spline kernel of the DG basis functions for regular triangulations. The method was extended to 2D by Ryan et al. in [15]. The reconstruction operator of the previously mentioned authors provided an order of accuracy of $2N + 1$ for basis functions with order N. They did not include the reconstruction operator in the numerical scheme during time advancement but only applied it at the end of time-stepping. Therefore they called this reconstruction operator a *post-processing technique* for the discontinuous Galerkin method.

In this paper we follow a different approach. We first will use the reconstruction operator in each time step during time advancement and second we will apply an operator working only on equally spaced Cartesian grids, but with an order of accuracy of $3N + 2$. When applying these approaches for aeroacoustical computations over very long distances where time accuracy of the method is crucial, we think that even using a $2N + 1$ accurate postprocessor after the last time step can not recover all the information which is lost during the wave propagation process due to numerical damping of the original DG method with N-th order basis functions. A method which in each time step reconstructs a numerical solution of order $3N+2$ and also performs time integration of this order of accuracy should result in less damping due to a lower numerical viscosity.

In the following sections we will derive the reconstructed ADER discontinuous Galerkin scheme and show the corresponding convergence rates obtained in numerical experiments. We further demonstrate that the use of the reconstruction operator in each time step in fact reduces numerical dissipation and dispersion considerably when applied to wave propagation problems over very long distances.

3.1 Derivation of the reconstructed ADER-DG schemes

In this section we consider discontinuous Galerkin methods on equally spaced Cartesian grids only, where the 2D basis functions of order N are chosen to be tensor products of N-th order 1D Legendre polynomials. Hence the number of degrees of freedom for one element is $n_d = (N+1)^2$. The reference square Q_E is defined in $\xi - \eta$ space as $\left[-\frac{1}{2}, \frac{1}{2}\right] \times \left[\frac{1}{2}, \frac{1}{2}\right]$.

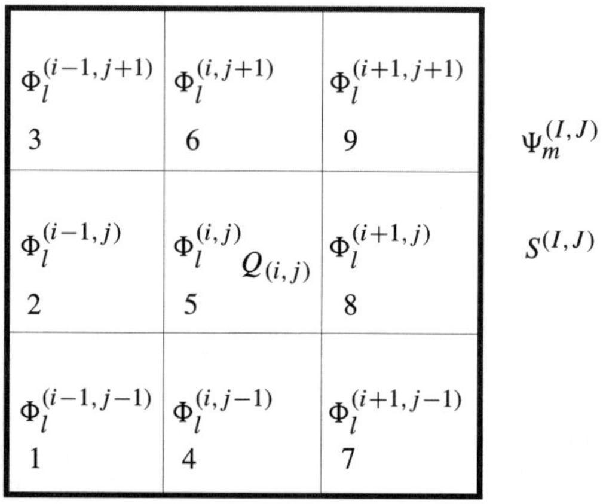

$\Phi_l^{(i-1,j+1)}$ 3	$\Phi_l^{(i,j+1)}$ 6	$\Phi_l^{(i+1,j+1)}$ 9
$\Phi_l^{(i-1,j)}$ 2	$\Phi_l^{(i,j)}$ $Q_{(i,j)}$ 5	$\Phi_l^{(i+1,j)}$ 8
$\Phi_l^{(i-1,j-1)}$ 1	$\Phi_l^{(i,j-1)}$ 4	$\Phi_l^{(i+1,j-1)}$ 7

$\Psi_m^{(I,J)}$

$S^{(I,J)}$

Figure 5. Stencil used for reconstruction of the higher order basis functions $\Psi_m^{(I,J)}$ on the stencil $S^{(I,J)}$ drawn in bold using the basis functions $\Phi_l^{(i+ii,j+jj)}$ of the elements $Q^{(i+ii,j+jj)}$ with $-1 \leq ii, jj \leq 1$.

In order to obtain a higher accurate approximation of the numerical solution, we now construct a suitable reconstruction operator which acts on a stencil $S^{(I,J)}$ consisting of a central element $Q^{(i,j)}$ and the adjacent cells $Q^{(i+ii,j+jj)}$ with $-1 \leq ii, jj \leq 1$ as shown in Figure 5. Let the numerical solution in each square of the stencil $S^{(I,J)}$ originally be represented by

$$u_p^{(i+ii,j+jj)} = \hat{u}_{pl}^{(i+ii,j+jj)} \Phi_l^{(i+ii,j+jj)} \tag{3.1}$$

and let the reconstructed solution in the stencil be defined as

$$w_p^{(I,J)} = \hat{w}_{pm}^{(I,J)} \Psi_m^{(I,J)} \tag{3.2}$$

with $\Psi_m^{(I,J)} \in W$ some higher order basis functions of a higher order polynomial space $W \subset L^2$. Our strategy for obtaining the degrees of freedom $\hat{w}_{pm}^{(I,J)}$ of the reconstructed solution is the following: We perform a L^2 projection of the still unknown reconstructed solution (3.2) onto the basis functions $\Phi_l^{(i+ii,j+jj)}$ of each element in the stencil $S^{(I,J)}$ and require the result of the projection to be equal to the corresponding degree of freedom $\hat{u}_{pl}^{(i+ii,j+jj)}$

$$\hat{u}_{pl}^{(i+ii,j+jj)} = \frac{\langle \Phi_n^{(i+ii,j+jj)}, \hat{w}_{pm}^{(I,J)} \Psi_m^{(I,J)} \rangle^{(ii,jj)}}{\langle \Phi_n^{(i+ii,j+jj)}, \Phi_l^{(i+ii,j+jj)} \rangle^{(ii,jj)}} \tag{3.3}$$

where $\langle .\,, . \rangle^{(ii,jj)}$ denotes the L^2 inner product over the element $Q^{(i+ii,j+jj)} \in S^{(I,J)}$. The division means again multiplication by the inverse of the matrix. Equation (3.3) is a generalization of typical conservative reconstruction formulae used in the finite volume context which require integral conservation of the cell averages, see for instance [11]. (3.3) constitutes furthermore an equation system which defines the unknowns $\hat{w}_{pm}^{(I,J)}$. Because of our choice for the basis functions $\Phi_l^{(i+ii,j+jj)}$ to be 2D tensor products of Legendre polynomials of degree N and given the stencil consisting of 9 elements, the number of rows of the system (3.3) is equal to $9 \cdot (N+1)^2 = [(3N+2)+1]^2$. So if we choose the higher order reconstruction basis functions $\Psi_m^{(I,J)}$ to be 2D tensor products of Legendre polynomials of the order $3N+2$, the system matrix becomes square and it can also be shown that there exists a unique solution for (3.3):

$$\hat{w}_{pm}^{(I,J)} = \left(\frac{\langle \Phi_n^{(i+ii,j+jj)}, \Psi_m^{(I,J)} \rangle^{(ii,jj)}}{\langle \Phi_n^{(i+ii,j+jj)}, \Phi_l^{(i+ii,j+jj)} \rangle^{(ii,jj)}} \right)^{-1} \hat{u}_{pl}^{(i+ii,j+jj)}. \tag{3.4}$$

If we use the reconstructed solution in all integrals of (2.14) except of the first volume integral, we obtain the following semi-discrete form of the reconstructed discontinuous Galerkin scheme

$$\begin{aligned}
&\frac{\partial}{\partial t} \hat{u}_{pl}^{(i,j)} |J| \int_{Q_E} \Phi_k \Phi_l d\xi d\eta \\
&+ \sum_{\nu=1}^{4} T_{pq}^{\nu} \frac{1}{2} (A_{qr}^T + |A_{qr}^T|)(T_{rs}^{\nu})^{-1} \hat{w}_{sl}^{(I,J)} |S_\nu| \int_0^1 \Phi_k^{(i,j)}(\chi_\nu) \Psi_l^{(I,J)}(\chi_\nu) d\chi_\nu \\
&+ \sum_{\nu=1}^{4} T_{pq}^{\nu} \frac{1}{2} (A_{qr}^T - |A_{qr}^T|)(T_{rs}^{\nu})^{-1} \hat{w}_{sl}^{(I,J)} |S_\nu| \int_0^1 \Phi_k^{(i,j)}(\chi_\nu) \Psi_l^{(K_\nu,L_\nu)}(\chi_\nu) d\chi_\nu \\
&- A_{pq}^* \hat{w}_{ql}^{(I,J)} |J| \int_{Q_E} \frac{\partial \Phi_k}{\partial \xi} \Psi_l d\xi d\eta - B_{pq}^* \hat{w}_{ql}^{(I,J)} |J| \int_{Q_E} \frac{\partial \Phi_k}{\partial \eta} \Psi_l d\xi d\eta = 0
\end{aligned} \tag{3.5}$$

where (K_ν, L_ν) denotes the index pair of a neighbor stencil $S^{(K_\nu, L_\nu)}$ whose central element is adjacent to the ν-th edge of element (i, j). The edges have the length $|S_\nu|$, they are parametrized by $0 \leq \chi_\nu \leq 1$ and are numbered counterclockwise starting at the right edge. The fact that the test functions are chosen out of the space V (and also that the basis functions are taken out of V in the first volume integral in order to obtain a quadratic and invertible mass matrix) has the consequence that the numerical solution is projected back onto the normal basis functions $\Phi_k^{(i,j)} \in V$ after time advancement. Similar to equations (2.17)–(2.21) we can define and pre-compute some matrices which are now *rectangular* matrices, except of the mass matrix which remains square:

$$M_{kl} = \int_{Q_E} \Phi_k^{(i,j)} \Phi_l^{(i,j)} d\xi d\eta, \tag{3.6}$$

$$F_{kl}^{\nu+} = \int_0^1 \Phi_k^{(i,j)}(\chi_\nu) \Psi_l^{(I,J)}(\chi_\nu) d\chi_\nu, \tag{3.7}$$

$$F_{kl}^{\nu-} = \int_0^1 \Phi_k^{(i,j)}(\chi_\nu) \Psi_l^{(K_\nu,L_\nu)}(\chi_\nu) d\chi_\nu, \tag{3.8}$$

$$K_{kl}^{\xi} = \int_{Q_E} \frac{\partial \Phi_k^{(i,j)}}{\partial \xi} \Psi_l^{(I,J)} d\xi d\eta, \tag{3.9}$$

$$K_{kl}^{\eta} = \int_{Q_E} \frac{\partial \Phi_k^{(i,j)}}{\partial \eta} \Psi_l^{(I,J)} d\xi d\eta. \tag{3.10}$$

In order to obtain the fully discretized version of (3.5) by using the ADER approach, we proceed in the same way as for the regular ADER-DG scheme, but this time feeding the Lax–Wendroff procedure up to order $3N + 2$ with the reconstructed numerical solution. We finally obtain

$$I_{plqm}(\Delta t) = \frac{\langle \Psi_n^{(I,J)}, \sum_{k=0}^{3N+2} \frac{\Delta t^{(k+1)}}{(k+1)!} (-1)^k \left(A_{pq}^* \partial_\xi + B_{pq}^* \partial_\eta \right)^k \Psi_m^{(I,J)}(\xi), \rangle^{(0,0)}}{\langle \Psi_n^{(I,J)}, \Psi_l^{(I,J)} \rangle^{(0,0)}} \tag{3.11}$$

and

$$\int_0^{\Delta t} \hat{w}_{pl}(t) dt = I_{plqm}(\Delta t) \hat{w}_{qm}(0). \tag{3.12}$$

So the fully discrete reconstructed ADER-DG scheme after integration of (3.5) in time is

$$\begin{aligned}
&\left[\left(\hat{u}_{pl}^{(i,j)}\right)^{n+1} - \left(\hat{u}_{pl}^{(i,j)}\right)^{n}\right] |J|\, M_{kl} \\
&+ \frac{1}{2}\sum_{\nu=1}^{4} T_{pq}^{j}\left(A_{qr}^{T} + |A_{qr}^{T}|\right)(T_{rs}^{j})^{-1}|S_j|\, F_{kl}^{\nu^+} \cdot I_{slmn}(\Delta t)\left(\hat{w}_{mn}^{(I,J)}\right)^{n} \\
&+ \frac{1}{2}\sum_{\nu=1}^{4} T_{pq}^{j}\left(A_{qr}^{T} - |A_{qr}^{T}|\right)(T_{rs}^{j})^{-1}|S_j|\, F_{kl}^{\nu^-} \cdot I_{slmn}(\Delta t)\left(\hat{w}_{mn}^{(K_\nu,L_\nu)}\right)^{n} \\
&- A_{pq}^{*}\, |J|\, K_{kl}^{\xi} \cdot I_{qlmn}(\Delta t)\left(\hat{w}_{mn}^{(I,J)}\right)^{n} - B_{pq}^{*}\, |J|\, K_{kl}^{\eta} \cdot I_{qlmn}(\Delta t)\left(\hat{w}_{mn}^{(I,J)}\right)^{n} = 0.
\end{aligned} \tag{3.13}$$

Since spatial and temporal discretization are performed with polynomials of degree $3N+2$ we expect the scheme to be of the order of accuracy $3N+3$ in space and time.

3.2 Numerical results for the linearized Euler equations

In order to verify numerically the design-order $3N+3$, we perform a similar numerical experiment as described in Section 2.3, i.e. advection of a Gaussian density fluctuation (2.35) with $\sigma = 5$ over one period of advection ($t = 100$) on a Cartesian grid with 100×100 cells. The computational domain has the extents $[0, 100] \times [0, 100]$, 4-periodic boundary conditions and the mean flow is constantly $\rho_0 = 1$, $u_0 = v_0 = 1$, $p_0 = \frac{1}{\gamma}$, $\gamma = 1.4$. Tables 17 to 19 confirm clearly that the design-order of $3N + 3$ has been reached very well.

The next numerical example is chosen to illustrate the wave propagation properties of the reconstructed ADER-DG scheme over very long distances. We want to demonstrate numerically that the use of the reconstruction operator in each time step together with the resulting high order ADER time discretization reduce numerical dissipation and dispersion errors considerably. Therefore we perform a long-time simulation of the previous example over 100 periods ($t = 10000$). The results for the cell averages obtained with the different DG methods are depicted in Figure 6. We can clearly see that both second order methods, RK-DG and ADER-DG with P1 elements, introduce a high amount of numerical dissipation and dispersion. This can not be seen after one period, but becomes obvious after 100 periods of advection where the time discretization error becomes significant. Figure 6 shows clearly that the use of a reconstruction operator as a *postprocessor* on the result of the ADER-DG P1 or RK-DG P1 method would only be able to smoothen the spatial discretization error but not the one introduced by time discretization in this long-time simulation where information is lost irreversibly due to time discretization errors. The reconstructed ADER-DG P1 scheme with reconstruction in each time step is able to retrieve almost the exact solution.

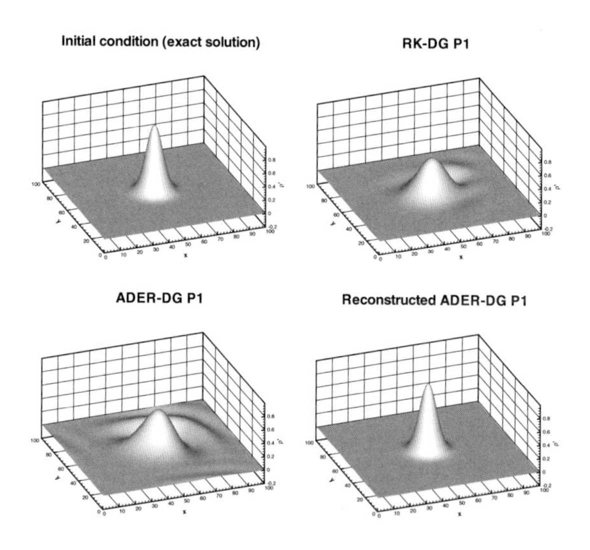

Figure 6. Cell average at $t = 0$ and after 100 periods of advection ($t = 10000$) for different DG methods.

4 ADER-DG schemes for nonlinear hyperbolic systems

In the discontinuous Galerkin framework the numerical solution is represented in form of piecewise polynomials which may be discontinuous at the element boundaries. This particular data representation poses generalized Riemann problems called GRP_N at the element interfaces where the initial condition on the left and on the right hand side of the interface are polynomials of degree N. The polynomials are in general separated by a discontinuity at the interface. So, a quite natural way to perform high order time integration of the discontinuous Galerkin scheme seems to be the application of the ADER approach of Toro et al. [21], [22], [19], [23] which constructs a solution to this generalized Riemann problem for small $\tau > 0$ where $\tau = t - t^n$ is the time t relative to the current time level n.

The ADER-DG strategy is in the nonlinear case different from the Lax–Wendroff DG approach presented in [14] where the numerical solution at the next time level $n+1$ is the direct result of a Taylor series in time expanded around time level n and where the space derivatives are obtained by a DG discretization. In the ADER-DG approach, the Lax–Wendroff procedure is performed in each Gaussian quadrature point in space and then the flux is integrated by a Gaussian quadrature rule of suitable order in time. The space derivatives on the boundaries are obtained by solving a generalized Riemann problem of order N whereas the space derivatives in the interior of the element are directly obtained by deriving the numerical solution given by (2.2) with respect to x and y. Since all basis functions are defined in the reference triangle T_E in $\xi - \eta$ space, the polynomials are first converted to the physical $x - y$ space because the $\xi - \eta$ derivatives are not compatible between two triangles.

4.1 Derivation of nonlinear ADER-DG schemes

The nonlinear hyperbolic systems which we consider have the form

$$u_t + \operatorname{div} F(u) = 0 \quad \text{with } F = [f, g] \tag{4.1}$$

where f and g are the nonlinear fluxes in x and y direction, respectively. Multiplication with a test function $\Phi_k \in V \subset L^2$, integration over a triangle $T^{(m)}$, introduction of the numerical solution u_h, and integration by parts finally yield

$$|J| \int_{T_E} \Phi_k \, (u_h)_t \, d\xi d\eta + \int_{\partial T^{(m)}} \Phi_k F^h \vec{n} dS - |J| \int_{T_E} F(u_h) \operatorname{grad} (\Phi_k) \, d\xi d\eta = 0 \tag{4.2}$$

where $F^h\vec{n}$ denotes again the numerical flux over the element boundary. The numerical solution u_h is again defined by (2.2). In the nonlinear case, the first integral yields the mass matrix as in the linear case, but we now approximate the second and the third integral appearing in (4.2) with Gaussian quadrature rules of suitable order, see [3], [4], [5], [6], where N_S^{GP} is the number of Gaussian quadrature points needed for the boundary integral and N_V^{GP} the number of Gaussian points needed for the volume integral. The weights are ω_n and the quadrature points are χ_n on the boundary and (ξ_n, η_n) in the volume. For general multidimensional integration formulae see [18]. We obtain:

$$\int_{\partial T^{(m)}} \Phi_k F^h \vec{n} dS \approx \sum_{j=1}^{3} |S_j| \sum_{n=1}^{N_S^{\mathrm{GP}}} \omega_n \Phi_k \, (\chi_n) \, F^h \big(u^{(m)}(\chi_n), u^{(k_j)}(\chi_n)\big) \vec{n}_j, \tag{4.3}$$

$$\int_{T_E} F(u_h) \operatorname{grad} (\Phi_k) \, d\xi d\eta \approx \sum_{n=1}^{N_V^{\mathrm{GP}}} \omega_n F \, (u_h(\xi_n, \eta_n)) \operatorname{grad} (\Phi_k \, (\xi_n, \eta_n)) \,. \tag{4.4}$$

If we want to perform high order time integration of (4.2) we first need the Lax–Wendroff procedure for the nonlinear hyperbolic system (4.1). In the nonlinear case, the procedure gets much more complicated than for linear hyperbolic systems since the chain rule has to be applied. This introduces derivatives $A'(u) = \frac{\partial A(u)}{\partial u}$ and $B'(u) = \frac{\partial B(u)}{\partial u}$ of the Jacobians $A(u) = \frac{\partial f}{\partial u}$ and $B(u) = \frac{\partial g}{\partial u}$ of the fluxes. We formally illustrate the procedure on the example for determining the first and second time derivatives of u by using (4.1) repeatedly, the general case can be treated by a computer algebra system. Starting with (4.1) and introducing the Jacobians $A(u)$ and

$B(u)$ we get by repeated differentiation of (4.1) with respect to x and t:

$$u_t = -A(u)u_x - B(u)u_y, \tag{4.5}$$
$$u_{tt} = -A(u)u_{xt} - (A'(u)u_t)u_x - B(u)u_{yt} - (B'(u)u_t)u_y, \tag{4.6}$$
$$u_{tx} = -A(u)u_{xx} - (A'(u)u_x)u_x - B(u)u_{yx} - (B'(u)u_x)u_y, \tag{4.7}$$
$$u_{ty} = -A(u)u_{xy} - (A'(u)u_y)u_x - B(u)u_{yy} - (B'(u)u_y)u_y. \tag{4.8}$$

Equation (4.6) can be expressed completely in terms of pure space derivatives using equations (4.5), (4.7) and (4.8). One might imagine that this procedure gets quite cumbersome for higher orders of accuracy, but thanks to computer algebra systems, it can be done in an automatic manner. In the general case the k-th time derivative can be expressed as a nonlinear function G_k of all (mixed) space derivatives up to order k, so

$$\frac{\partial^k u}{\partial t^k} = G_k\left(\frac{\partial^p u}{\partial x^q \partial y^q}\right) \quad \text{for all } 0 \le p \le k,\ q + r = p. \tag{4.9}$$

The essential part of the nonlinear ADER-DG schemes is now the solution of the generalized Riemann problem GRP_N posed by the DG polynomials $u_h^{(m)}(x, y)$ and $u_h^{(k_j)}(x, y)$ on the cell interface j at time $\tau = 0$ in each Gaussian quadrature point χ_n on the boundary:

$$\text{PDE}: \quad u_t + \text{div}\, F(u) = 0,$$
$$\text{IC}: \quad u_h(x, y, 0) = \begin{cases} u_h^{(m)}(x, y) & \text{if } (x, y) \in T^{(m)}, \\ u_h^{(k_j)}(x, y) & \text{if } (x, y) \in T^{(k_j)}. \end{cases} \tag{4.10}$$

The solution of GRP_N is first determined by the solution $u_h^{\text{RP}_0}(\chi_n)$ of a conventional (piecewise constant data) nonlinear Riemann problem RP_0 of the boundary extrapolated values $u_h^{(m)}(\chi_n)$ and $u_h^{(k_j)}(\chi_n)$ on the left and right hand side, respectively, in the quadrature point χ_n,

$$\text{PDE}: \quad u_t + \text{div}\, F(u) = 0,$$
$$\text{IC}: \quad u_h(x, y, 0) = \begin{cases} u_h^{(m)}(\chi_n) & \text{if } (x, y) \in T^{(m)}, \\ u_h^{(k_j)}(\chi_n) & \text{if } (x, y) \in T^{(k_j)}. \end{cases} \tag{4.11}$$

and second it is determined by the set of solutions $u_h^{\text{RP}_{(q,r)}}(\chi_n)$ of a sequence of linearized conventional Riemann problems $\text{RP}_{(q,r)}$ for all space derivatives $u^{(q,r)}$ of u with $1 \le p \le N$ and $q + r = p$. Linearization is performed around the solution

$u_h^{\mathrm{RP}_0}(\chi_n)$:

$$
\begin{aligned}
\mathrm{PDE}: \quad & u_t^{(q,r)} + A\big(u_h^{\mathrm{RP}_0}(\chi_n)\big)u_x^{(q,r)} + B\big(u_h^{\mathrm{RP}_0}(\chi_n)\big)u_y^{(q,r)} = 0, \\
\mathrm{IC}: \quad & u_h^{(q,r)}(x,y,0) = \begin{cases} \dfrac{\partial^p u_h^{(m)}}{\partial x^q \partial y^r}(\chi_n) & \text{if } (x,y) \in T^{(m)}, \\ \dfrac{\partial^p u_h^{(k_j)}}{\partial x^q \partial y^r}(\chi_n) & \text{if } (x,y) \in T^{(k_j)}. \end{cases}
\end{aligned}
\tag{4.12}
$$

The solution of the sequence of linearized Riemann problems $\mathrm{RP}_{(q,r)}$ can be calculated in a very efficient way since the Jacobians and their left and right eigenvectors remain the same for all space derivatives and thus only need to be calculated once.

The solution of the generalized Riemann problem as a function of relative time τ is finally given by a Taylor series where time derivatives have been replaced by space derivatives according to (4.9):

$$
u_h^{\mathrm{GRP}_N}(\chi_n, \tau) = u_h^{\mathrm{RP}_0}(\chi_n) + \sum_{k=1}^{N} \frac{\tau^k}{k!} G_k\big(u_h^{\mathrm{RP}_{(q,r)}}(\chi_n)\big) \tag{4.13}
$$

for all $0 \leq p \leq k$, $q + r = p$. The time dependent approximate solution of GRP_N (4.13) can now be plugged into the approximation of the numerical flux of the nonlinear DG scheme (4.3). For the temporal approximation of the volume integral (4.4) the space derivatives can directly be obtained by deriving equation (2.2) at the Gaussian points (x_n, y_n) since the polynomials are differentiable inside each triangle $T^{(m)}$ and equation (4.9) can be applied directly:

$$
u_h^{\mathrm{ADER}}(x_n, y_n, \tau) = \sum_{k=0}^{N} \frac{\tau^k}{k!} G_k\left(\frac{\partial^p u_h^{(m)}(x_n, y_n, 0)}{\partial x^q \partial y^r}\right) \tag{4.14}
$$

for all $0 \leq p \leq k$, $q + r = p$. With these ingredients, the fluxes in equations (4.3) and (4.4) can be integrated in time using a Gaussian quadrature rule with N integration points τ_l and the associated weights α_l in the interval $[0, \Delta t]$:

$$
\begin{aligned}
&\int_0^{\Delta t} \left(\int_{\partial T^{(m)}} \Phi_k F^h \vec{n}\, dS \right) d\tau \\
&\quad \approx \sum_{l=1}^{N} \alpha_l \sum_{j=1}^{3} |S_j| \sum_{n=1}^{N_S^{\mathrm{GP}}} \omega_n \Phi_k(\chi_n)\, F\big(u_h^{\mathrm{GRP}_N}(\chi_n, \tau_l)\big)\vec{n}_j,
\end{aligned}
\tag{4.15}
$$

$$
\begin{aligned}
&\int_0^{\Delta t} \left(\int_{T_E} F(u_h)\, \mathrm{grad}(\Phi_k) \right. d\xi\, d\eta) d\tau \\
&\quad \approx \sum_{l=1}^{N} \alpha_l \sum_{n=1}^{N_V^{\mathrm{GP}}} \omega_n F\big(u_h^{\mathrm{ADER}}(\xi_n, \eta_n, \tau_l)\big)\, \mathrm{grad}(\Phi_k(\xi_n, \eta_n)).
\end{aligned}
\tag{4.16}
$$

Integration of (4.2) in time and inserting equations (4.15) and (4.16) finally yields the fully discrete ADER discontinuous Galerkin scheme.

We emphasize that in equation (4.15) the flux function F is the *physical* flux, which is evaluated by using the solution of the GRP_N at time τ_l. It is therefore a high order generalization of Godunov's scheme if the exact Riemann solver is used for the solution of RP_0. Any other approximate Riemann solver might be used as well if it is able to deliver the approximate solution $u_h^{\mathrm{RP}_0}(\chi_n)$ of the conventional nonlinear Riemann problem at the interface. For an exhaustive overview of Riemann solvers see [20].

4.2 Numerical results for the nonlinear Euler equations

Example 1. We first consider the smooth two-dimensional example of an isentropic vortex given by Hu and Shu in [11]. The initial condition is

$$\begin{pmatrix} \rho \\ u \\ v \\ p \end{pmatrix} = \begin{pmatrix} 1+\delta\rho \\ 1+\delta u \\ 1+\delta v \\ 1+\delta p \end{pmatrix}. \tag{4.17}$$

The velocity perturbation of the vortex is

$$\begin{pmatrix} \delta u \\ \delta v \end{pmatrix} = \frac{\epsilon}{2\pi} e^{\frac{1-r^2}{2}} \begin{pmatrix} -(y-5) \\ (x-5) \end{pmatrix}. \tag{4.18}$$

There is no perturbation in the entropy ($S = \frac{p}{\rho^\gamma}$), so

$$\delta S = 0 \tag{4.19}$$

and the temperature perturbation is

$$\delta T = -\frac{(\gamma-1)\,\epsilon^2}{8\gamma\pi^2} e^{1-r^2} \tag{4.20}$$

with $r^2 = (x-5)^2 + (y-5)^2$ and the vortex strength $\epsilon = 5$. The computational domain has the extents $[0, 10] \times [0, 10]$ and 4-periodic boundary conditions are imposed. After one period of $t = 10$, the exact solution is given by the initial condition. The numerical convergence rates for the nonlinear ADER-DG schemes up to sixth order of accuracy are given in Tables 20–24 The errors presented are those for the pressure p. Computations have been performed on a Cartesian grid. In all examples, we used Godunov's exact Riemann solver to solve the GRPs on the element boundaries.

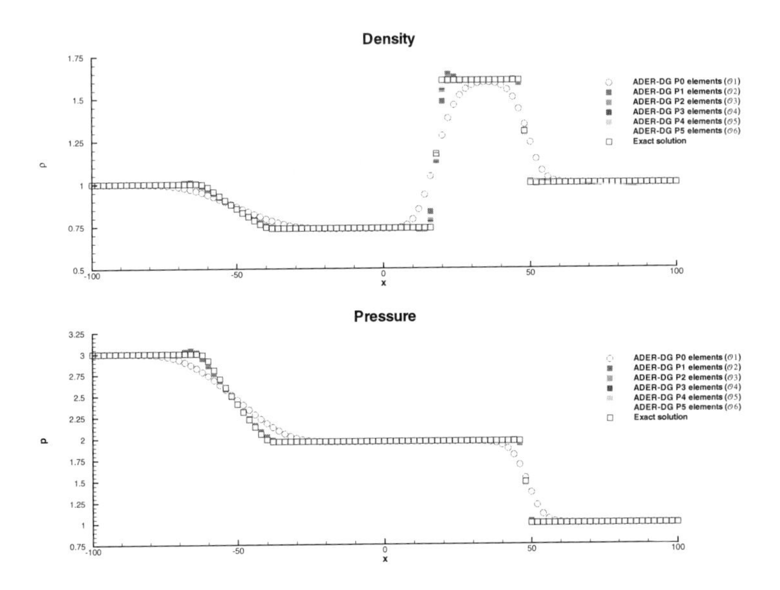

Figure 7. Numerical results for $\mathcal{O}1$–$\mathcal{O}6$ ADER-DG schemes for Example 2 at $t = 30$.

Example 2. We now consider a mild one-dimensional Riemann problem whose initial condition is given by

$$(\rho, u, p) = \begin{cases} (1, 0, 1) & \text{if } x \leq 0, \\ (1, 0, 3) & \text{if } x > 0. \end{cases} \tag{4.21}$$

The solution at $t = 30$ is plotted in Figure 7, a zoom into the same solution is shown in Figure 8. We emphasize that in the illustrated examples, we did not use any limiter at all, except that we revert to first order in the code if the density or the pressure become negative. The robustness of the DG method even without the use of limiters has already been pointed out by Atkins et al. in [1] which is clearly confirmed by our results. In Figures 7 and 8 we can see some mild overshoots of the numerical solution for P1 and P2 elements in the vicinity of the shock and the contact discontinuity and at the end of the rarefaction fan. But altogether we note that the solution gets less oscillatory if the order of the basis polynomials is increased. For P5 elements, we almost get a monotone solution, very close to the exact one. Nevertheless we underline the fact that the method produces oscillatory results if no limiter is used. The reason why we did not use a limiter was to show the behavior of the method for very high orders of basis polynomials since the aim of the ADER approach is to obtain schemes of *arbitrary* high order of accuracy. For this reason, the most promising limiters seem to be the HWENO limiters recently proposed by Qiu and Shu in [13].

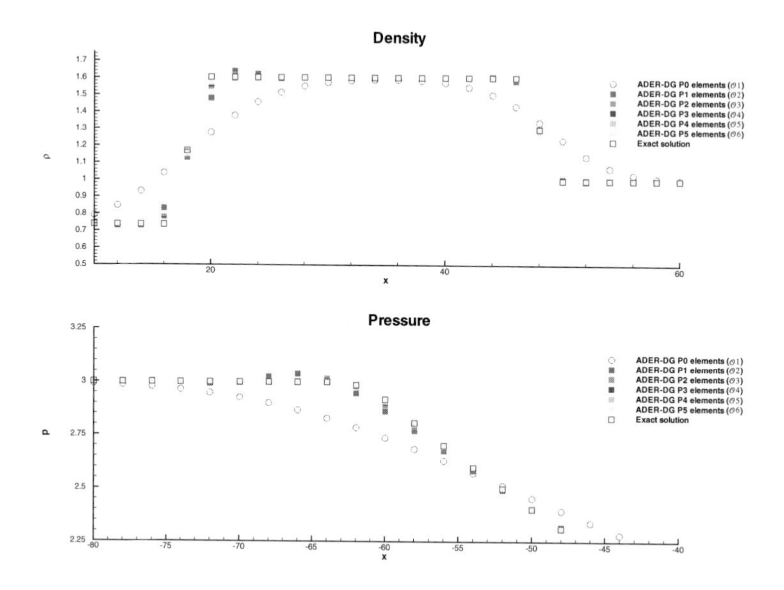

Figure 8. Zoom into the numerical results for $\mathcal{O}1$–$\mathcal{O}6$ ADER-DG schemes for Example 2 at $t = 30$.

Example 3. We now consider the Riemann problem proposed by Toro in [20] as test case number 5 which exhibits strong discontinuities and whose initial condition is given by

$$(\rho, u, p) = \begin{cases} (5.99924, \quad 19.5975, \quad 460.895) & \text{if } x \leq 0, \\ (5.99242, \quad -6.19633, \quad 46.0950) & \text{if } x > 0. \end{cases} \tag{4.22}$$

The numerical results for ADER-DG schemes with P1 to P5 elements and the exact solution are shown in Figure 9. This time we also included the numerical solution obtained by an ADER-DG-P1 scheme using the TVBM limiter proposed by Cockburn et al. [4] with the TVBM constant set to $M = 50$. The pressure jump of the shock wave is much higher than the one in the previous example but also in this case the ADER-DG method behaves quite robustly even without the use of a limiter. Figure 9 shows clearly that also in this example the oscillations decrease if higher order basis functions are used.

5 Conclusions

In the previous sections we developed an arbitrary high order discontinuous Galerkin scheme for linear and nonlinear hyperbolic conservation laws. In the linear case, the

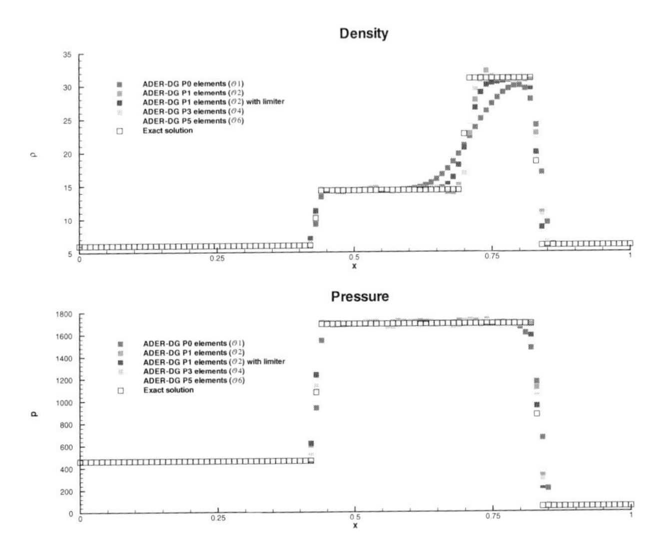

Figure 9. Numerical results for $\mathcal{O}1$, $\mathcal{O}2$, $\mathcal{O}4$ and $\mathcal{O}6$ ADER-DG schemes for Toro's test case 5 (Example 3) at $t = 0.035$.

scheme can be coded such that the order of accuracy becomes only a numerical parameter which can be arbitrarily set to any positive integer number. We have shown numerical convergence results up to tenth order of accuracy in space and time on Cartesian and regular triangular grids. Concerning convergence rates of the discontinuous Galerkin method on very distorted grids we refer to the literature, see e.g. [1]. We have also shown computations of the locally linearized Euler equations with non-uniform mean flow and discontinuities in the solution. The discontinuities can be treated very robustly with the ADER-DG schemes. We found that with increasing the order of the basis functions, the oscillations near jumps in the numerical solution decrease. Computations of the linearized Euler equations with non-uniform mean flow have been performed on regular and irregular triangular meshes. The CPU time was about 40% to 50% lower than the one of a quadrature-free Runge–Kutta DG method of equal order and equal CFL number. Concerning CPU time we remark that when solving linear systems with constant Jacobians on Cartesian grids, an entire time step of the ADER-DG scheme needs the same CPU time as a single Runge–Kutta substage since the tensor $I_{plqm}(\Delta t)$ can be precomputed for the entire domain and can be multiplied to the stiffness and flux matrices in advance. In this case, the CPU time can be reduced by 75% using an ADER-DG P3 scheme compared to a fourth order quadrature-free Runge–Kutta DG P3 scheme.

Independently of this, the memory required for the entire ADER-DG time integration always remains the same as for explicit first order Euler time-stepping since there is no need to store intermediate Runge–Kutta stages or previous time steps as e.g. for an Adams–Bashforth type time integration scheme. Beside the computation of (2.31) our scheme (2.33) looks very similar in coding like one substage of a quadrature-free Runge–Kutta discontinuous Galerkin scheme [1].

In the nonlinear case, our computations have shown that the method produces also less oscillatory results if the order of the basis functions is increased. However, the numerical solution can not be monotone without limiters, so in order to get rid of spurious oscillations in the ADER-DG scheme for very high order of accuracy, in our future research we will try the use of HWENO limiters [13] which are in our opinion very promising for this application.

Due to its explicit single-step nature and the extremely local stencil (high order accurate discretization in space and time is performed by using only the element and its direct side neighborhood), the ADER-DG scheme should furthermore be an ideal candidate for parallelization and vectorization on modern supercomputers. Further research will also concern hp-adaptivity since the order of accuracy in space *and time* may vary locally with ADER-DG schemes.

6 Acknowledgments

The research presented in this paper was financed by the EU–supported network HYKE, no. HPRN-CT-2002-00282, during the CEMRACS 2003 in Luminy, France, and by Deutsche Forschungsgemeinschaft (DFG) in the framework of research group FOR 508, “Noise Generation in Turbulent Flows”.

A Examples for basis functions and matrices for the ADER-DG method on triangular meshes

A.1 P1 elements

The basis functions [12] are orthogonal and read for P1 elements as follows:

$$\Phi_0 = 1, \;\; \Phi_1 = 2\xi + \eta - 1, \;\; \Phi_2 = 3\eta - 1. \tag{A.1}$$

The mass matrix is

$$M_{kl} = \begin{pmatrix} \frac{1}{2} & 0 & 0 \\ 0 & \frac{1}{12} & 0 \\ 0 & 0 & \frac{1}{4} \end{pmatrix}, \tag{A.2}$$

the stiffness matrices are

$$K^{\xi}_{kl} = \begin{pmatrix} 0 & 0 & 0 \\ 1 & 0 & 0 \\ 0 & 0 & 0 \end{pmatrix} \qquad K^{\eta}_{kl} = \begin{pmatrix} 0 & 0 & 0 \\ \frac{1}{2} & 0 & 0 \\ \frac{3}{2} & 0 & 0 \end{pmatrix} \tag{A.3}$$

and finally the flux matrices for all three edges are

$$\begin{aligned} F^{1,0}_{kl} &= \begin{pmatrix} 1 & 0 & -1 \\ 0 & \frac{1}{3} & 0 \\ -1 & 0 & 1 \end{pmatrix} \qquad F^{1,1}_{kl} = \begin{pmatrix} 1 & 0 & -1 \\ 0 & -\frac{1}{3} & 0 \\ -1 & 0 & 1 \end{pmatrix} \\ F^{1,2}_{kl} &= \begin{pmatrix} 1 & \frac{1}{2} & \frac{1}{2} \\ 0 & \frac{1}{6} & -\frac{1}{2} \\ -1 & -\frac{1}{2} & -\frac{1}{2} \end{pmatrix} \qquad F^{1,3}_{kl} = \begin{pmatrix} 1 & -\frac{1}{2} & \frac{1}{2} \\ 0 & \frac{1}{6} & \frac{1}{2} \\ -1 & \frac{1}{2} & -\frac{1}{2} \end{pmatrix}, \end{aligned} \tag{A.4}$$

$$\begin{aligned} F^{2,0}_{kl} &= \begin{pmatrix} 1 & \frac{1}{2} & \frac{1}{2} \\ \frac{1}{2} & \frac{1}{3} & 0 \\ \frac{1}{2} & 0 & 1 \end{pmatrix} \qquad F^{2,1}_{kl} = \begin{pmatrix} 1 & 0 & -1 \\ \frac{1}{2} & \frac{1}{6} & -\frac{1}{2} \\ \frac{1}{2} & -\frac{1}{2} & -\frac{1}{2} \end{pmatrix} \\ F^{2,2}_{kl} &= \begin{pmatrix} 1 & \frac{1}{2} & \frac{1}{2} \\ \frac{1}{2} & \frac{1}{6} & \frac{1}{2} \\ \frac{1}{2} & \frac{1}{2} & -\frac{1}{2} \end{pmatrix} \qquad F^{2,3}_{kl} = \begin{pmatrix} 1 & -\frac{1}{2} & \frac{1}{2} \\ \frac{1}{2} & -\frac{1}{3} & 0 \\ \frac{1}{2} & 0 & 1 \end{pmatrix}, \end{aligned} \tag{A.5}$$

$$\begin{aligned} F^{3,0}_{kl} &= \begin{pmatrix} 1 & -\frac{1}{2} & \frac{1}{2} \\ -\frac{1}{2} & \frac{1}{3} & 0 \\ \frac{1}{2} & 0 & 1 \end{pmatrix} \qquad F^{3,1}_{kl} = \begin{pmatrix} 1 & 0 & -1 \\ -\frac{1}{2} & \frac{1}{6} & \frac{1}{2} \\ \frac{1}{2} & \frac{1}{2} & -\frac{1}{2} \end{pmatrix} \\ F^{3,2}_{kl} &= \begin{pmatrix} 1 & \frac{1}{2} & \frac{1}{2} \\ -\frac{1}{2} & -\frac{1}{3} & 0 \\ \frac{1}{2} & 0 & 1 \end{pmatrix} \qquad F^{3,3}_{kl} = \begin{pmatrix} 1 & -\frac{1}{2} & \frac{1}{2} \\ -\frac{1}{2} & \frac{1}{6} & -\frac{1}{2} \\ \frac{1}{2} & -\frac{1}{2} & -\frac{1}{2} \end{pmatrix}. \end{aligned} \tag{A.6}$$

A.2 P2 elements

In the P2 case, the basis functions are

$$\begin{aligned} &\Phi_0 = 1, && \Phi_1 = 2, \xi + \eta - 1, \\ &\Phi_2 = 6, \xi^2 + \eta^2 + 6\xi\eta - 6\xi - 2\eta + 1, && \Phi_3 = 3, \eta - 1, \\ &\Phi_4 = 5, \eta^2 + 10\xi\eta - 2\xi - 6\eta + 1, && \Phi_5 = 10, \eta^2 - 8\eta + 1. \end{aligned} \tag{A.7}$$

The mass matrix is

$$M_{kl} = \begin{pmatrix} \frac{1}{2} & 0 & 0 & 0 & 0 & 0 \\ 0 & \frac{1}{12} & 0 & 0 & 0 & 0 \\ 0 & 0 & \frac{1}{30} & 0 & 0 & 0 \\ 0 & 0 & 0 & \frac{1}{4} & 0 & 0 \\ 0 & 0 & 0 & 0 & \frac{1}{18} & 0 \\ 0 & 0 & 0 & 0 & 0 & \frac{1}{6} \end{pmatrix}, \tag{A.8}$$

the stiffness matrices are

$$K^{\xi}_{kl} = \begin{pmatrix} 0 & 0 & 0 & 0 & 0 & 0 \\ 1 & 0 & 0 & 0 & 0 & 0 \\ 0 & \frac{1}{2} & 0 & 0 & 0 & 0 \\ 0 & 0 & 0 & 0 & 0 & 0 \\ \frac{2}{3} & 0 & 0 & \frac{5}{6} & 0 & 0 \\ 0 & 0 & 0 & 0 & 0 & 0 \end{pmatrix} \qquad K^{\eta}_{kl} = \begin{pmatrix} 0 & 0 & 0 & 0 & 0 & 0 \\ \frac{1}{2} & 0 & 0 & 0 & 0 & 0 \\ \frac{1}{3} & \frac{1}{4} & 0 & -\frac{1}{12} & 0 & 0 \\ \frac{3}{2} & 0 & 0 & 0 & 0 & 0 \\ \frac{1}{3} & \frac{5}{12} & 0 & \frac{5}{12} & 0 & 0 \\ -\frac{2}{3} & 0 & 0 & \frac{5}{3} & 0 & 0 \end{pmatrix} \tag{A.9}$$

and finally the flux matrices for all three edges are

$$F^{1,0}_{kl} = \begin{pmatrix} 1 & 0 & 0 & -1 & 0 & 1 \\ 0 & \frac{1}{3} & 0 & 0 & -\frac{1}{3} & 0 \\ 0 & 0 & \frac{1}{5} & 0 & 0 & 0 \\ -1 & 0 & 0 & 1 & 0 & -1 \\ 0 & -\frac{1}{3} & 0 & 0 & \frac{1}{3} & 0 \\ 1 & 0 & 0 & -1 & 0 & 1 \end{pmatrix} \qquad F^{1,1}_{kl} = \begin{pmatrix} 1 & 0 & 0 & -1 & 0 & 1 \\ 0 & -\frac{1}{3} & 0 & 0 & \frac{1}{3} & 0 \\ 0 & 0 & \frac{1}{5} & 0 & 0 & 0 \\ -1 & 0 & 0 & 1 & 0 & -1 \\ 0 & \frac{1}{3} & 0 & 0 & -\frac{1}{3} & 0 \\ 1 & 0 & 0 & -1 & 0 & 1 \end{pmatrix}$$

$$F^{1,2}_{kl} = \begin{pmatrix} 1 & \frac{1}{2} & \frac{1}{3} & \frac{1}{2} & \frac{1}{3} & \frac{1}{3} \\ 0 & \frac{1}{6} & \frac{1}{6} & -\frac{1}{2} & -\frac{1}{6} & -\frac{1}{3} \\ 0 & 0 & \frac{1}{30} & 0 & -\frac{1}{6} & \frac{1}{3} \\ -1 & -\frac{1}{2} & -\frac{1}{3} & -\frac{1}{2} & -\frac{1}{3} & -\frac{1}{3} \\ 0 & -\frac{1}{6} & -\frac{1}{6} & \frac{1}{2} & \frac{1}{6} & \frac{1}{3} \\ 1 & \frac{1}{2} & \frac{1}{3} & \frac{1}{2} & \frac{1}{3} & \frac{1}{3} \end{pmatrix} \qquad F^{1,3}_{kl} = \begin{pmatrix} 1 & -\frac{1}{2} & \frac{1}{3} & \frac{1}{2} & -\frac{1}{3} & \frac{1}{3} \\ 0 & \frac{1}{6} & -\frac{1}{6} & \frac{1}{2} & -\frac{1}{6} & \frac{1}{3} \\ 0 & 0 & \frac{1}{30} & 0 & \frac{1}{6} & \frac{1}{3} \\ -1 & \frac{1}{2} & -\frac{1}{3} & -\frac{1}{2} & \frac{1}{3} & -\frac{1}{3} \\ 0 & -\frac{1}{6} & \frac{1}{6} & -\frac{1}{2} & \frac{1}{6} & -\frac{1}{3} \\ 1 & -\frac{1}{2} & \frac{1}{3} & \frac{1}{2} & -\frac{1}{3} & \frac{1}{3} \end{pmatrix}, \tag{A.10}$$

$$F_{kl}^{2,0} = \begin{pmatrix} 1 & \frac{1}{2} & \frac{1}{3} & \frac{1}{2} & \frac{1}{3} & \frac{1}{3} \\ \frac{1}{2} & \frac{1}{3} & \frac{1}{4} & 0 & \frac{1}{12} & 0 \\ \frac{1}{3} & \frac{1}{4} & \frac{1}{5} & -\frac{1}{12} & 0 & 0 \\ \frac{1}{2} & 0 & -\frac{1}{12} & 1 & \frac{5}{12} & \frac{2}{3} \\ \frac{1}{3} & \frac{1}{12} & 0 & \frac{5}{12} & \frac{1}{3} & 0 \\ \frac{1}{3} & 0 & 0 & \frac{2}{3} & 0 & 1 \end{pmatrix} \quad F_{kl}^{2,1} = \begin{pmatrix} 1 & 0 & 0 & -1 & 0 & 1 \\ \frac{1}{2} & \frac{1}{6} & 0 & -\frac{1}{2} & -\frac{1}{6} & \frac{1}{2} \\ \frac{1}{3} & \frac{1}{6} & \frac{1}{30} & -\frac{1}{3} & -\frac{1}{6} & \frac{1}{3} \\ \frac{1}{2} & -\frac{1}{2} & 0 & -\frac{1}{2} & \frac{1}{2} & \frac{1}{2} \\ \frac{1}{3} & -\frac{1}{6} & -\frac{1}{6} & -\frac{1}{3} & \frac{1}{6} & \frac{1}{3} \\ \frac{1}{3} & -\frac{1}{3} & \frac{1}{3} & -\frac{1}{3} & \frac{1}{3} & \frac{1}{3} \end{pmatrix}$$

$$F_{kl}^{2,2} = \begin{pmatrix} 1 & \frac{1}{2} & \frac{1}{3} & \frac{1}{2} & \frac{1}{3} & \frac{1}{3} \\ \frac{1}{2} & \frac{1}{6} & \frac{1}{12} & \frac{1}{2} & \frac{1}{4} & \frac{1}{3} \\ \frac{1}{3} & \frac{1}{12} & \frac{1}{30} & \frac{5}{12} & \frac{1}{6} & \frac{1}{3} \\ \frac{1}{2} & \frac{1}{2} & \frac{5}{12} & -\frac{1}{2} & -\frac{1}{12} & -\frac{1}{3} \\ \frac{1}{3} & \frac{1}{4} & \frac{1}{6} & -\frac{1}{12} & \frac{1}{6} & -\frac{1}{3} \\ \frac{1}{3} & \frac{1}{3} & \frac{1}{3} & -\frac{1}{3} & -\frac{1}{3} & \frac{1}{3} \end{pmatrix} \quad F_{kl}^{2,3} = \begin{pmatrix} 1 & -\frac{1}{2} & \frac{1}{3} & \frac{1}{2} & -\frac{1}{3} & \frac{1}{3} \\ \frac{1}{2} & -\frac{1}{3} & \frac{1}{4} & 0 & -\frac{1}{12} & 0 \\ \frac{1}{3} & -\frac{1}{4} & \frac{1}{5} & -\frac{1}{12} & 0 & 0 \\ \frac{1}{2} & 0 & -\frac{1}{12} & 1 & -\frac{5}{12} & \frac{2}{3} \\ \frac{1}{3} & -\frac{1}{12} & 0 & \frac{5}{12} & -\frac{1}{3} & 0 \\ \frac{1}{3} & 0 & 0 & \frac{2}{3} & 0 & 1 \end{pmatrix}, \tag{A.11}$$

$$F_{kl}^{3,0} = \begin{pmatrix} 1 & -\frac{1}{2} & \frac{1}{3} & \frac{1}{2} & -\frac{1}{3} & \frac{1}{3} \\ -\frac{1}{2} & \frac{1}{3} & -\frac{1}{4} & 0 & \frac{1}{12} & 0 \\ \frac{1}{3} & -\frac{1}{4} & \frac{1}{5} & -\frac{1}{12} & 0 & 0 \\ \frac{1}{2} & 0 & -\frac{1}{12} & 1 & -\frac{5}{12} & \frac{2}{3} \\ -\frac{1}{3} & \frac{1}{12} & 0 & -\frac{5}{12} & \frac{1}{3} & 0 \\ \frac{1}{3} & 0 & 0 & \frac{2}{3} & 0 & 1 \end{pmatrix} \quad F_{kl}^{3,1} = \begin{pmatrix} 1 & 0 & 0 & -1 & 0 & 1 \\ -\frac{1}{2} & \frac{1}{6} & 0 & \frac{1}{2} & -\frac{1}{6} & -\frac{1}{2} \\ \frac{1}{3} & -\frac{1}{6} & \frac{1}{30} & -\frac{1}{3} & \frac{1}{6} & \frac{1}{3} \\ \frac{1}{2} & \frac{1}{2} & 0 & -\frac{1}{2} & -\frac{1}{2} & \frac{1}{2} \\ -\frac{1}{3} & -\frac{1}{6} & \frac{1}{6} & \frac{1}{3} & \frac{1}{6} & -\frac{1}{3} \\ \frac{1}{3} & \frac{1}{3} & \frac{1}{3} & -\frac{1}{3} & -\frac{1}{3} & \frac{1}{3} \end{pmatrix}$$

$$F_{kl}^{3,2} = \begin{pmatrix} 1 & \frac{1}{2} & \frac{1}{3} & \frac{1}{2} & \frac{1}{3} & \frac{1}{3} \\ -\frac{1}{2} & -\frac{1}{3} & -\frac{1}{4} & 0 & -\frac{1}{12} & 0 \\ \frac{1}{3} & \frac{1}{4} & \frac{1}{5} & -\frac{1}{12} & 0 & 0 \\ \frac{1}{2} & 0 & -\frac{1}{12} & 1 & \frac{5}{12} & \frac{2}{3} \\ -\frac{1}{3} & -\frac{1}{12} & 0 & -\frac{5}{12} & -\frac{1}{3} & 0 \\ \frac{1}{3} & 0 & 0 & \frac{2}{3} & 0 & 1 \end{pmatrix} \quad F_{kl}^{3,3} = \begin{pmatrix} 1 & -\frac{1}{2} & \frac{1}{3} & \frac{1}{2} & -\frac{1}{3} & \frac{1}{3} \\ -\frac{1}{2} & \frac{1}{6} & -\frac{1}{12} & -\frac{1}{2} & \frac{1}{4} & -\frac{1}{3} \\ \frac{1}{3} & -\frac{1}{12} & \frac{1}{30} & \frac{5}{12} & -\frac{1}{6} & \frac{1}{3} \\ \frac{1}{2} & -\frac{1}{2} & \frac{5}{12} & -\frac{1}{2} & \frac{1}{12} & -\frac{1}{3} \\ -\frac{1}{3} & \frac{1}{4} & -\frac{1}{6} & \frac{1}{12} & \frac{1}{6} & \frac{1}{3} \\ \frac{1}{3} & -\frac{1}{3} & \frac{1}{3} & -\frac{1}{3} & \frac{1}{3} & \frac{1}{3} \end{pmatrix}. \tag{A.12}$$

B Examples for basis functions and matrices for the reconstructed ADER-DG method on Cartesian grids

B.1 P0 elements

The original basis function for P0 elements is trivially:

$$\Phi_0 = 1. \tag{B.1}$$

The reconstructed basis functions for P0 elements are

$$\begin{aligned} \Psi_0 &= 1, & \Psi_1 &= \xi, & \Psi_2 &= \xi^2 - \frac{1}{3}, \\ \Psi_3 &= \eta, & \Psi_4 &= \xi\eta, & \Psi_5 &= \left(\xi^2 - \frac{1}{3}\right)\eta, \\ \Psi_6 &= \eta^2 - \frac{1}{3}, & \Psi_7 &= \xi\left(\eta^2 - \frac{1}{3}\right), & \Psi_8 &= \left(\xi^2 - \frac{1}{3}\right)\left(\eta^2 - \frac{1}{3}\right). \end{aligned} \tag{B.2}$$

The mass matrix is

$$M_{kl} = \left(1 \right), \tag{B.3}$$

the stiffness matrices are

$$\begin{aligned} K^{\xi}_{kl} &= \left(0\ 0\ 0\ 0\ 0\ 0\ 0\ 0\ 0 \right) \\ K^{\eta}_{kl} &= \left(0\ 0\ 0\ 0\ 0\ 0\ 0\ 0\ 0 \right), \end{aligned} \tag{B.4}$$

and finally the flux matrices for all four edges are

$$\begin{aligned} \frac{1}{2}F^{1+}_{kl} &= \left(\tfrac{1}{2}\ \ \tfrac{1}{2}\ \tfrac{1}{3}\ 0\ 0\ 0\ 0\ 0\ 0 \right) \\ \frac{1}{2}F^{1-}_{kl} &= \left(\tfrac{1}{2}\ -\tfrac{1}{2}\ \tfrac{1}{3}\ 0\ 0\ 0\ 0\ 0\ 0 \right) \\ \frac{1}{2}F^{3+}_{kl} &= \left(\tfrac{1}{2}\ \ \tfrac{1}{2}\ \tfrac{1}{3}\ 0\ 0\ 0\ 0\ 0\ 0 \right) \\ \frac{1}{2}F^{3-}_{kl} &= \left(\tfrac{1}{2}\ -\tfrac{1}{2}\ \tfrac{1}{3}\ 0\ 0\ 0\ 0\ 0\ 0 \right), \end{aligned} \tag{B.5}$$

$$\begin{aligned} \frac{1}{2}F^{2+}_{kl} &= \left(\tfrac{1}{2}\ 0\ 0\ \ \tfrac{1}{2}\ 0\ 0\ \tfrac{1}{3}\ 0\ 0 \right) \\ \frac{1}{2}F^{2-}_{kl} &= \left(\tfrac{1}{2}\ 0\ 0\ -\tfrac{1}{2}\ 0\ 0\ \tfrac{1}{3}\ 0\ 0 \right) \\ \frac{1}{2}F^{4+}_{kl} &= \left(\tfrac{1}{2}\ 0\ 0\ \ \tfrac{1}{2}\ 0\ 0\ \tfrac{1}{3}\ 0\ 0 \right) \\ \frac{1}{2}F^{4-}_{kl} &= \left(\tfrac{1}{2}\ 0\ 0\ -\tfrac{1}{2}\ 0\ 0\ \tfrac{1}{3}\ 0\ 0 \right). \end{aligned} \tag{B.6}$$

B.2 P1 elements

The original basis functions for P1 elements are

$$\begin{aligned} \Phi_0 &= 1, & \Phi_1 &= \xi, \\ \Phi_2 &= \eta, & \Phi_3 &= \xi\eta. \end{aligned} \tag{B.7}$$

The reconstructed basis functions for P1 elements are 2D tensor products of the following fifth order Legendre polynomials, so there is a total number of 36 reconstructed degrees of freedom

$$
\begin{aligned}
L_0(\xi) &= 1, & L_1(\xi) &= \xi, \\
L_2(\xi) &= \xi^2 - \frac{1}{3}, & L_3(\xi) &= \xi^3 - \frac{3}{5}\xi, \\
L_4(\xi) &= \xi^4 - \frac{6}{7}\xi^2 + \frac{3}{35}, & L_5(\xi) &= \xi^5 - \frac{10}{9}\xi^3 + \frac{5}{21}\xi.
\end{aligned} \tag{B.8}
$$

The mass matrix is

$$
M_{kl} = \begin{pmatrix} 1 & 0 & 0 & 0 \\ 0 & \frac{1}{3} & 0 & 0 \\ 0 & 0 & \frac{1}{3} & 0 \\ 0 & 0 & 0 & \frac{1}{9} \end{pmatrix}, \tag{B.9}
$$

the stiffness matrices are

$$
\begin{aligned}
K^{\xi}_{kl} &= \begin{pmatrix} 0 & 0 & 0 & 0 & 0 & 0 & 0 & 0 & \dots & 0 \\ 2 & 0 & 0 & 0 & 0 & 0 & 0 & 0 & \dots & 0 \\ 0 & 0 & 0 & 0 & 0 & 0 & 0 & 0 & \dots & 0 \\ 0 & 0 & 0 & 0 & 0 & 0 & \frac{2}{3} & 0 & \dots & 0 \end{pmatrix} \\
K^{\eta}_{kl} &= \begin{pmatrix} 0 & 0 & 0 & 0 & 0 & 0 & 0 & 0 & \dots & 0 \\ 0 & 0 & 0 & 0 & 0 & 0 & 0 & 0 & \dots & 0 \\ 2 & 0 & 0 & 0 & 0 & 0 & 0 & 0 & \dots & 0 \\ 0 & \frac{2}{3} & 0 & 0 & 0 & 0 & 0 & 0 & \dots & 0 \end{pmatrix}
\end{aligned} \tag{B.10}
$$

and finally the flux matrices e.g. in ξ-direction are

$$
\frac{1}{2}F^{1^+}_{kl} = \begin{pmatrix} \frac{1}{2} & \frac{1}{2} & \frac{1}{3} & \frac{1}{5} & \frac{4}{35} & \frac{4}{63} & 0 & 0 & 0 & 0 & 0 & 0 & 0 \dots 0 \\ \frac{1}{2} & \frac{1}{2} & \frac{1}{3} & \frac{1}{5} & \frac{4}{35} & \frac{4}{63} & 0 & 0 & 0 & 0 & 0 & 0 & 0 \dots 0 \\ 0 & 0 & 0 & 0 & 0 & 0 & \frac{1}{6} & \frac{1}{6} & \frac{1}{9} & \frac{1}{15} & \frac{4}{105} & \frac{4}{189} & 0 \dots 0 \\ 0 & 0 & 0 & 0 & 0 & 0 & \frac{1}{6} & \frac{1}{6} & \frac{1}{9} & \frac{1}{15} & \frac{4}{105} & \frac{4}{189} & 0 \dots 0 \end{pmatrix}
$$

$$
\frac{1}{2}F^{1^-}_{kl} = \begin{pmatrix} \frac{1}{2} & -\frac{1}{2} & \frac{1}{3} & -\frac{1}{5} & \frac{4}{35} & -\frac{4}{63} & 0 & 0 & 0 & 0 & 0 & 0 & 0 \dots 0 \\ \frac{1}{2} & -\frac{1}{2} & \frac{1}{3} & -\frac{1}{5} & \frac{4}{35} & -\frac{4}{63} & 0 & 0 & 0 & 0 & 0 & 0 & 0 \dots 0 \\ 0 & 0 & 0 & 0 & 0 & 0 & \frac{1}{6} & -\frac{1}{6} & \frac{1}{9} & -\frac{1}{15} & \frac{4}{105} & -\frac{4}{189} & 0 \dots 0 \\ 0 & 0 & 0 & 0 & 0 & 0 & \frac{1}{6} & -\frac{1}{6} & \frac{1}{9} & -\frac{1}{15} & \frac{4}{105} & -\frac{4}{189} & 0 \dots 0 \end{pmatrix}
$$

$$\frac{1}{2}F_{kl}^{3+} = \begin{pmatrix} \frac{1}{2} & \frac{1}{2} & \frac{1}{3} & \frac{1}{5} & \frac{4}{35} & \frac{4}{63} & 0 & 0 & 0 & 0 & 0 & 0 & 0\dots 0 \\ -\frac{1}{2} & -\frac{1}{2} & -\frac{1}{3} & -\frac{1}{5} & -\frac{4}{35} & -\frac{4}{63} & 0 & 0 & 0 & 0 & 0 & 0 & 0\dots 0 \\ 0 & 0 & 0 & 0 & 0 & 0 & \frac{1}{6} & \frac{1}{6} & \frac{1}{9} & \frac{1}{15} & \frac{4}{105} & \frac{4}{189} & 0\dots 0 \\ 0 & 0 & 0 & 0 & 0 & 0 & -\frac{1}{6} & -\frac{1}{6} & -\frac{1}{9} & -\frac{1}{15} & -\frac{4}{105} & -\frac{4}{189} & 0\dots 0 \end{pmatrix}$$

$$\frac{1}{2}F_{kl}^{3-} = \begin{pmatrix} \frac{1}{2} & -\frac{1}{2} & \frac{1}{3} & -\frac{1}{5} & \frac{4}{35} & -\frac{4}{63} & 0 & 0 & 0 & 0 & 0 & 0 & 0\dots 0 \\ -\frac{1}{2} & \frac{1}{2} & -\frac{1}{3} & \frac{1}{5} & -\frac{4}{35} & \frac{4}{63} & 0 & 0 & 0 & 0 & 0 & 0 & 0\dots 0 \\ 0 & 0 & 0 & 0 & 0 & 0 & \frac{1}{6} & -\frac{1}{6} & \frac{1}{9} & -\frac{1}{15} & \frac{4}{105} & -\frac{4}{189} & 0\dots 0 \\ 0 & 0 & 0 & 0 & 0 & 0 & -\frac{1}{6} & \frac{1}{6} & -\frac{1}{9} & \frac{1}{15} & -\frac{4}{105} & \frac{4}{189} & 0\dots 0 \end{pmatrix}. \tag{B.11}$$

C Numerical convergence tables

In the following tables N_G denotes the number of grid cells in each dimension and N_d is the total number of degrees of freedom. The next three columns contain the error measured in different norms followed by three columns with the corresponding convergence rates. The error norms have been calculated using Gaussian integration formulae with an order of accuracy which is four times higher than the one of the numerical scheme.

Table 3. Numerical convergence rates for ADER-DG $\mathcal{O}2$ scheme (P1) on the Cartesian grid.

ADER-DG $\mathcal{O}2$							
N_G	N_d	L_∞	L_1	L_2	$\mathcal{O}_{L_\infty}$	$\mathcal{O}_{L_1}$	$\mathcal{O}_{L_2}$
100	30,000	1.5660E-01	8.2060E+00	6.4463E-01			
150	67,500	7.2868E-02	3.4193E+00	2.8628E-01	1.9	2.2	2.0
200	120,000	4.0392E-02	1.8397E+00	1.5835E-01	2.1	2.2	2.1
300	270,000	1.7306E-02	7.8121E-01	6.8851E-02	2.1	2.1	2.1

Table 4. Numerical convergence rates for ADER-DG $\mathcal{O}3$ scheme (P2) on the Cartesian grid.

ADER-DG $\mathcal{O}3$							
N_G	N_d	L_∞	L_1	L_2	$\mathcal{O}_{L_\infty}$	$\mathcal{O}_{L_1}$	$\mathcal{O}_{L_2}$
50	15,000	6.7613E-02	2.8419E+00	2.4326E-01			
75	33,750	1.9147E-02	7.5917E-01	6.7480E-02	3.1	3.3	3.2
100	60,000	7.6597E-03	3.0250E-01	2.6902E-02	3.2	3.2	3.2
150	135,000	2.1437E-03	8.5185E-02	7.5264E-03	3.1	3.1	3.1

Table 5. Numerical convergence rates for ADER-DG $\mathcal{O}4$ scheme (P3) on the Cartesian grid.

ADER-DG $\mathcal{O}4$							
N_G	N_d	L_∞	L_1	L_2	$\mathcal{O}_{L_\infty}$	$\mathcal{O}_{L_1}$	$\mathcal{O}_{L_2}$
25	6.250	7.8971E-02	4.0460E+00	2.8840E-01			
50	25.000	4.5254E-03	1.7027E-01	1.4428E-02	4.1	4.6	4.3
75	56.250	6.9786E-04	2.6585E-02	2.2745E-03	4.6	4.6	4.6
100	100.000	2.0581E-04	7.5912E-03	6.4561E-04	4.2	4.4	4.4

Table 6. Numerical convergence rates for ADER-DG $\mathcal{O}5$ scheme (P4) on the Cartesian grid.

ADER-DG $\mathcal{O}5$							
N_G	N_d	L_∞	L_1	L_2	$\mathcal{O}_{L_\infty}$	$\mathcal{O}_{L_1}$	$\mathcal{O}_{L_2}$
25	9,375	7.7497E-03	6.2479E-01	4.1506E-02			
50	37,500	2.3475E-04	1.0357E-02	7.8186E-04	5.0	5.9	5.7
75	84,375	2.6994E-05	1.1818E-03	9.1796E-05	5.3	5.4	5.3
100	150,000	6.1512E-06	2.6783E-04	2.1121E-05	5.1	5.2	5.1

Table 7. Numerical convergence rates for ADER-DG $\mathcal{O}6$ scheme (P5) on the Cartesian grid.

ADER-DG $\mathcal{O}6$							
N_G	N_d	L_∞	L_1	L_2	$\mathcal{O}_{L_\infty}$	$\mathcal{O}_{L_1}$	$\mathcal{O}_{L_2}$
25	13,125	1.0116E-03	7.4786E-02	4.6516E-03			
50	52,500	1.6321E-05	1.0351E-03	7.0422E-05	6.0	6.2	6.0
75	118,125	1.5928E-06	7.6535E-05	5.4450E-06	5.7	6.4	6.3
100	210,000	2.6423E-07	1.2825E-05	9.1726E-07	6.2	6.2	6.2

Table 8. Numerical convergence rates for ADER-DG $\mathcal{O}7$ scheme (P6) on the Cartesian grid.

ADER-DG $\mathcal{O}7$							
N_G	N_d	L_∞	L_1	L_2	$\mathcal{O}_{L_\infty}$	$\mathcal{O}_{L_1}$	$\mathcal{O}_{L_2}$
25	17,500	7.6424E-05	5.9554E-03	3.6798E-04			
50	70,000	2.1902E-06	9.0208E-05	5.7881E-06	5.1	6.0	6.0
75	157,500	1.4254E-07	7.2541E-06	4.8813E-07	6.7	6.2	6.1
100	280,000	1.8566E-08	1.0555E-06	7.2583E-08	7.1	6.7	6.6

Table 9. Numerical convergence rates for ADER-DG $\mathcal{O}8$ scheme (P7) on the Cartesian grid.

ADER-DG $\mathcal{O}8$							
N_G	N_d	L_∞	L_1	L_2	$\mathcal{O}_{L_\infty}$	$\mathcal{O}_{L_1}$	$\mathcal{O}_{L_2}$
25	22,500	4.4134E-05	1.2390E-03	8.0753E-05			
50	90,000	1.6532E-07	7.0635E-06	4.8826E-07	8.1	7.5	7.4
75	202,500	9.6372E-09	3.6892E-07	2.5577E-08	7.0	7.3	7.3
100	360,000	8.8004E-10	4.5762E-08	3.1527E-09	8.3	7.3	7.3

Table 10. Numerical convergence rates for ADER-DG $\mathcal{O}9$ scheme (P8) on the Cartesian grid.

ADER-DG $\mathcal{O}9$							
N_G	N_d	L_∞	L_1	L_2	$\mathcal{O}_{L_\infty}$	$\mathcal{O}_{L_1}$	$\mathcal{O}_{L_2}$
10	4,500	6.3325E-03	6.1045E-01	2.9457E-02			
25	28,125	3.2383E-06	2.1242E-04	1.2898E-05	8.3	8.7	8.4
50	112,500	2.6208E-08	8.9930E-07	6.5512E-08	6.9	7.9	7.6
75	253,125	6.7687E-10	3.0629E-08	2.3645E-09	9.0	8.3	8.2

Table 11. Numerical convergence rates for ADER-DG $\mathcal{O}10$ scheme (P9) on the Cartesian grid.

ADER-DG $\mathcal{O}10$							
N_G	N_d	L_∞	L_1	L_2	$\mathcal{O}_{L_\infty}$	$\mathcal{O}_{L_1}$	$\mathcal{O}_{L_2}$
10	5,500	2.3342E-03	1.6519E-01	9.1034E-03			
25	34,375	7.0918E-07	4.5060E-05	2.8242E-06	8.8	9.0	8.8
50	137,500	1.0632E-09	5.4547E-08	4.1396E-09	9.4	9.7	9.4
75	309,375	2.6470E-11	8.8686E-10	7.1228E-11	9.1	10.2	10.0

Table 12. Numerical convergence rates for ADER-DG $\mathcal{O}2$ scheme (P1) on the regular unstructured mesh.

ADER-DG $\mathcal{O}2$							
N_G	N_d	L_∞	L_1	L_2	$\mathcal{O}_{L_\infty}$	$\mathcal{O}_{L_1}$	$\mathcal{O}_{L_2}$
25	7,500	1.0143E-01	1.7988E+01	8.4504E-01			
40	19,200	4.3561E-02	7.4431E+00	3.5315E-01	1.8	1.9	1.9
50	30,000	2.9151E-02	4.8430E+00	2.2921E-01	1.8	1.9	1.9
75	67,500	1.3540E-02	2.2258E+00	1.0488E-01	1.9	1.9	1.9

Table 13. Numerical convergence rates for ADER-DG $\mathcal{O}4$ scheme (P3) on the regular unstructured mesh.

ADER-DG $\mathcal{O}4$							
N_G	N_d	L_∞	L_1	L_2	$\mathcal{O}_{L_\infty}$	$\mathcal{O}_{L_1}$	$\mathcal{O}_{L_2}$
10	4,000	2.6614E-02	3.6940E+00	1.7035E-01			
20	16,000	2.1150E-03	2.2322E-01	1.1262E-02	3.7	4.0	3.9
40	64,000	1.7118E-04	1.4971E-02	7.7531E-04	3.6	3.9	3.9
60	144,000	3.7768E-05	3.1228E-03	1.6343E-04	3.7	3.9	3.8

Table 14. Numerical convergence rates for ADER-DG $\mathcal{O}6$ scheme (P5) on the regular unstructured mesh.

ADER-DG $\mathcal{O}6$							
N_G	N_d	L_∞	L_1	L_2	$\mathcal{O}_{L_\infty}$	$\mathcal{O}_{L_1}$	$\mathcal{O}_{L_2}$
5	2,100	2.0367E-02	4.6454E+00	1.6683E-01			
10	8,400	8.1399E-04	9.7671E-02	4.5373E-03	4.6	5.6	5.2
20	33,600	2.3346E-05	1.6271E-03	8.3467E-05	5.1	5.9	5.8
30	75,600	2.3879E-06	1.4918E-04	7.8982E-06	5.6	5.9	5.8

Table 15. Numerical convergence rates for ADER-DG $\mathcal{O}8$ scheme (P7) on the regular unstructured mesh.

ADER-DG $\mathcal{O}8$							
N_G	N_d	L_∞	L_1	L_2	$\mathcal{O}_{L_\infty}$	$\mathcal{O}_{L_1}$	$\mathcal{O}_{L_2}$
5	3,600	2.3401E-03	4.1436E-01	1.5853E-02			
10	14,400	2.2099E-05	2.3774E-03	1.1356E-04	6.7	7.4	7.1
15	32,400	1.8084E-06	1.0472E-04	5.3533E-06	6.2	7.7	7.5
20	57,600	1.9437E-07	1.0761E-05	5.5335E-07	7.8	7.9	7.9

Table 16. Numerical convergence rates for ADER-DG $\mathcal{O}10$ scheme (P9) on the regular unstructured mesh.

ADER-DG $\mathcal{O}9$							
N_G	N_d	L_∞	L_1	L_2	$\mathcal{O}_{L_\infty}$	$\mathcal{O}_{L_1}$	$\mathcal{O}_{L_2}$
4	3,520	1.1908E-03	2.2494E-01	7.5866E-03			
8	14,080	7.1357E-06	4.6672E-04	2.1936E-05	7.4	8.9	8.4
12	31,680	1.4006E-07	9.2523E-06	4.4864E-07	9.7	9.7	9.6
16	56,320	1.2676E-08	5.4173E-07	2.81E-08	8.4	9.9	9.6

Table 17. Numerical convergence rates for the reconstructed ADER-DG $\mathcal{O}3$ scheme (P0) on the Cartesian grid.

Reconstructed ADER-DG $\mathcal{O}3$							
N_G	N_d	L_∞	L_1	L_2	$\mathcal{O}_{L_\infty}$	$\mathcal{O}_{L_1}$	$\mathcal{O}_{L_2}$
50	2,500	3.4005E-02	1.1944E+01	3.5249E-01			
100	10,000	4.6946E-03	1.5548E+00	4.7308E-02	2.9	2.9	2.9
200	40,000	5.9522E-04	1.9543E-01	5.9760E-03	3.0	3.0	3.0
300	90,000	1.7677E-04	5.7934E-02	1.7725E-03	3.0	3.0	3.0

Table 18. Numerical convergence rates for the reconstructed ADER-DG $\mathcal{O}6$ scheme (P1) on the Cartesian grid.

Reconstructed ADER-DG $\mathcal{O}6$							
N_G	N_d	L_∞	L_1	L_2	$\mathcal{O}_{L_\infty}$	$\mathcal{O}_{L_1}$	$\mathcal{O}_{L_2}$
25	2,500	1.3612E-01	1.1667E+01	6.4359E-01			
50	10,000	7.1692E-03	3.1377E-01	2.5343E-02	4.2	5.2	4.7
75	22,500	6.3138E-04	2.3395E-02	2.0454E-03	6.0	6.4	6.2
100	40,000	9.9688E-05	3.7224E-03	3.2837E-04	6.4	6.4	6.4

Table 19. Numerical convergence rates for the reconstructed ADER-DG $\mathcal{O}9$ scheme (P2) on the Cartesian grid.

Reconstructed ADER-DG $\mathcal{O}9$							
N_G	N_d	L_∞	L_1	L_2	$\mathcal{O}_{L_\infty}$	$\mathcal{O}_{L_1}$	$\mathcal{O}_{L_2}$
25	5,625	6.2255E-03	3.5897E-01	2.4752E-02			
50	22,500	2.0513E-05	7.2158E-04	6.3804E-05	8.2	9.0	8.6
75	50,625	4.6286E-07	1.7171E-05	1.5291E-06	9.4	9.2	9.2
100	90,000	3.4620E-08	1.2158E-06	1.0849E-07	9.0	9.2	9.2

Table 20. Numerical convergence rates for the nonlinear ADER-DG $\mathcal{O}2$ scheme (P1) on the Cartesian grid.

Nonlinear ADER-DG $\mathcal{O}2$							
N_G	N_d	L_∞	L_1	L_2	$\mathcal{O}_{L_\infty}$	$\mathcal{O}_{L_1}$	$\mathcal{O}_{L_2}$
30	2,700	5.0503E-02	2.3753E-01	6.5840E-02			
50	7,500	1.6149E-02	6.3506E-02	1.78E-02	2.2	2.6	2.6
75	16,875	6.1842E-03	2.1920E-02	6.1867E-03	2.4	2.6	2.6
100	30,000	3.0862E-03	1.0467E-02	2.9395E-03	2.4	2.6	2.6

Table 21. Numerical convergence rates for the nonlinear ADER-DG $\mathcal{O}3$ scheme (P2) on the Cartesian grid.

Nonlinear ADER-DG $\mathcal{O}3$							
N_G	N_d	L_∞	L_1	L_2	$\mathcal{O}_{L_\infty}$	$\mathcal{O}_{L_1}$	$\mathcal{O}_{L_2}$
10	600	5.1832E-02	3.8129E-01	7.9602E-02			
20	2,400	1.3945E-02	4.2144E-02	1.0735E-02	1.9	3.2	2.9
40	9,600	8.9570E-04	3.2167E-03	8.2627E-04	4.0	3.7	3.7
60	21,600	2.5259E-04	7.9205E-04	2.0026E-04	3.1	3.5	3.5

Table 22. Numerical convergence rates for the nonlinear ADER-DG $\mathcal{O}4$ scheme (P3) on the Cartesian grid.

Nonlinear ADER-DG $\mathcal{O}4$							
N_G	N_d	L_∞	L_1	L_2	$\mathcal{O}_{L_\infty}$	$\mathcal{O}_{L_1}$	$\mathcal{O}_{L_2}$
10	1,000	3.8283E-02	1.0800E-01	2.2654E-02			
15	2,250	3.8778E-03	1.8743E-02	3.94E-03	5.6	4.3	4.3
20	4,000	1.1365E-03	4.7590E-03	9.5647E-04	4.3	4.8	4.9
30	9,000	1.4403E-04	6.4132E-04	1.3806E-04	5.1	4.9	4.8

Table 23. Numerical convergence rates for the nonlinear ADER-DG $\mathcal{O}5$ scheme (P4) on the Cartesian grid.

Nonlinear ADER-DG $\mathcal{O}5$							
N_G	N_d	L_∞	L_1	L_2	$\mathcal{O}_{L_\infty}$	$\mathcal{O}_{L_1}$	$\mathcal{O}_{L_2}$
10	1,500	7.9368E-03	2.4853E-02	5.4678E-03			
15	3,375	5.2042E-04	2.5758E-03	4.3588E-04	6.7	5.6	6.2
20	6,000	1.1981E-04	5.9624E-04	9.3568E-05	5.1	5.1	5.3
25	9,375	4.4390E-05	1.6765E-04	2.8067E-05	4.4	5.7	5.4

Table 24. Numerical convergence rates for the nonlinear ADER-DG $\mathcal{O}6$ scheme (P5) on the Cartesian grid.

Nonlinear ADER-DG $\mathcal{O}6$							
N_G	N_d	L_∞	L_1	L_2	$\mathcal{O}_{L_\infty}$	$\mathcal{O}_{L_1}$	$\mathcal{O}_{L_2}$
5	525	3.6304E-02	2.1472E-01	4.2824E-02			
10	2,100	6.7181E-04	6.5338E-03	8.6322E-04	5.8	5.0	5.6
15	4,725	8.3470E-05	7.1073E-04	9.7455E-05	5.1	5.5	5.4
20	8,400	2.1026E-05	9.5374E-05	1.41E-05	4.8	7.0	6.7

References

[1] H. Atkins and C.-W. Shu, Quadrature-free implementation of the discontinuous Galerkin method for hyperbolic equations, *AIAA Journal* 36 (1998), 775–782.

[2] J. C. Butcher, The Numerical Analysis of Ordinary Differential Equations: Runge–Kutta and General Linear Methods, Wiley, Chichester 1987.

[3] B. Cockburn, C.-W. Shu, TVB Runge-Kutta local projection discontinuous Galerkin finite element method for conservation laws II: general framework, *Math. Comp.* 52 (1989), 411–435.

[4] B. Cockburn, S.-Y. Lin, C.-W. Shu, TVB Runge-Kutta local projection discontinuous Galerkin finite element method for conservation laws III: one dimensional systems, *J. Comput. Phys.* 84 (1989), 90–113.

[5] B. Cockburn, S. Hou, C.-W. Shu, The Runge-Kutta local projection discontinuous Galerkin finite element method for conservation laws IV: the multidimensional case, *Math. Comp.* 54 (1990), 545–581

[6] B. Cockburn, C.-W. Shu, The Runge-Kutta discontinuous Galerkin method for conservation laws V: multidimensional systems, *J. Comput. Phys.* 141 (1998), 199–224

[7] B. Cockburn, M. Luskin, C.-W. Shu and E. Suli, Enhanced accuracy by post-processing for finite element methods for hyperbolic equations, *Math. Comp.* 72 (2003), 577–606.

[8] R. Courant, E. Isaacson, M. Rees, On the solution of nonlinear hyperbolic differential equations by finite differences, *Comm. Pure Appl. Math.* 5 (1952), 243–255.

[9] M. Dumbser, On the Improvement of Efficiency and Storage Requirements of the Discontinuous Galerkin Method for Aeroacoustics, *PAMM* 3 (2003), 426–427.

[10] M. Dumbser, C.-D. Munz, ADER Discontinuous Galerkin Schemes for Aeroacoustics, *Comptes Rendus Mécanique*, to appear.

[11] C. Hu, C.-W. Shu, Weighted essentially non-oscillatory schemes on triangular meshes, *J. Comput. Phys.* 150 (1999), 97–127.

[12] G. E. Karniadakis, S. J. Sherwin, *Spectral/hp Element Methods in CDF*, Numer. Math. Sci. Comput., Oxford University Press, New York 1999.

[13] J. Qiu, C.-W. Shu, Hermite WENO schemes and their application as limiters for Runge-Kutta discontinuous Galerkin method: one-dimensional case, *J. Comput. Phys.* 193 (2003), 115–135.

[14] J. Qiu, M. Dumbser, C.-W. Shu, The Discontinuous Galerkin Method with Lax-Wendroff Type Time Discretizations, *Comput. Methods Appl. Mech. Engrg.*, to appear.

[15] J. K. Ryan, C.-W. Shu, H.L. Atkins, Extension of a post-processing technique for the discontinuous Galerkin method for hyperbolic equations with applications to an aeroacoustic problem, *SIAM J. Sci. Comput.*, to appear.

[16] T. Schwartzkopff, C.-D. Munz, E. F. Toro, ADER: A high-order approach for linear hyperbolic systems, *J. Sci. Comput.* 17 (2002), 231–240.

[17] T. Schwartzkopff, M. Dumbser, C.-D. Munz, Fast high order ADER schemes for linear hyperbolic equations, *J. Comput. Phys.* 197 (2004), 532–39 .

[18] A. H. Stroud, *Approximate Calculation of Multiple Integrals*, Prentice-Hall Ser. Automatic Comput., Prentice-Hall Inc., Englewood Cliffs, New Jersey, 1971.

[19] V. A. Titarev, E. F. Toro, ADER: Arbitrary High Order Godunov Approach, *J. Sci. Comput.* 17 (2002), 609–618.

[20] E. F. Toro, *Riemann Solvers and Numerical Methods for Fluid Dynamics*, Springer-Verlag, Berlin 1999.

[21] E. F. Toro, R. C. Millington and L. A. M. Nejad, Towards very high–order Godunov schemes, In *Godunov Methods: Theory and Applications* (E. F. Toro, ed.), Kluwer Academic/Plenum Publishers, New York 2001, 905–937

[22] E. F. Toro, V. A. Titarev, Very-High Order Godunov-Type Schemes for Non-Linear Scalar Conservation Laws, *Proceedings of the ECCOMAS Computational Fluid Dynamics Conference*, Swansea, UK, 2001.

[23] E. F. Toro, V. A. Titarev, Solution of the generalized Riemann problem for advection-reaction equations, *Proc. Roy. Soc. London* 458 (2002), 271–281.

The multiple pressure variables method for fluid dynamics and aeroacoustics at low Mach numbers

C.-D. Munz[1], *M. Dumbser*[1] *and M. Zucchini*[2]

[1] *Institut für Aerodynamik und Gasdynamik, Pfaffenwaldring 21*
70550 Stuttgart, Germany
email: munz@iag.uni-stuttgart.de
michael.dumbser@iag.uni-stuttgart.de

[2] *Robert Bosch GmbH, Corporate Research and Development*
Department of Applied Physics, 70839 Gerlingen, Germany
email: marco.zucchini@de.bosch.com

Abstract. The incompressible limit of a compressible fluid, when the Mach number tends to zero, is a singular limit due to the fact that the governing equations change their type. The propagation rate of the pressure waves become infinite. Based on asymptotic results the multiple pressure variables (MPV-) method is reviewed as it may be used for the construction of numerical methods in this low Mach number regime. The MPV-schemes formally converge to incompressible methods, if the Mach number tends to zero. The insight into the behavior in the incompressible limit is used to introduce a perturbation method to calculate the noise generation and propagation in the flow region. In this approach the fluid flow is calculated by an incompressible method. The introduction of a perturbation about this incompressible solution leads to a system describing the generation and propagation of the compressible corrections to the incompressible flow field. The wave propagation model are the linearized Euler equations and the noise generation is given by source terms for these equations. This relates very well to the EIF-approach (Expansion about Incompressible Flow) as proposed by Hardin and Pope. The new part is a scaling motivated by the MPV-method and $M = 0$ limit. By that a linearization is introduced and the most important terms in the sources can easily be identified. For a simple one-dimensional problem with compression from the boundary the perturbation method is validated. Numerical results are also presented for the noise generation by a rotating vortex pair and for the applied problem of the noise generation of a plane jet.

1 Introduction

When the Mach number tends to zero, the compressible equations converge to their incompressible counterpart. Within this limit the pressure waves become infinitely fast with respect to the fluid velocity and an immediate pressure equalization will take place. Hence, local gradients in the velocity field can not longer generate large pressure gradients. Subsequently no large density gradient is generated. Due to that fact that wave velocities become infinitely large this incompressible limit is mathematically a rather subtle one. In the inviscid case the pure hyperbolic compressible equations become hyperbolic-elliptic. Using asymptotic analysis the incompressible limit of a compressible flow has been considered by Klainerman and Majda for the isentropic case and they gave a mathematically rigorous justification. Later Klein extended the formal asymptotic considerations to the non-isentropic case and to multiple scales. He also gives a short survey and references about the previous work in this flow regime.

Due to this singular limit the numerical simulation of fluid flow in the low Mach number regime becomes difficult. The numerical schemes designed for highly compressible flow such as finite volume shock-capturing schemes run into problems. For non-stationary problems usually explicit time integration methods are used. When the Mach number tends to zero, the wave speeds associated with the propagation of the pressure waves become infinite and the time step restriction for explicit numerical schemes becomes very restrictive. Preconditioning methods have been proposed to eliminate the stiffness in the low Mach number regime by reducing the fast wave speeds. For non-stationary problems a dual time-stepping has to be introduced which strongly reduces the efficiency of the methods. But also implicit methods that are not affected by time step restrictions proportional to the Mach number may fail at low Mach numbers. They may produce physically wrong solutions. Beside the increasing stiffness of the equations the other difficulty is the pressure term itself. For Mach number zero the pressure is not longer a thermodynamic but a pure hydrodynamic variable. The incompressible equations do not depend on any equation of state. The numerical method has to be consistent with that limit behavior. We will explain this fact in this paper in more detail. After the introduction of a non-dimensionalization of the governing equations in Section 2 we give a review about the incompressible limit of a compressible flow in Section 3. We also consider the case when temperature gradients and heat conduction occurs and derive the zero Mach number equations.

The main difficulty in the calculation of sound generated by fluid flow at low Mach numbers is the occurrence of quite different scales. While the fluid flow may be affected by small fluid structures containing large energy, such as small vortices in a turbulent flow, the acoustic waves are phenomena of low energy with long wavelengths that may travel over long distances. These different scales and different physical behaviors of fluid flow and sound propagation lead often to the situation that the numerical methods can not solve both together. The inherent numerical dissipation of the CFD-solver necessary to capture the strong gradients in the fluid flow may damp out the acoustic waves within a small distance or the numerical errors produce more noise than the physics. In the case, when the compressible effects only slightly influence the flow, the better way seems to be to solve the incompressible equations to obtain the flow field. The noise propagation is solved by acoustic equations with source terms given from the flow field. This is nowadays called hybrid approach and includes acoustic analogies and perturbation methods. In this paper we consider the perturbation approach as first proposed by Hardin and Pope. We use the low Mach number asymptotic results to introduce a scaling of the approach and to consider the case with heat conduction and compression from the boundary.

2 Non-dimensionalization in the low Mach number regime

For moderate and high Mach numbers the non-dimensionalization of the compressible Navier–Stokes equations is usually performed using the basic reference values: length, density and fluid velocity. All other reference values are determined from these via their functional relationship. If the Mach number as the quotient of the fluid and sound velocities tends to zero and hence the sound velocity tends to infinity with respect to the fluid velocity, then the choice of different basic reference values for the two different velocities becomes favorable. We do this by using four basic reference values

$$x_{\mathrm{ref}},\ \rho_{\mathrm{ref}},\ v_{\mathrm{ref}} \quad \text{and} \quad p_{\mathrm{ref}} \tag{2.1}$$

for length, density, fluid velocity and in addition the pressure, respectively. We emphasize that the number

$$M = \frac{v_{\mathrm{ref}}}{c_{\mathrm{ref}}}, \tag{2.2}$$

called the global Mach number, characterizes the compressibility of the fluid flow. Note that the speed of sound is $c = \sqrt{(\gamma p)/\rho}$. Here, we simply take $c_{\mathrm{ref}} := \sqrt{p_{\mathrm{ref}}/\rho_{\mathrm{ref}}}$.

The compressible Navier–Stokes equations for a perfect gas, non-dimensionalized in this way, read in conservation form as

$$\rho_t + \nabla \cdot (\rho v) = 0, \tag{2.3}$$

$$v_t + \nabla((\rho v) \circ v) + \frac{1}{M^2} \nabla p = \frac{1}{\mathrm{Re}} \nabla \cdot \tau, \tag{2.4}$$

$$e_t + \nabla \cdot (v(e + p)) = \frac{1}{\mathrm{Re}} \nabla \cdot (\tau v) - \frac{1}{\mathrm{Pe}} \nabla \cdot q, \tag{2.5}$$

where ρ, v, p, e, τ and q denote the density, velocity, pressure, total energy per unit volume, the viscous stress tensor and the heat flux, respectively. These equations represent the conservation of mass, momentum and energy. They are closed by an equation of state $p = p(\rho, \epsilon)$ with the specific internal energy ϵ. We assume the equation of state of a perfect gas to be valid:

$$p = (\gamma - 1)\rho\epsilon = \rho T, \tag{2.6}$$

where T denotes the temperature and the non-dimensionalization is performed such that the reference values satisfy

$$\frac{p_{\text{ref}}}{\rho_{\text{ref}}} = RT_{\text{ref}} = (c_p - c_v)T_{\text{ref}} = \frac{\gamma - 1}{\gamma} c_p T_{\text{ref}}. \tag{2.7}$$

The total energy consists of the internal, the kinetic and the potential energy:

$$e = \rho\epsilon + \frac{M^2}{2}\rho v^2. \tag{2.8}$$

The viscous stress tensor τ is given in the usual way for a Newtonian fluid and depends linearly on the velocity space derivatives. The heat flux is for simplicity assumed to be

$$q = \kappa \nabla T, \tag{2.9}$$

with a constant heat conduction coefficient κ.

The Mach number M, the Reynolds number Re, and the Peclet number Pe appear in the equations as global dimensionless characteristic quantities being measures for compressibility, for viscosity, and heat conduction, respectively.

It is often convenient to rewrite the compressible Navier–Stokes equations in primitive variables:

$$\rho_t + \nabla \cdot (\rho\, v) = 0, \tag{2.10}$$

$$v_t + (v \cdot \nabla)v + \frac{1}{\rho M^2}\nabla p = \frac{1}{\rho \,\text{Re}}\nabla \cdot \tau, \tag{2.11}$$

$$p_t + v \cdot \nabla p + \gamma p \nabla \cdot v = -\frac{\gamma}{\text{Pe}}\nabla \cdot q + \frac{M^2(\gamma - 1)}{\text{Re}}(\nabla \cdot (\tau\, v) - v\nabla \cdot \tau). \tag{2.12}$$

In the following considerations we first neglect for simplification viscous effects, gravity, and the influence of the heat flux. As governing equations then we have the Euler equations written in the primitive variables as

$$\rho_t + \nabla \cdot (\rho\, v) = 0, \tag{2.13}$$

$$v_t + (v \cdot \nabla)v + \frac{1}{\rho M^2}\nabla p = 0, \tag{2.14}$$

$$p_t + v \cdot \nabla p + \gamma p \nabla \cdot v = 0. \tag{2.15}$$

The factor $1/M^2$ in (2.14), (2.11), or (2.4) clearly shows the singular behavior of the incompressible limit. The term $\nabla p/M^2$ has to converge to the gradient of the "incompressible" pressure. This pressure then does not have the role of a thermodynamic variable anymore. More insight into this limit behavior is obtained by the asymptotic considerations in the next section.

3 The incompressible limit of a compressible fluid flow

The incompressible limit of a compressible fluid flow is characterized by a velocity of sound that tends to infinity with respect to the velocity of the fluid flow. This means that the pressure waves become very fast, an immediate pressure equalization takes place and the pressure becomes nearly constant. Hence, gradients in the fluid velocity cannot longer generate strong pressure gradients that lead to changes in density: The fluid flow becomes incompressible in the limit.

As the small parameter M appears explicitly in the governing system of partial differential equations and tends to zero in the incompressible limit, an asymptotic series in powers of the Mach number is a good candidate to get insight into the physical behavior. We assume that the solution of the system can be represented by the following expansion:

$$\rho(x,t) = \rho^{(0)} + M\rho^{(1)} + M^2\rho^{(2)} + O(M^3), \tag{3.1}$$

$$v(x,t) = v^{(0)} + Mv^{(1)} + M^2v^{(2)} + O(M^3), \tag{3.2}$$

$$p(x,t) = p^{(0)} + Mp^{(1)} + M^2p^{(2)} + O(M^3). \tag{3.3}$$

These expansions are substituted into the equations (2.10)–(2.12). The unknown expansion functions $\rho^{(i)}$, $v^{(i)}$, $p^{(i)}$ with $i = 0, 1, 2, \ldots$ depend on the space variable x and the time variable t. Collecting all terms multiplied by equal powers of the Mach number M and separately equating these to zero gives a hierarchy of asymptotic equations for the various expansion functions.

The goal of the following analysis is to discuss the asymptotic equations and by this to get insight in the behavior of the system for small and zero Mach number. In this section we investigate especially the equations for the expansion functions at leading order. The equations of the incompressible limit $M = 0$ is our main interest in this section.

In the velocity equations terms of order M^{-2} and M^{-1} appear and thus we obtain as first conditions

$$\nabla p^{(0)}(x,t) = 0, \quad \nabla p^{(1)}(x,t) = 0, \tag{3.4}$$

from which it follows that $p^{(0)}$ as well as $p^{(1)}$ depend on time only:

$$p^{(0)} = p^{(0)}(t), \quad p^{(1)} = p^{(1)}(t). \tag{3.5}$$

Next, we consider the pressure equation of leading order and use (3.3) to obtain the equation

$$p_t^{(0)} + \gamma p^{(0)}\nabla\cdot v^{(0)} = \frac{\gamma-1}{\mathrm{Pe}}\nabla\cdot q^{(0)}. \tag{3.6}$$

This equation is integrated over the whole domain Ω of the fluid flow:

$$p_t^{(0)} \int_{\Omega} d\Omega + \gamma\, p^{(0)} \int_{\partial\Omega} v^{(0)} \cdot n\, dS = \frac{\gamma - 1}{\mathrm{Pe}} \int_{\partial\Omega} q^{(0)} \cdot n\, dS. \tag{3.7}$$

Here, we applied the Gaussian theorem, where n denotes the unit normal, and used (3.5).

Let us first consider the case, when heat conduction can be neglected and no compression from the boundary $\partial\Omega$ occurs, i.e.,

$$\int_{\partial\Omega} v^{(0)} \cdot n dS = 0, \tag{3.8}$$

often called the incompressibility constraint. No volume flux through the boundary of Ω appears. In this case, the integrated pressure equation (3.7) immediately gives that $p^{(0)}$ is constant in time. Without heat conduction it follows then from (3.6) that the velocity field has zero divergence. Together with the zeroth order continuity and velocity equations we get the basic equations

$$\rho_t^{(0)} + v^{(0)} \cdot \nabla \rho^{(0)} = 0, \tag{3.9}$$

$$v_t^{(0)} + (v^{(0)} \cdot \nabla) v^{(0)} + \frac{1}{\rho^{(0)}} \nabla p^{(2)} = \frac{1}{\rho^{(0)}\mathrm{Re}} \nabla \tau^{(0)}, \tag{3.10}$$

$$\nabla \cdot v^{(0)} = 0. \tag{3.11}$$

These are the equations of incompressible fluid flow with variable density. If the density is constant initially, then it is constant for all times. In this case the density equation cancels out.

We shortly discuss the role of the pressure in this singular limit. Two pressure terms $p^{(0)}$ and $p^{(2)}$ survive in the limit $M \to 0$, the pressure term $p^{(1)}$ in the pressure expansion has the same behavior as $p^{(0)}$ and is suppressed in the following. The leading order pressure $p^{(0)}$ becomes constant in space and time. It satisfies the equation of state for $M = 0$ and may be called the thermodynamic pressure. The second order pressure term $p^{(2)}$ is the agent to balance the inertial forces and to establish the divergence free condition of the velocity. It is called the hydrodynamic pressure. In the velocity equations (2.4), (2.11) and (2.14) we formally get

$$\frac{1}{M^2} \nabla p \to \nabla p^{(2)} \quad \text{for } M \to 0. \tag{3.12}$$

This corresponds quite well to the physical situation. The pressure wave propagation becomes infinite with respect to the fluid velocity and a fast pressure equalization takes place. This means that the pressure in the compressible equations tends to the thermodynamic background pressure $p^{(0)}$ in the limit. Due to the fast pressure waves velocity gradients in the flow cannot longer generate large pressure gradients and no density changes occur. According to (2.8) the total energy equals the internal energy. The velocity fluctuations are very small and do not contribute to the total energy.

In the case of outer compression and heat conduction the divergence-free constraint (3.11) must be replaced according to the leading order pressure equation (3.6) by

$$\nabla \cdot v^{(0)} = -\frac{1}{\gamma\, p^{(0)}} p_t^{(0)} + \frac{\gamma}{\mathrm{Pe}\, p^{(0)}} \nabla \cdot q^{(0)}. \tag{3.13}$$

The continuity equation at leading order now reads

$$\rho_t^{(0)} + v^{(0)} \cdot \nabla \rho^{(0)} = -\rho^{(0)} \nabla \cdot v^{(0)}. \tag{3.14}$$

The velocity equation (3.10) remains unchanged.

The system (3.14), (3.10), and (3.13) is closed by the evolution equation for $p^{(0)}$ which is derived from integrated pressure equation (3.7) as

$$p_t^{(0)} = -\frac{\gamma\, p^{(0)}}{|\Omega|} \int_{\partial\Omega} v^{(0)} \cdot n\, dS + \frac{\gamma}{\mathrm{Pe}\, |\Omega|} \int_{\partial\Omega} q^{(0)} \cdot n\, dS. \tag{3.15}$$

The situation is now as follows. Compression from the boundary, when the incompressibility constraint (3.8) is not valid, leads to a background pressure which is constant in space but a function in time. In this case and further neglecting heat conduction, the divergence of the velocity is constant but non-zero in space and a function in time. The density varies in time as given by (3.14).

A rigorous mathematical justification of such an analysis under several constraints on the initial data is given by Klainerman and Majda in [8] and Majda and Sethian in [12]. For a barotropic flow $p = p(\rho)$ they show that compressible flow in fact converges to incompressible flow as the Mach number vanishes, if the pressure is uniform initially, except for perturbations of $O(M^2)$ and a single characteristic length scale governs the initial data. Klein [9] extended the formal asymptotic analysis to the compressible equation of a perfect gas with variable entropy and to initial data with long-wavelength pressure perturbations of order $O(M)$ introducing multiple length scales.

4 The multiple pressure variables method for weakly compressible flow

As shown in the previous section, the relationship between compressible and incompressible equations is difficult due to the fact that the equations change their type and wave velocities become infinite. All explicit methods necessarily run into trouble due to the CFL-stability condition. The time step necessarily tends to zero and the computational efforts increase when the Mach number converges to zero. Beside this stiffness of the equations in the zero Mach number limit the other problem that occurs is an accuracy problem. All compressible methods based on the total compressible pressure may run into accuracy problems for small Mach numbers, because they have to capture the hydrodynamic pressure that tends to zero with the power M^2 with respect to the thermodynamic pressure term. Hence, rounding errors may strongly influence the numerical results. In practical calculations it is observed that unphysical results are produced by unmodified compressible methods. Several methods were designed to handle low Mach number fluid flow. These are extensions or modifications of numerical methods either for the incompressible equations, often called pressure-based methods or for the compressible equations, often called density-based methods.

The density-based methods often use a preconditioning of the equations. The principle idea has its origin in a method for the incompressible equations: Chorin proposed in [2] to replace the $\nabla \cdot v = 0$ equation in the incompressible equation by

$$p_t + c^2 \nabla \cdot v = 0 \tag{4.1}$$

with a constant c. By doing this, a direct pressure velocity coupling has been introduced and the whole system becomes hyperbolic in the inviscid case. The infinite propagation rate of the pressure is replaced by the finite constant rate c. Hence, the equations are modified by some sort and Chorin called it the method of artificial compressibility. In the steady state the time derivatives vanish and the modified equations equal to the stationary incompressible Navier–Stokes equations.

For the compressible equations the multiplication of the vector containing the time derivatives by a matrix P^{-1} is performed to introduce a preconditioning of the equations. A survey about preconditioning is given by Turkel, Fiterman, and van Leer [22], Darmofal and van Leer [3] and Weiss and Smith [23]. It turned out that some of the coefficients of the preconditioning matrix should be proportional to M^2. Otherwise the numerical dissipation of the schemes dominate the results at low Mach numbers. This was confirmed by Guillard and Viozat [5] using the low Mach number asymptotic results. They showed that the numerical dissipation is proportional to M only and hence increases with M in comparison to the hydrodynamic pressure. Guillard and Viozat then proposed an implicit scheme with the preconditioning of the numerical dissipation only. Using dual time stepping this method of preconditioning may be extended to the simulation of time accurate problems, but needs much computational effort.

The so called pressure correction methods for incompressible flow have their origins in the SIMPLE method (Semi-implicit method for pressure linked equations) of Patankar and Spalding [16] for the incompressible Navier–Stokes equations. A detailed description is given in the book of Patankar [15]. This method became very common for the numerical simulation of practical problems for incompressible fluid flow. This incompressible pressure correction method has been extended to low Mach number compressible fluid flow by Karki and Patankar [7] in the case of a homentropic equation of state. The idea in their approach is to replace the divergence-free condition by the full continuity equation. The density is replaced via the homentropic equation to obtain a pressure correction equation of the state $p = p(\rho)$.

This concept has been extended to general fluid flow with density variations by several authors (see [17], [19]). Again the elliptic pressure equation is obtained from the continuity equation in which the density is eliminated via the equation of state. In this case, of course, an unknown variable remains, because the pressure is a function of both the density and the internal energy. The usual way is to handle this problem iteratively. In each iteration cycle a pressure correction is calculated at first by the same procedure as in the homentropic case with an internal energy fixed at the old iteration level. After applying the pressure and velocity correction, the transport equation for the internal energy is solved in an explicit way. Due to the fact that these schemes have an implicit time approximation at least for all the pressure terms the stiffness in the equations is eliminated, but the accuracy problem may still be present, because the full pressure is used.

To handle the stiffness as well as the accuracy problem, Klein and Munz proposed in [10] and [14] to introduce the decomposition of the pressure terms into the numerical modeling and called this the multiple pressure variables approach. They introduced a numerical approximation of the pressure decomposition

$$p = p^{(0)} + M^2 p^{(2)}. \tag{4.2}$$

We assume that the Mach number M is small such that $p^{(0)}$ becomes constant in space. In this case the thermodynamic pressure can be obtained as the pressure p averaged over the whole computational domain

$$p^{(0)} = \bar{\bar{p}} := \frac{1}{|V|}\int_V p(x)dx. \tag{4.3}$$

The pressure $p^{(2)}$ is then given according to the consistency with (4.2) by

$$M^2 p^{(2)} = p - p^{(0)}. \tag{4.4}$$

Multiple pressure variables are introduced by applying the averaging operator in (4.3) to the discrete values. The basic philosophy of the MPV-approach is indicated by discussing the approximation of the term $\nabla p/M^2$ via the decomposition

$$\frac{1}{M^2}\nabla p = \frac{1}{M^2}\nabla p^{(0)} + \nabla p^{(2)}. \tag{4.5}$$

By construction, the thermodynamic pressure term is constant in space, so $\nabla p^{(0)} = 0$. The idea in the multiple pressure variables approach is now to replace the approximation of $\nabla p/M^2$ in the velocity equations by an approximation of (4.5) where the singularity is eliminated and which converges towards $\nabla p^{(2)}$.

If this pressure decomposition is introduced into the compressible Navier–Stokes equations in their primitive formulation, the velocity and the pressure equations read as

$$v_t + (v\cdot\nabla)v + \frac{1}{\rho}\nabla p^{(2)} = \frac{1}{\rho\,\mathrm{Re}}\nabla\cdot\tau, \tag{4.6}$$

$$\begin{aligned} p_t^{(2)} + v\cdot\nabla p^{(2)} + \gamma p\nabla\cdot v = {} & -p_t^{(0)} - \frac{\gamma}{\mathrm{Pe}}\nabla\cdot q \\ & + \frac{M^2(\gamma-1)}{\mathrm{Re}}(\nabla\cdot(\tau\, v) - v\nabla\cdot\tau). \end{aligned} \tag{4.7}$$

The density equation remains unchanged. The pressure term $p^{(0)}$ is constant in space. The time development is directly given from (3.15) as

$$p_t^{(0)} = -\frac{\gamma\, p^{(0)}}{|\Omega|} \int_{\partial\Omega} v \cdot n\, dS + \frac{\gamma}{\mathrm{Pe}\, |\Omega|} \int_{\partial\Omega} q \cdot n\, dS. \tag{4.8}$$

This system is equivalent to the compressible Navier–Stokes equations in their primitive formulation for any value of M. The advantage is that when M tends to zero, all terms multiplied by M^2 vanish. Furthermore, the total pressure p equals $p^{(0)}$ and becomes constant in space and ∇p becomes zero. If we have no compression from the boundaries, then $p_t^{(0)} = 0$, and (4.7) formally converges to the divergence-free constraint of the incompressible Navier–Stokes equations.

If a numerical method is applied to this formulation, it formally converges to an incompressible method. Of course, the stiffness of that limit equations has to be handled by the corresponding numerical approximation, but the accuracy problem is eliminated by that re-scaling. It becomes obvious from these considerations that it is necessary to introduce some sort of pressure decomposition to obtain the correct limit. The incompressible pressure correction methods have been extended to the weakly compressible regime in [14] by applying the basic ideas to the system (2.10), (4.6)–(4.8). A similar procedure introducing such a technique of re-scaling has been proposed by Bijl and Wesseling in [1] and by Demirdzic et al. in [4]. The MPV-method may also be applied to the conservation form of the compressible Navier–Stokes equations. Again, the pressure term $p^{(2)}$ becomes a primary variable. Within the formulation of the numerical method the conservation property is preserved.

5 Acoustic perturbation equations based on the MPV-approach

In the previous section, we considered the numerical approximation of the fluid flow at low Mach numbers. The generation of the sound by the flow is mathematically modeled by the compressible equations. Hence, a compressible flow calculation should give the acoustic noise, too. But due to the fact that in the low Mach number regime the amplitudes of the acoustic pressure fluctuations are much smaller than those of the hydrodynamic pressure, there is often no chance to capture the aeroacoustic noise within a flow calculation in an appropriate way. Numerical errors may produce even more noise than the physics. If the fluid flow is nearly incompressible and not influenced by sound waves, the numerical simulation of the flow and the sound may be calculated separately.

Here, the calculation of acoustic wave propagation is based on acoustic wave equations with source terms determined by the fluid flow. In an acoustic analogy as introduced by Sir Lighthill [11] the acoustic wave propagation is determined by the simple linear wave equation in a non-moving medium as valid in the far field with source terms determined in the flow region. In or in the vicinity of the flow filed a perturbation method seems to be a better approach. Here, the interaction of the sound propagation with the flow field is taken into account. In this section we reconsider the EIF approach proposed by Hardin and Pope [6]. The asymptotic results are used to introduce a scaling into these equations to identify the main terms. We start with the inviscid equations and then switch to those with viscosity and heat conduction.

Using the scaling motivated by the asymptotic considerations above we introduce the following perturbation ansatz:

$$\rho = \rho^{(0)} + M^2 \left(\rho_{\text{inc}} + \rho'\right), \tag{5.1}$$

$$v = {}' v_{\text{inc}} + M\, v', \tag{5.2}$$

$$p = p^{(0)} + M^2 \left(p_{\text{inc}} + p'\right). \tag{5.3}$$

The functions $\rho^{(0)}$ and $p^{(0)}$ again denote the thermodynamic part of the density and pressure. We first consider the simplest case and assume the flow to be isentropic: $p = p(\rho)$ and the temperature as well as the background density to be constant. The variables v_{inc} and p_{inc} denote the incompressible solution as given by the incompressible Euler equations. The function ρ_{inc} does not appear in the incompressible equations and is introduced as the change of density by the hydrodynamic pressure and is defined to be

$$p_{\text{inc}} = c_0^2 \rho_{\text{inc}}\,, \tag{5.4}$$

where c_0 denotes the background sound velocity. We will discuss this term later in more detail. The prime quantities are introduced as the compressible corrections of the hydrodynamic flow field.

This ansatz is substituted into the compressible Euler equations (2.13)–(2.15). After some reformulation we obtain

$$\rho'_t + v_{\text{inc}} \nabla \cdot \rho' + \frac{\rho^{(0)}}{M} \nabla v' \;=\; -\left((\rho_{\text{inc}})_t + v_{\text{inc}} \nabla \rho_{\text{inc}}\right), \tag{5.5}$$

$$v'_t + \nabla \cdot \left(v_{\text{inc}} \circ v' + v' \circ v_{\text{inc}}\right) + \frac{1}{M\,\rho^{(0)}} \nabla p' \;=\; 0, \tag{5.6}$$

$$p'_t + v_{\text{inc}} \nabla p' + \frac{\gamma p^{(0)}}{M} \nabla \cdot v' \;=\; -\left((p_{\text{inc}})_t + v_{\text{inc}} \nabla p_{\text{inc}}\right). \tag{5.7}$$

The system on the left hand side are linearized Euler equations. It states a linear hyperbolic system of linear wave equations with the wave speeds $v_{\text{inc}} \pm c_0/M$ and v_{inc}. On the right hand side source terms appear that are determined by the incompressible solution itself. There appear some additional terms, especially with products of primed quantities and their derivatives. But, because they are of less order and have a factor M in front, they are neglected within these considerations.

The density equation and the pressure equation may be combined to give

$$\left(p' - c_0^2 \rho'\right)_t + v^{(0)} \nabla \left(p' - c_0^2 \rho'\right) = 0, \tag{5.8}$$

showing that the wave propagation is isentropic. In this case the linearized system may be reduced to two equations. If additionally $v_{\text{inc}} \equiv 0$ in the whole domain, then the usual linear acoustic wave equation is obtained by combining both equations:

$$p'_{tt} - \frac{\gamma p^{(0)}}{\rho^{(0)} M^2} \Delta p' = -(p_{\text{inc}})_{tt}. \tag{5.9}$$

The perturbation ansatz corresponds to that of Hardin and Pope. They also considered the nonlinear terms and nonlinear wave propagation equations. We only introduced the scaling with respect to the Mach number as motivated by the asymptotics. But, this gives information about the most important terms for small Mach numbers and introduces the linearization by neglecting the higher order terms. There was a discussion about the introduction of the change of the density due to the hydrodynamic pressure as proposed by Hardin and Pope [6] and also considered here in the equation (5.4). The density equation (5.5) clearly shows that this is necessary to get in the isentropic case the proper relation (5.8). We will also show results for a simple example that support these considerations. If viscous terms, heat conduction and compression from the boundary may occur, then the incompressible solution in the perturbation ansatz has to be replaced by that solution for the equations at $M = 0$ taking these effects into account.

We consider the case of fluid flow with viscosity and compression from the boundary. The different terms $\rho^{(0)}$, $p^{(0)}$, $v^{(2)}$, and $p^{(2)}$ in the perturbation ansatz have now to be the solutions of the equations at $M = 0$ with viscosity and compression from the boundary. The corresponding system describing the acoustic wave propagation now reads as

$$\rho'_t + \nabla \cdot \left(v^{(0)} \rho'\right) + \frac{\rho^{(0)}}{M} \nabla v' = -\left(\rho_t^{(2)} + \nabla \cdot v^{(0)} \rho^{(2)}\right), \tag{5.10}$$

$$v'_t + \nabla \cdot \left(v^{(0)} \circ v' + v' \circ v^{(0)}\right) + \frac{1}{\rho^{(0)} M} \nabla p' = \frac{1}{\rho^{(0)} \mathrm{Re}} \nabla \cdot \tau^{(1)}, \tag{5.11}$$

$$\begin{aligned} p'_t + v^{(0)} \nabla p' + \frac{1}{M} \gamma p^{(0)} \nabla v' \\ = -\left(p_t^{(2)} + v^{(0)} \nabla p^{(2)} + \gamma p^{(2)} \nabla \cdot v^{(0)} + \gamma p' \nabla \cdot v^{(0)}\right). \end{aligned} \tag{5.12}$$

This system is more complicated due to the fact that the terms with $\nabla \cdot v^{(0)}$ do not cancel out and $\rho^{(0)}$ is a function of time. The expression $\tau^{(1)}$ denotes the $\mathcal{O}(M)$ terms of viscosity.

Other approaches using asymptotic results for low Mach number flows are based on matched asymptotic expansions and are given by Ting and Miksis [21] and Slimon et al. [20].

6 Numerical results

6.1 A one-dimensional non-trivial example

One-dimensional examples are often good first test problems to validate the numerical method proposed. In the case of the noise generation by an incompressible flow field in one space dimension means that the flow becomes trivial with constant velocity. But, if we allow compression from the boundary the situation changes. The equations for Mach number $M = 0$ allow a non-trivial solution that may be calculated analytically.

In our example considered, the boundary values for the velocity are given by

$$v^{(0)}(a,t) = 0.0025\sin(t), \qquad v^{(0)}(b,t) = 0.0025\sin(t+\pi). \tag{6.1}$$

The difference of these values does not vanish. Hence, the incompressibility constraint (3.8) is not satisfied and compression from the boundary occurs. We have the following properties of the flow field: the thermodynamic pressure $p^{(0)}$ is constant in space, but a function of time, the background density $\rho^{(0)}$ is constant in space, but a function of time, the incompressible velocity $v^{(0)}$ is a linear function in space, and the hydrodynamic pressure $p^{(2)}$ is a parabola in space. According to the excitation from the boundary values (6.1) the thermodynamic pressure is given by (4.8) that simplifies in our case to

$$p_t^{(0)} = -\frac{\gamma\, p^{(0)}}{|I|}\big(v^{(0)}(b,t) - v^{(0)}(a,t)\big), \tag{6.2}$$

where I denotes the spatial interval $I = [a,b] = [0,\pi]$. Hence, the x-derivative of the leading order velocity is given by

$$v_x^{(0)} = -\frac{1}{\gamma\, p^{(0)}}\, p_t^{(0)}. \tag{6.3}$$

The hydrodynamic pressure is calculated from the velocity equation

$$v_t^{(0)} + v^{(0)} v_x^{(0)} + \frac{1}{\rho^{(0)}}\, p_x^{(2)} = 0 \tag{6.4}$$

and the density from the continuity equation:

$$\rho_t^{(0)} = -\rho^{(0)} v_x^{(0)}. \tag{6.5}$$

In Figure 1 the thermodynamic pressure is constant in space and the sum with the hydrodynamic pressure is shown at the fixed time $t = 0.5$. This zero Mach number solution is used to calculate the source terms on the right hand side of the acoustic perturbation equations (5.10)–(5.12) using a second order Discontinuous Galerkin method with 100 grid points and CFL $= \frac{1}{6}$. These numerical results are shown in Figure 2. The numerical values for the acoustic pressure fluctuations as function of space for the fixed time $t = 0.5$ are plotted as circles. For comparison the full Euler equations with the corresponding boundary values and boundary conditions are

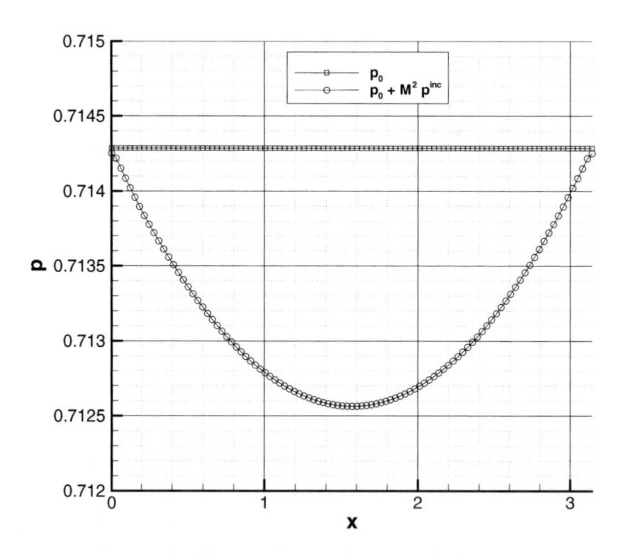

Figure 1. The incompressible solution at $t = 0.5$ of the 1D example.

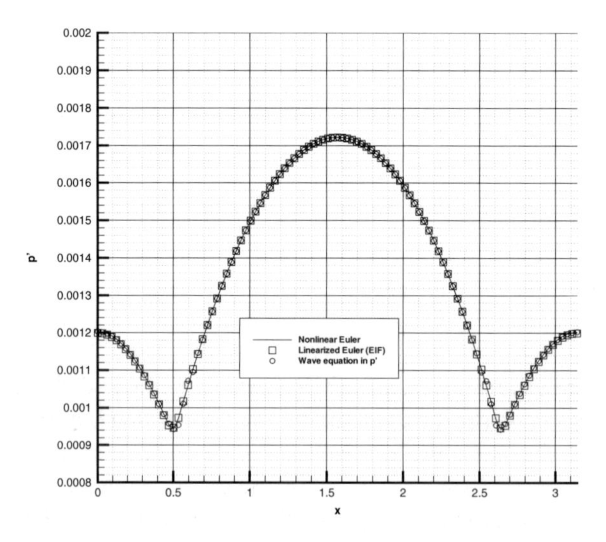

Figure 2. The pressure fluctuation $M^2 p'$ at $t = 0.5$ of the 1D example.

numerically solved using a second order Discontinuous Galerkin compressible method also with 100 grid points at CFL $= \frac{1}{6}$. The difference of the compressible pressure and the hydrodynamic pressure term from the Mach number zero solution is shown as a straight line within this plot. The numerical solutions produced by the different approaches are almost identical. The same conclusion is obtained, of course, if the hydrodynamic pressure and the acoustic pressure is summed-up and is compared directly to the pressure obtained from the compressible calculation (see Figure 3).

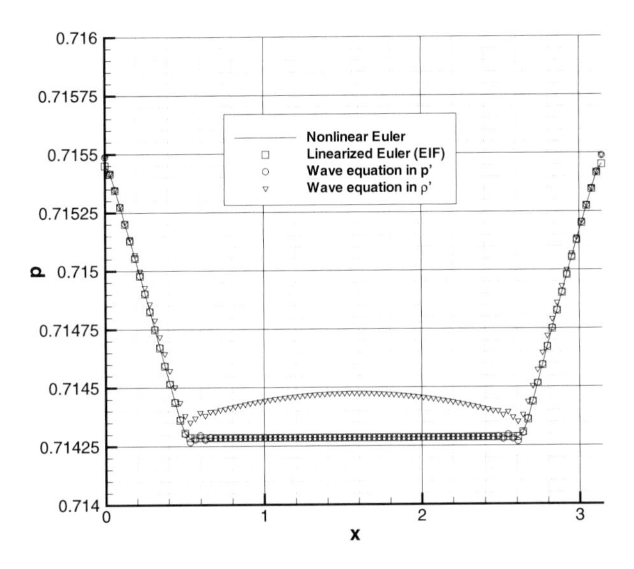

Figure 3. The full compressible pressure $p = p_0 + M^2(p_{\text{inc}} + p')$ at $t = 0.5$ of the 1D example.

In this problem the values of the velocity are small. Hence, the influence of the flow field to the acoustic wave propagation should be small or even negligible. Instead of solving the system of the linearized Euler equations the reduced model as given by the simple wave equation (5.9) assuming zero background flow should produce a good approximation. The numerical values obtained by this model are additionally plotted into the figures Figure 2 and Figure 3. Some very small differences are visible especially at the kinks of the pressure in the Figure 3. There was some discussion in the literature about the term of density change due to the hydrodynamic pressure. The scaled expansion about the incompressible flow indicates that this density term is necessary. Neglecting this term then in the isentropic case, it turns out that the acoustic wave propagation would be non-isentropic. Figure 3 shows also the numerical results where this term is neglected. Here, we solve directly a wave equation for the density fluctuations with the source term $-[\rho^{(0)}(v^{(0)})^2]_{xx}$.

6.2 The co-rotating vortex pair

A typical two-dimensional test case for the validation of a computational aeroacoustic code is the so-called rotating quadrupole. It is generated by a pair of co-rotating vortices of strength $\kappa = \Gamma/(2\pi)$ each, where Γ is the circulation. They are placed at a distance of $2r_0$ and thus each vortex induces a velocity $q = \Gamma/(4\pi r_0)$ on the other. This causes the vortices to rotate around their common midpoint.

For this setting, the exact solution of the potential theory for the incompressible and inviscid flow field as well as the solution of the acoustic far field equations by matched asymptotic expansions can be determined analytically [13]. Therefore, this example is a good test case for the algorithm validation in 2D.

We used the perturbation ansatz about incompressible flow in (5.5)–(5.7) to couple the flow field to the acoustic field. With the analytic solution of the incompressible and inviscid flow field the acoustic source terms on the right hand side are determined. The propagation of the sound waves is then calculated numerically by solving the linearized Euler equations. The numerical results are compared to the analytical acoustic solution. We picked up the case with a vortex distance of $2r_0 = 2.0$ and a "rotating" Mach number of $M = 0.095$ which leads to a rotation period of $T = 2\pi$. The computational domain is set to be $[-400, 400] \times [-400, 400]$ and is discretized by 320×320 grid points, i.e. with grid spacing of $\Delta x = 2.5$. The numerical computations were performed with a finite volume ADER-scheme of second and fourth order in space and time. These schemes are well adopted to linear hyperbolic systems and possess low dispersion and dissipation errors, for further details see [18]. Figure 4 shows the contour plot of the acoustic pressure. The acoustic source is located in the center and the acoustic waves spiral out of the origin.

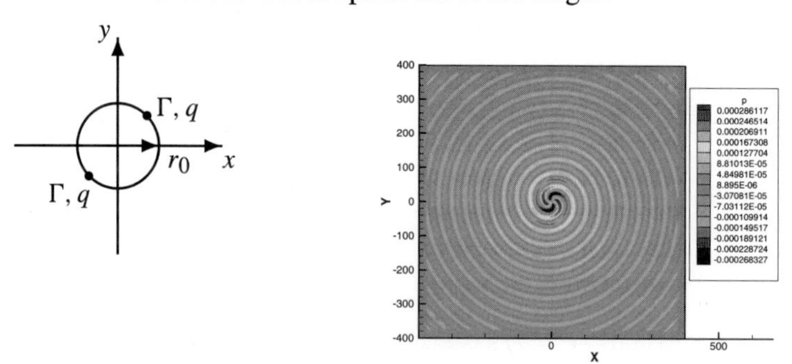

Figure 4. Initial setting and contour plot of the acoustic pressure.

Figure 5 compares the results of the numerical simulation to the analytical solution. Two points on the x-axes have been chosen at distance $x = 150$ and $x = 300$ from the origin. At these points, the acoustic pressure is plotted as function of time. The end time $t \approx 44$ corresponds to about 7 rotational periods. Since the analytic solution is periodic in time and the source terms are calculated from the analytic solution, the amplitude remains constant, no damping of the amplitude should appear within the whole time interval. Figure 5 indicates that there is some difference between the amplitudes of the numerical wave and the analytical one, but it does not increase from point $x = 150$ to $x = 300$. Using the fourth order scheme for the calculation of the sound waves (figure 5), the phase and the dissipation errors are small over this distance. A high order scheme is necessary for computational aeroacoustics to get a good wave propagation over long distances.

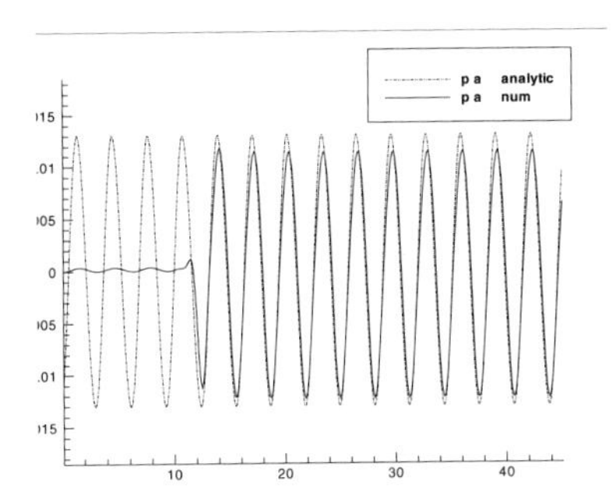

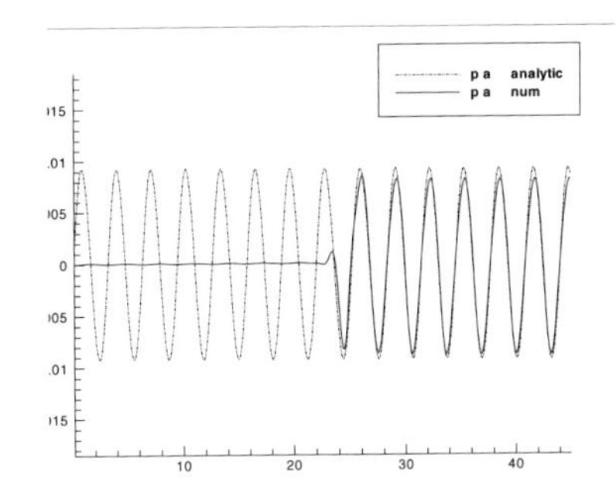

Figure 5. Time development of the acoustic pressure at the fixed points (150,0) and (300,0). Exact solution and numerical solution obtained with a 4th order scheme.

6.3 Noise from a free jet

The scaled perturbation approach has also been applied to a high Reynolds number subsonic free jet (Re $= 2 \cdot 10^5$) with Mach number $M = 0.3$. We assume in the following calculations the jet to be plane and investigated the practical situation with a 2D numerical approach. The simulation is split into two steps:

First: Perform a large eddy incompressible simulation (LES) with the commercial software *Ansys CFX* 5.6 to calculate the unsteady velocity and pressure fields. Second: These data are then used to determine the source terms of the acoustic perturbation equations and to calculate and the propagation of the sound waves with linearized Euler equations (LEE).

The computational domains for the fluid calculation and the noise propagation have been defined to fulfill different requirements: The smallest element size of the CFD model should resolve the turbulence of the shear layers in an appropriate way. The CAA domain is chosen to be larger in order to show the low frequency sound waves. The size of the grid elements is chosen to resolve acoustic waves up to 3000 Hz. This means that the elements of the acoustic grid may be chosen larger than those of the CFD calculation to save computational effort. See Tables 1 and 2 for more details and Figure 6 for an overview of the computational domains.

Table 1. CFD Model Data.

Method	Finite Volume, incompressible Navier–Stokes, DES
Jet speed	103 m/s with block profile
Jet outlet dimension	D=103 mm; L= 60mm
Element quantity	200 K
Element form and size	Hexahedron 1–10 mm
Domain size	60 cm · 120 cm · 1 mm
Advection scheme	implicit scheme, high resolution
Convergence criteria	RMS error = 10^{-5}
Time step	20 μs
Export time step	0.1 ms

Table 2. CAA Model Data.

Method	Discontinuous Galerkin finite elements, LEE
Element quantity	20 K
Element form and size	Triangular 10–30 mm
Domain size	150 · 200 cm
Gauss points per element	3
Time step	~ 7 μs

Because of the different meshes an interpolation of the data has to be introduced. Consequently a stand alone export user function was developed to export all the data from CFX, that are needed for the scaled EIF-Method. A build-in Fortran routine was developed to interpolate the data on the coarser CAA mesh. This interface works with any mesh type and mesh type combination because only coordinates of the nodes are requested regardless of the topology (structured or unstructured). Figure 7 gives an example of mesh combinations.

More detailed: Each CAA element has three or more Gaussian integration points, at which input data (pressure and velocity field) should be known. The first step is to search at each Gaussian point for the three nearest nodes of the CFD mesh, to calculate a weight factor according to the relative distance and to save it in computer memory. This occurs just once at the beginning of the acoustic calculation. During the calculation, the input sources are linearly interpolated using these factors at each time step.

Because of the stability criterion for the explicit second order Discontinuous Galerkin method, the LEE time steps are much smaller then the export time steps from LES. A linear time interpolation of LES export data allows for smooth calculation preventing a higher LES export rate, which is not needed for the investigated acoustic frequencies and which would require more disk capacity.

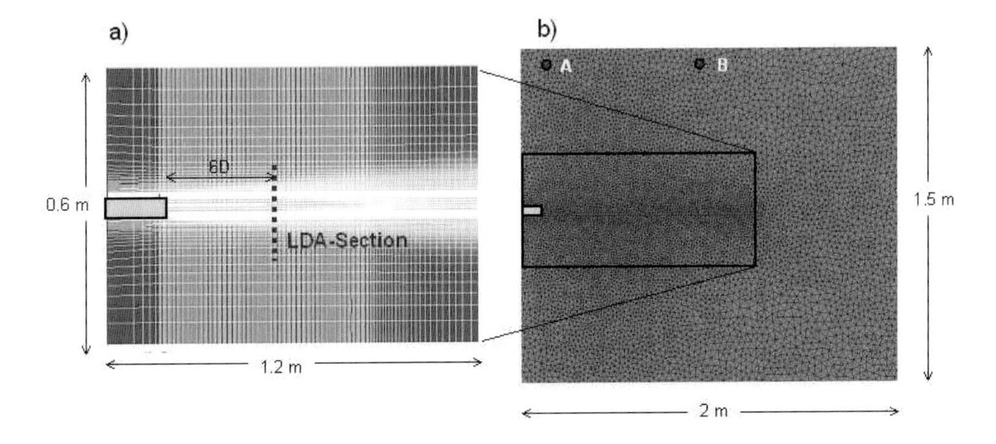

Figure 6. Calculation domains: a) CFD-Mesh.The velocity field on the dotted line is compared with experiment (Laser Doppler Anemometry). b) CAA-Mesh. The pressure spectrum at the dots is compared with literature data.

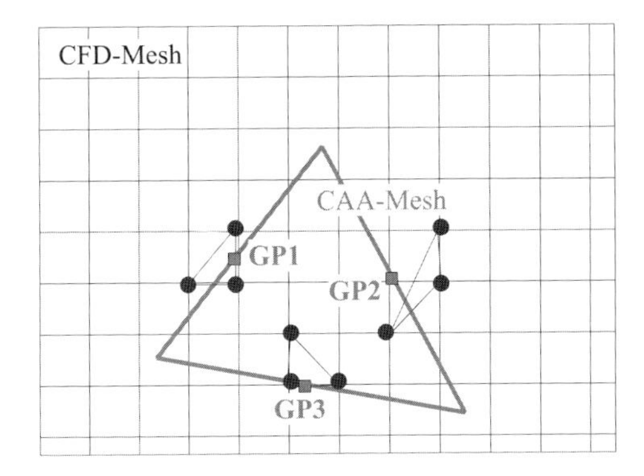

Figure 7. Example of Mesh-interpolation: the three nodes of CFD-Mesh (blue dots) nearest to the Gaussian points of the CAA-Mesh (red squares) are used for the data interpolation.

In the computation of the jet a fully developed turbulent flow field should occur before recording pressure and velocity fields. The recording time was only 35 ms with 10 KHz sampling rate to shorten the CPU time and to achieve qualitative statistical results.

A comparison was made with experimental data of a 3D axial-symmetrical free jet with the same boundary conditions (speed, outlet profile velocity and initial turbulence). Figure 8 shows a good agreement of the CFD results with the measurement by 3D Laser Doppler Anemometry (LDA) concerning integral quantities: mean velocity and turbulent kinetic energy (TKE) profiles.

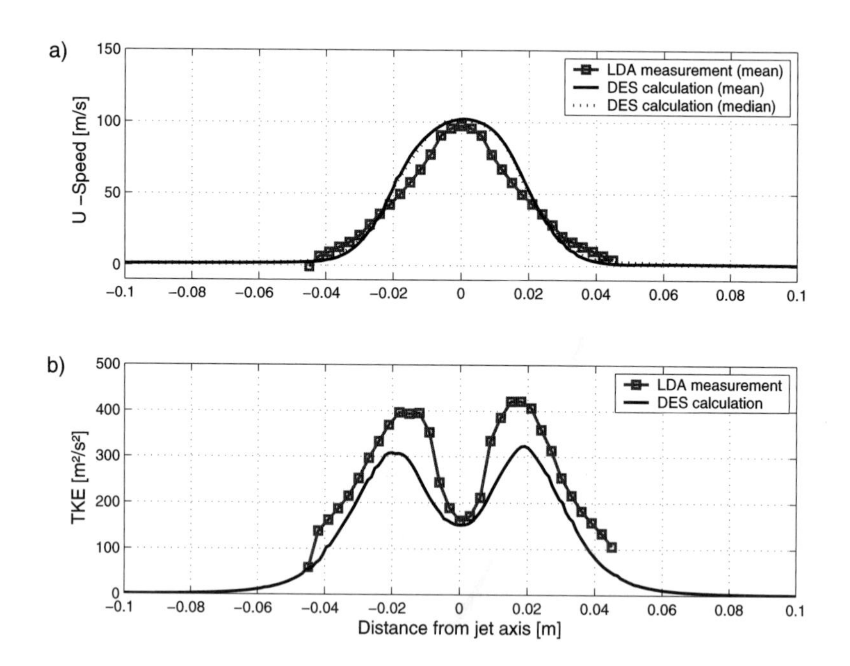

Figure 8. Comparison between CFD results and LDA measurements on a line situated six diameters away (= 180 mm) from the jet outlet. a) Speed on jet axis direction. b) Turbulent kinetic energy (TKE).

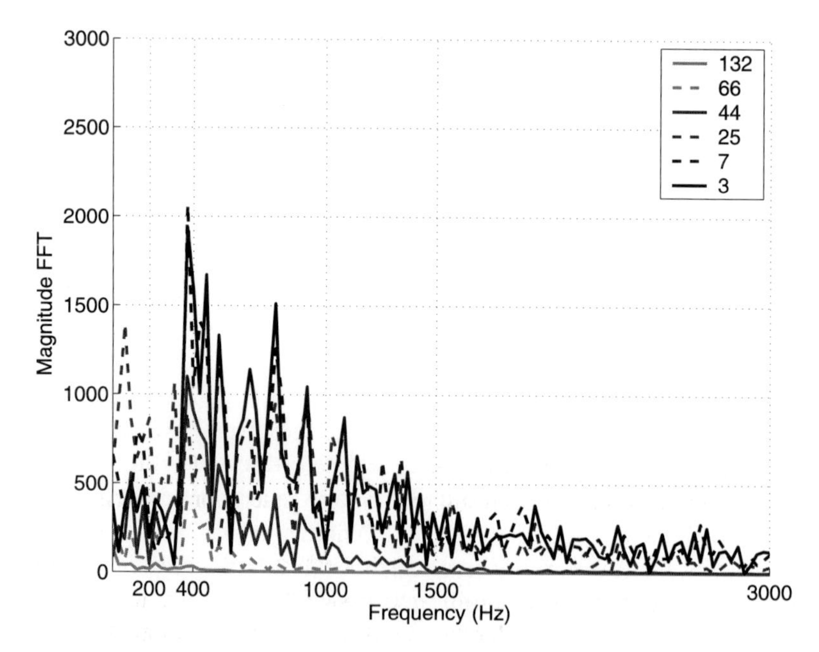

Figure 9. Velocity spectra on a line situated six diameters away (= 180 mm) from the jet outlet with increasing distance (mm) from the jet axis.

To evaluate the frequency content of the 2D LES unsteady simulation, a Fourier analysis was done (See Figure 6.3). A dominant value around 400 Hz was found. We anticipate that this correlates well to the generated acoustic main frequency.

Figure 10 gives a snapshot of the perturbated pressure field and of the source intensity at a given instant. The pressure at observation points A and B (see Figure 6 for location) is analyzed with a fast Fourier transform.

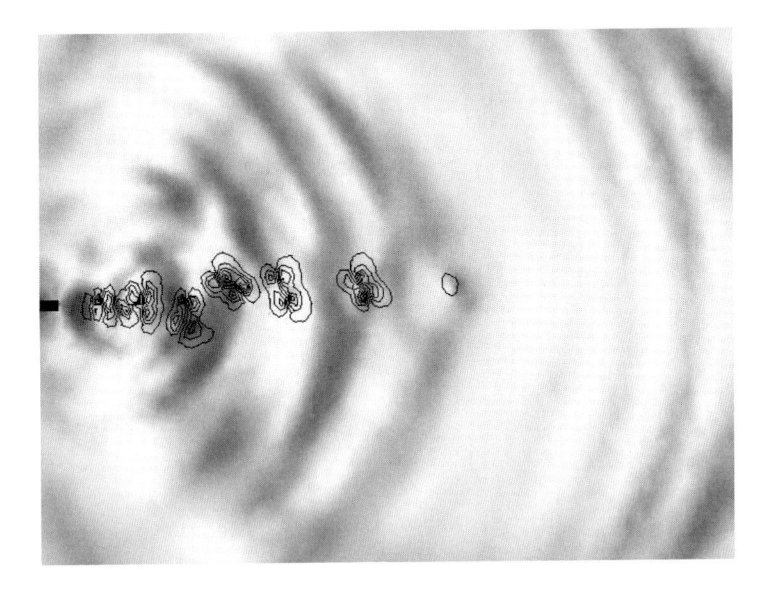

Figure 10. Instantaneous pressure field (colored background) and instantaneous source intensity distribution (iso-lines).

Figure 11 shows the spectrum profile and the maximum at 400 Hz (Strouhal number = 0.2). This matches very well with the experimental data from the literature. The calculated overall sound pressure levels (138–141 dB), however, are much too high, but, the cause should be found in the intrinsic nature of a two dimensional simulation. In fact the pressure fluctuation in LES and the acoustic pressure perturbations have in two space dimension a much bigger intensity than in a three dimensional free space and decay more slowly: the SPL-decay-law for a point source in 2D far field is $\mathrm{SPL}_2 = \mathrm{SPL}_1 - 10 \cdot \lg\left(\frac{d_2}{d_1}\right)$ instead of $\mathrm{SPL}_2 = \mathrm{SPL}_1 - 20 \cdot \lg\left(\frac{d_2}{d_1}\right)$ for 3D far field.

We conclude that the numerical result for the noise generation of a jet is qualitatively confirmed from literature data. However, the 2D simulation intrinsically overestimates pressure intensity. Furthermore, the physical time (35 ms) simulated in this study may be too short for accurate acoustic results and should be considered just as a feasibility demonstration. In order to get smoother spectra we should continue the calculation for a physical time of at least 300 ms. The present calculations already needed 500 CPU hours on a Pentium processor with a clock frequency of 1.8 GHz (250h for LES and 250h for LEE). The calculations could be performed on parallel CPUs allowing a reduction of the computation time.

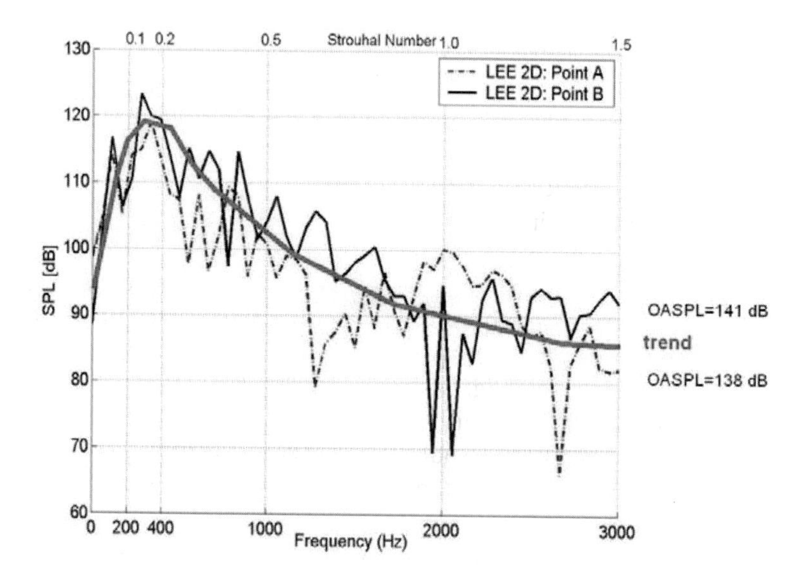

Figure 11. Spectrum of sound pressure plotted against two scales (frequency and Strouhal number). The irregular shape is due to the small number of calculated time steps. A clear trend curve is anyhow visible which matches literature data.

7 Conclusions

The perturbation methods seem to close the gap between the calculation of acoustic wave propagation into the far field and the simulation of the fluid flow generating the sound. Here, the sound generation as well as its propagation near the flow is captured in the time domain in a similar way as the simulation of the flow is performed. This has been proposed by Hardin and Pope in [6]. We followed this direction in this paper, but scaled the different terms in the perturbation ansatz by powers of the Mach number as motivated by the multiple pressure variables method and a low Mach number asymptotic expansion. For the pressure we propose the perturbation ansatz

$$p = p^{(0)} + M^2(p^{\mathrm{inc}} + p'), \tag{7.1}$$

where p' denotes the acoustic pressure fluctuations and $p^{(0)}$ and p^{inc} denote the background thermodynamic pressure and the hydrodynamic pressure, respectively.

This hybrid approach separating the flow and acoustic calculation is attractive in the low Mach number regime. A good incompressible flow simulation is combined with a wave propagation method. If noise will be calculated as generated by turbulent flow, then the techniques considered can be applied only in the case that the flow simulation captures the basic non-stationary flow. Hence, a direct numerical simulation (DNS) should be used to simulate the flow field. Assuming that the noise generated by large eddies has larger energies than the small scale vortices, a large eddy simulation or detached eddy simulation of the turbulent flow may also be successful to capture the main features of the physical problem well. But, these techniques for simulating turbulent flow are still in their beginning for applications to real problems. Numerical methods for wave propagation in the time domain are designed to preserve all important wave properties. Typical approximations used in solving fluid flow are not favorable in this situation. They inherently posses too much numerical dissipation. This, of course, enables them to capture large local gradients in the flow, not necessary in numerical wave propagation. High order finite difference, compact schemes or the ADER finite volume schemes do this job much better.

If the interaction between flow and acoustic waves become important and have to be modeled numerically, it seems to be the only general way to do direct numerical simulation of the compressible equations. In this case the flow solver has to capture the propagation of the acoustic fluctuations in an appropriate way at least over a small distance. The solver based on the compressible Navier–Stokes equations should only be applied to the very inner region where nonlinear effects play an important role. Into the direction of the far field, next a region is connected where the locally linearized Euler equations are a good mathematical model to resolve the physical situation. The next lower level in the mathematical modeling seems to be the linear acoustics. In this heterogenous domain decomposition the acoustic fluctuations are exchanged at the boundaries only. This strategy, of course, will work only if appropriate boundary and transmission conditions are found which do not generate artificial noise that is larger than the physical one.

8 Acknowledgments

We gratefully acknowledge that this work was supported by Deutsche Forschungsgemeinschaft (DFG) in the framework of the French-German research group FOR 508, *Noise Generation in Turbulent Flows* and by the EU financed network HYKE, no. HPRN-CT-2002-00282, during the CEMRACS 2003 in Luminy, France. The authors thank Roland Fortenbach, who performed the calculations of the rotating vortices and thank Stéphanie Salmon who calculated the 1D example solving the linear second order wave equation with source terms.

References

[1] H. Bijl and P. Wesseling, A unified method for computing incompressible and compressible flows in boundary-fitted Coordinates, *J. Comput Phys.* 141 (1998), 153–173.

[2] A. J. Chorin, A numerical method for solving incompressible viscous flow problems, *J. Comput. Phys.* 2 (1967), 12–26.

[3] D. L. Darmofal and B. van Leer, Local preconditioning: Manipulating mother nature to fool father time, in *Computing the Future II: Advances and Prospects in Computational Aerodynamics* (M. Hafez and D. Caughey, eds.), John Wiley and Sons, 1998, 30.

[4] I. Demirdzic, Z. Lilek, and M. Peric, A collocated finite volume method for predicting flows at all speeds, *Int. J. Numer. Meth. Fluids* 16 (1993), 1029–1042.

[5] H. Guillard and C. Viozat, On the behavior of upwind schemes in the low Mach number limit, *Computers and Fluids* 28 (1999), 63–86.

[6] J. Hardin and D. Pope, An acoustic/viscous splittting technique for computational aeroacoustics, *Theoretical and Computational Fluid Dynamics* 6 (1994), 323–340.

[7] K. C. Karki and S. V. Patankar, Pressure based calculation procedure for viscous flows at all speeds in arbitrary configurations, *AIAA Journal* 27 (1989), 1167–1174.

[8] S. Klainerman and A. Majda, Compressible and incompressible fluids, *Comm. Pure Appl. Math* XXXV (1982), 629–651.

[9] R. Klein, Semi-implicit extension of a godunov-type scheme based on low Mach number asymptotics I: One dimensional flow, *J. Comput. Phys.* 121 (1995), 213–237.

[10] R. Klein and C.-D. Munz, The multiple pressure variable (MPV) approach for the numerical approximation of weakly compressible fluid flow, in *Proceedings of "Numerical Modelling in Continuum Mechanics"* (M. Feistauer, R. Rannacher, and K. Kozel, eds.), Charles University Prag, 1995, 123–133.

[11] M. Lighthill, On sound generated aerodynamically - 1. General theory, *Proc. Roy. Soc. London* 211 (1952), 564–587.

[12] A. Majda and J. Sethian, The derivation and numerical solution of the equations for zero Mach number combustion, *Combust. Sci. and Tech.* 42 (1985), 185–205.

[13] B. E. Mitchell, S. K. Lele, and P. Moin, Direct computation of the sound from a compressible co-rotating vortex pair, *J. Fluid Mech.* 285 (1995), 181–202.

[14] C. Munz, S. Roller, K. Geratz, and R. Klein, The extension of incompressible flow solvers to the weakly compressible regime, *Computers and Fluids* (2003).

[15] S. Patankar, *Numerical Heat Transfer and Fluid Flow*, McGraw Hill, New York 1980.

[16] S. V. Patankar and D. B. Spalding, A calculation procedure for heat, mass, and momentum transfer in three-dimensional parabolic flow, *Int. J. Heat Mass Transfer* 15 (1972), 1787–1806.

[17] C. Rhie, Pressure based Navier–Stokes solver using the multi grid method, *AIAA Journal* 27 (1989), 1017–1018.

[18] T. Schwartzkopff, M. Dumbser, C.-D. Munz, Fast high order ADER schemes for linear hyperbolic equations, *J. Comput. Phys.* 197 (2004), 532–39.

[19] W. Shyy, Elements of pressure-based computational algorithms for complex fluid flow and heat transfer, *Adv. Heat Transfer* 24 (1994), 191–275.

[20] S. Slimon, M. Soteriou, and D. Davis, Development of computational aeroacoustics equations for subsonic flows using a mach number expansion approach, *J. Comput. Phys.* 159 (2000), 377–406.

[21] L. Ting and J. Miksis, On vortical flow and sound generation, *Soc. Indust. Appl. Math.* 50 (1990), 521–536.

[22] E. Turkel, A. Fiterman, and B. van Leer, Preconditioning and the limit of the compressible to the incompressible flow equations for finite difference schemes, in *Computing the Future: Advances and Prospects for Computational Aerodynamics* (M. Hafez and D. Caughey, eds.), John Wiley and Sons, 1994, 215–234.

[23] J. Weiss and W. Smith, Preconditioning applied to variable and constant density flows, *AIAA Journal* 33 (1995), 2050–2057.